Handbook of Affinity Chromatography

CHROMATOGRAPHIC SCIENCE SERIES

A Series of Monographs

Editor: JACK CAZES
Sanki Laboratories, Inc.
Mount Laurel, New Jersey

1. Dynamics of Chromatography, *J. Calvin Giddings*
2. Gas Chromatographic Analysis of Drugs and Pesticides, *Benjamin J. Gudzinowicz*
3. Principles of Adsorption Chromatography: The Separation of Nonionic Organic Compounds, *Lloyd R. Snyder*
4. Multicomponent Chromatography: Theory of Interference, *Friedrich Helfferich and Gerhard Klein*
5. Quantitative Analysis by Gas Chromatography, *Josef Novák*
6. High-Speed Liquid Chromatography, *Peter M. Rajcsanyi and Elisabeth Rajcsanyi*
7. Fundamentals of Integrated GC-MS (in three parts), *Benjamin J. Gudzinowicz, Michael J. Gudzinowicz, and Horace F. Martin*
8. Liquid Chromatography of Polymers and Related Materials, *Jack Cazes*
9. GLC and HPLC Determination of Therapeutic Agents (in three parts), *Part 1 edited by Kiyoshi Tsuji and Walter Morozowich, Parts 2 and 3 edited by Kiyoshi Tsuji*
10. Biological/Biomedical Applications of Liquid Chromatography, *edited by Gerald L. Hawk*
11. Chromatography in Petroleum Analysis, *edited by Klaus H. Altgelt and T. H. Gouw*
12. Biological/Biomedical Applications of Liquid Chromatography II, *edited by Gerald L. Hawk*
13. Liquid Chromatography of Polymers and Related Materials II, *edited by Jack Cazes and Xavier Delamare*
14. Introduction to Analytical Gas Chromatography: History, Principles, and Practice, *John A. Perry*
15. Applications of Glass Capillary Gas Chromatography, *edited by Walter G. Jennings*
16. Steroid Analysis by HPLC: Recent Applications, *edited by Marie P. Kautsky*
17. Thin-Layer Chromatography: Techniques and Applications, *Bernard Fried and Joseph Sherma*
18. Biological/Biomedical Applications of Liquid Chromatography III, *edited by Gerald L. Hawk*
19. Liquid Chromatography of Polymers and Related Materials III, *edited by Jack Cazes*
20. Biological/Biomedical Applications of Liquid Chromatography, *edited by Gerald L. Hawk*
21. Chromatographic Separation and Extraction with Foamed Plastics and Rubbers, *G. J. Moody and J. D. R. Thomas*

22. Analytical Pyrolysis: A Comprehensive Guide, *William J. Irwin*
23. Liquid Chromatography Detectors, *edited by Thomas M. Vickrey*
24. High-Performance Liquid Chromatography in Forensic Chemistry, *edited by Ira S. Lurie and John D. Wittwer, Jr.*
25. Steric Exclusion Liquid Chromatography of Polymers, *edited by Josef Janča*
26. HPLC Analysis of Biological Compounds: A Laboratory Guide, *William S. Hancock and James T. Sparrow*
27. Affinity Chromatography: Template Chromatography of Nucleic Acids and Proteins, *Herbert Schott*
28. HPLC in Nucleic Acid Research: Methods and Applications, *edited by Phyllis R. Brown*
29. Pyrolysis and GC in Polymer Analysis, *edited by S. A. Liebman and E. J. Levy*
30. Modern Chromatographic Analysis of the Vitamins, *edited by André P. De Leenheer, Willy E. Lambert, and Marcel G. M. De Ruyter*
31. Ion-Pair Chromatography, *edited by Milton T. W. Hearn*
32. Therapeutic Drug Monitoring and Toxicology by Liquid Chromatography, *edited by Steven H. Y. Wong*
33. Affinity Chromatography: Practical and Theoretical Aspects, *Peter Mohr and Klaus Pommerening*
34. Reaction Detection in Liquid Chromatography, *edited by Ira S. Krull*
35. Thin-Layer Chromatography: Techniques and Applications. Second Edition, Revised and Expanded, *Bernard Fried and Joseph Sherma*
36. Quantitative Thin-Layer Chromatography and Its Industrial Applications, *edited by Laszlo R. Treiber*
37. Ion Chromatography, *edited by James G. Tarter*
38. Chromatographic Theory and Basic Principles, *edited by Jan Åke Jönsson*
39. Field-Flow Fractionation: Analysis of Macromolecules and Particles, *Josef Janča*
40. Chromatographic Chiral Separations, *edited by Morris Zief and Laura J. Crane*
41. Quantitative Analysis by Gas Chromatography, Second Edition, Revised and Expanded, *Josef Novák*
42. Flow Perturbation Gas Chromatography, *N. A. Katsanos*
43. Ion-Exchange Chromatography of Proteins, *Shuichi Yamamoto, Kazuhiro Nakanishi, and Ryuichi Matsuno*
44. Countercurrent Chromatography: Theory and Practice, *edited by N. Bhushan Mandava and Yoichiro Ito*
45. Microbore Column Chromatography: A Unified Approach to Chromatography, *edited by Frank J. Yang*
46. Preparative-Scale Chromatography, *edited by Eli Grushka*
47. Packings and Stationary Phases in Chromatographic Techniques, *edited by Klaus K. Unger*
48. Detection-Oriented Derivatization Techniques in Liquid Chromatography, *edited by Henk Lingeman and Willy J. M. Underberg*
49. Chromatographic Analysis of Pharmaceuticals, *edited by John A. Adamovics*
50. Multidimensional Chromatography: Techniques and Applications, *edited by Hernan Cortes*
51. HPLC of Biological Macromolecules: Methods and Applications, *edited by Karen M. Gooding and Fred E. Regnier*

52. Modern Thin-Layer Chromatography, *edited by Nelu Grinberg*
53. Chromatographic Analysis of Alkaloids, *Milan Popl, Jan Fähnrich, and Vlastimil Tatar*
54. HPLC in Clinical Chemistry, *I. N. Papadoyannis*
55. Handbook of Thin-Layer Chromatography, *edited by Joseph Sherma and Bernard Fried*
56. Gas–Liquid–Solid Chromatography, *V. G. Berezkin*
57. Complexation Chromatography, *edited by D. Cagniant*
58. Liquid Chromatography–Mass Spectrometry, *W. M. A. Niessen and Jan van der Greef*
59. Trace Analysis with Microcolumn Liquid Chromatography, *Miloš Krejčí*
60. Modern Chromatographic Analysis of Vitamins: Second Edition, *edited by André P. De Leenheer, Willy E. Lambert, and Hans J. Nelis*
61. Preparative and Production Scale Chromatography, *edited by G. Ganetsos and P. E. Barker*
62. Diode Array Detection in HPLC, *edited by Ludwig Huber and Stephan A. George*
63. Handbook of Affinity Chromatography, *edited by Toni Kline*

ADDITIONAL VOLUMES IN PREPARATION

Handbook of Affinity Chromatography

edited by
Toni Kline

Bristol-Myers Squibb
Institute for Pharmaceutical Research
Princeton, New Jersey

Marcel Dekker, Inc. New York • Basel • Hong Kong

Library of Congress Cataloging-in-Publication Data

Handbook of affinity chromatography / edited by Toni Kline.
p. cm. — (Chromatographic science series : v. 63)
Includes bibliographical references and index.
ISBN 0-8247-8939-3 (acid-free)
1. Affinity chromatography. I. Kline, Toni.
II. Series: Chromatographic science : v. 63.
QP519.9.A35H36 1993
574.19'285—dc20 93-19412
CIP

The publisher offers discounts on this book when ordered in bulk quantities. For more information, write to Special Sales/Professional Marketing at the address below.

This book is printed on acid-free paper.

Marcel Dekker, Inc.
270 Madison Avenue, New York, New York 10016

Current printing (last digit):
10 9 8 7 6 5 4 3 2 1

PRINTED IN THE UNITED STATES OF AMERICA

Preface

Much of the concept for this handbook is explicitly presented in the title: it is intended to be a practical guide to be used in the laboratory. It is also intended to be used in the office *before* going into the laboratory, in the sense that it reflects the newer techniques and applications that affect the planning of both experimental tactics and research strategies. Traditional affinity chromatography relies on the high-affinity biospecific interactions as the means of separation. Thus proteins are purified on immobilized substrates/inhibitors (for enzymes), antibodies (for antigenic proteins), nucleic acids (for DNA- or RNA-binding proteins), or ligands (for receptors).

In recent years, the definition of biology itself has expanded under the influence of fully functional engineered proteins, abzymes, and other evidence of our ability to understand and exploit natural principles. In the same sense, "biospecific interactions" can now be taken as a composite of forces that can be manipulated at the molecular level. This enables proteins and molecules other than proteins to be affinity-selected according to fundamental properties of hydrophobicity, charge, and polarity. Similarly, the exact degree of affinity is now controlled, and weak affinity adsorbtion has become a versatile technique for separations.

Advances in the field have made it realistic to create a customized separation system unique to each problem without having to invent a whole new technology to support it. This handbook is intended to encourage researchers to do so, and it benefits from the contributors' own recent experiences in the field. The organization of this handbook reflects our intentions. The first section outlines the fundamental principles by which all interactions occur and presents the details that harness these forces into a functional machine. Since the success of any affinity purification is dependent on the success of every step, we have tried to provide the basis for efficiently conducting a precious substance with confidence through each step of the procedure. Section II presents some of the current scope of the technique directed toward purifications of several classes of proteins. In Section III, aspects are presented of the current research on affinity chromatography as a topic in its own right in the field of biorecognition.

Toni Kline

Contents

Contributors

Irwin Chaiken Biopharmaceuticals R&D, SmithKline Beecham, King of Prussia, Pennsylvania

Jean M. Egly Laboratoire de Génétique Moléculaire des Eucaryotes du CNRS, Unité 184 de Biologie Moléculaire et de Génie Génétique de l'INSERM, Institut de Chimie Biologique, Strasbourg, France

Lars G. Fägerstam Pharmacia Biosensor AB, Uppsala, Sweden

Felix Friedberg Department of Biochemistry and Molecular Biology, College of Medicine, Howard University, Washington, D.C.

Craig M. Jackson American Red Cross Blood Services, Detroit, Michigan

Lawrence M. Kauvar Terrapin Technologies, Inc., So. San Francisco, California

Toni Kline Bristol-Myers Squibb Institute for Pharmaceutical Research, Princeton, New Jersey

Per-Olof Larsson Pure and Applied Biochemistry, Chemical Center, University of Lund, Lund, Sweden

Vincent Moncollin Laboratoire de Génétique Moléculaire des Eucaryotes du CNRS, Unité 184 de Biologie Moléculaire et de Génie Génétique de l'INSERM, Institut de Chimie Biologique, Strasbourg, France

Sten Ohlson HyClone Laboratories, Inc., Logan, Utah

Daniel J. O'Shannessy Protein Biochemistry, SmithKline Beecham, King of Prussia, Pennsylvania

Malcolm G. P. Page Pharma Division, Preclinical Research, F. Hoffmann-La Roche Ltd., Basel, Switzerland

Allen R. Rhoads Department of Biochemistry and Molecular Biology, College of Medicine, Howard University, Washington, D.C.

Herbert Schott Institut für Organische Chemie, Universität Tübingen, Tübingen, Germany

Peter Toomik Institute of Molecular and Cell Biology, Tartu University, Tartu, Estonia

Richard Villems Institute of Molecular and Cell Biology, Tartu University, Tartu, Estonia

Donald J. Winzor Department of Biochemistry, University of Queensland, Brisbane, Queensland, Australia

David Zopf Neose Pharmaceuticals, Inc., Horsham, Pennsylvania

I

TECHNIQUES OF AFFINITY CHROMATOGRAPHY

1

Overview

Richard Villems and Peter Toomik

Tartu University, Tartu, Estonia

I. SOLID SUPPORT

The choice of support material may be crucial: for example, the binding capacity of immobilized Cibacron Blue F3GA dye differed up to 600-fold depending on the support used (1). For a review, see also Refs. 2 and 3.

A. Carbohydrate Polymers

The agarose molecule consists of alternately linked 1,3-bound β-D-galactopyranose and 2,4-bound 3,6-anhydro-α-L-galactopyranose residues (Figure 1) (4). Any secondary hydroxyl may be esterified with sulfuric acid (5) and D-galactose residue O-methylated (6). The degree of methylation depends on the raw material (7), and the content of most reactive hydroxyl groups is often unknown.

Although commercially available agarose in 1 to 12% gels is usually adequate, custom preparations may be made*[1] (8–10). Ligands should not be immobilized before gelling because it decreases both the gel strength and the gelling temperature (11). Cross-linking with epichlorohydrine or analogs,* 1,4-butanediol diglycidyl ether (12), or divinylsulfone (DVS) (13) increases both the chemical and thermal stability of the gel but may diminish the binding capacity [effects

[1]Asterisks indicate that the method is described in the Methods section of the chapter, pp. 41-46.

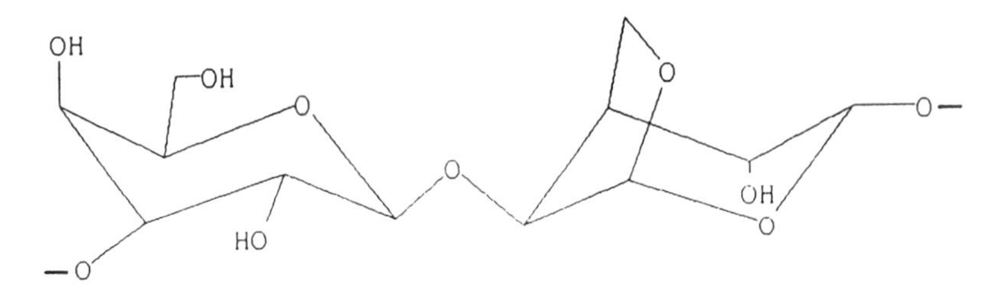

Figure 1 Repeating agarobiose unit.

other than loss of hydroxyls may have an influence (1)]. DVS increases the gel's rigidity to some orders of magnitude and enables the use of agarose for high-performance liquid chromatography (HPLC) (20). Alternatively, nonporous compressed agarose beads can be used (21,22).

Little is known about the structure of Pharmacia's two ultrarigid agarose gels: Superose and Sepharose Fast Flow. Superose is prepared by treatment with long-chain bi- and polyfunctional epoxides in a nonaqueous medium and then with "short" epoxides in aqueous solution (23). Another rigid agarose-based material, Macrosorb K4AX (containing kieselguhr as the rigid skeleton), is available from Sterling Organics (24).

Cross-linked agarose tolerates any solvent except strong acids and oxidizers, but may be damaged by some rare enzymes and must not be frozen or air dried. Mild acid hydrolysis increases the quantity of sterically available galactose residues and turns agarose into an excellent sorbent for galactose-binding proteins (25) (see also Table 1). Agarose gel activates complement system in human blood (26) and *must never be used in vivo* (e.g., for affinity removing of toxins from blood).

Cellulose is primarily of historical significance. In 1951, diazotized *p*-aminobenzylcellulose was used for immunoaffinity chromatography (27), but

Table 1 Protein Binding to Unmodified Agarose

Protein	Ref.
Agarase from *Littorina mandshurica*	14
Galactose oxidase	15
Lectins	16,17
α-Glycerophosphate dehydrogenase (NAD dependent)	18
Lysozyme	19

phospho- and DNA–cellulose are still used for DNA-related and other enzymes. Selection applications of phosphocellulose include:

DNA polymerase (28–31)
Restrictase EcoR1 (32)
Restrictase ThtI (33)
Restrictase AhaI, Aha II (34)
Restrictase Bst 31 (35), etc.
RNA polymerase (36,37)
Reversed transcriptase (38)
Alkaline phosphatase (39)

Of course, it is not always clear whether phospocellulose acts as an ion exchanger or as an affinity sorbent. The efficiency of recently developed affinity filters and polystyrene-cellulose copolymer-based sorbents is comparable to that of HPLC (40).

Dextran is a linear α-1,6-linked glucose polymer produced by *Leuconostoc mesenteroides*. Two types of gels, Sephadex (cross-linked by glyceryl bridges) and Sephacryl (allyldextran cross-linked by *N,N′*-methylenebisacrylamide) are used in the chromatography of biomolecules. Sephacryl S-1000 is especially suitable for affinity chromatography of very large molecules (e.g., factor VIII)* (41). At low pH Sephacryl S-200 adsorbs proteins (42).

Sephadex is used mainly as a glucose polymer (e.g., for the purification of lectins) (Table 2). For other applications it is mechanically too weak (G75 to G200) or has insufficient porosity (G10 to G75). Sephadex can be activated in aqueous media with non-cross-linking reagents, and the presence of *cis*-diol groups makes periodate oxidation more attractive than oxidation using agarose. Stults et al. (50) have shown that cross-linked Sepharose contains up to 38 μM/mL periodate-oxidizable groups and that vicinal diols present due to incomplete cross-linking are probably involved. A mixed dextran–agarose gel, Superdex is

Table 2 Protein Binding to Unmodified Sephadex

Protein	Ref.
Exoamylase from *Pseudomonas stutzeri* (3.8.1.60), G100	43
Glucoamylase, G200	44
Lectin from *Helix pomatia*, G200	45
Concanavalin A	
G50	46,47
G100	48
Lectin from *Vicia faba*, G150	49

designed for gel filtration and is still not used for affinity chromatography. A new polysaccharide gel, Vectrex (51), is recommended for the separation of biotinylated nucleic acids.

B. Synthetic Polymers

Polyacrylamide (PAA) was first used for protein chromatography in 1962 (52,53). It is synthesized by copolymerization of acrylamide and a cross-linking reagent (usually *N,N'*-methylenebisacrylamide). The primary disadvantage is the same as that of Sephadex: The gels are either soft or have small pores. PAA gel is resistant against enzymatic attacks and does not adsorb biomolecules; therefore, it is used occasionally despite its poor mechanical properties.

Ligands can be immobilized onto PAA gel by the following methods:

1. Entrapment, wherein PAA gel is polymerized from the mixture containing particles to be immobilized. The particles must be much larger than those to be separated and must tolerate polymerization catalysts (problems with SH groups of proteins may arise). Thus yeast cells are entrapped for purification of lectins (54).
2. Copolymerization with molecules bearing allyl, vinyl, acryl, and other groups. Various allyl glycosides (55) or polysaccharide allyl ethers (56) are typical examples, but even proteins have been copolymerized after acylation with acrylic acid chloroanhydride (57). Copolymerization with acrylic acid *N*-hydroxysuccinimidyl ester yields a reactive carrier material for immobilization via a NH_2 group* (58).
3. Substitution reactions at an amide group (e.g., reactions with other amines, dialdehydes, azides, etc.).

Hydroxyalkylmethacrylate gels (Spheron, Separon HEMA) are synthesized by copolymerization of 2-hydroxyethylmethacrylate and ethylene glycol dimethacrylate (59,60). The gel is chemically and mechanically more stable than PAA gel but is also more hydrophobic and thus inappropriate for many applications.

Trisacryl is produced by copolymerization of *N*-acryloyl-2-amino-2-hydroxymethyl-1,3-propane diol with diallyltartardiamide (Figure 2) (61) (for monomer synthesis, see Ref. 62). The product is hydrophilic, biologically inert, rigid, and macroporous. Despite its low working volume (it contains more than 30% solids) it is used for large-scale separations (63). Copolymerization with active group–bearing monomers also enables to prepare variety of affinity gels (for a review, see Ref. 64).

Hydrophilic vinyl polymers (Toyopearl, Toyo Soda, Japan; Fractogel, Merck, Germany) are remarkably stable but of low working volume. The achievable ligand concentration is up to 10 times higher than that of agarose, but the protein-binding capacity is lower (65) (probably due to the presence of small pores). The

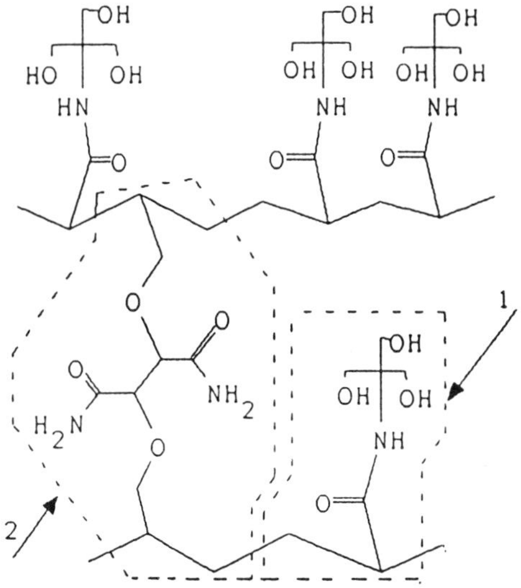

Figure 2 Partial structure of trisacryl gels. 1, *N*-(trishydroxymethyl)acrylamide residue; 2, diallyltartardiamide residue.

very popular HW 65 may not be the best choice for some proteins (66). Superfine and even fine-grade gels were used in the fast-performance liquid chromatography (FPLC) mode (67). Another vinyl polymer, TSK PW, is available for HPLC (68).

C. Mixed Gels

Polyacrylamide–agarose gels (Ultrogel AcA) were developed for gel filtration (69) and tested for affinity chromatography (70,71) but combine the disadvantages of both components: the thermolability of non-cross-linked agarose with the alkali sensitivity of acrylamide. The advantage is that AcA-type gels are easy to activate with glutaraldehyde (72).

D. Inorganic Materials

When extreme rigidity of the support is needed, as for HPLC (microspherical silica) and large-scale industrial applications, inorganic materials are used [irregular porous glass (73) or silica particles]. They are soluble at pH > 8, although a coating of zirconium will enhance their stability* (74,75). Otherwise, glass or silica may be coated with an inert polymers (76). Any friction between sorbent particles during column packing causes the exposition of nude silica and nonspecific sorption.

E. Magnetic Carriers

Magnetic sorbent particles (Magnogel AcA 44 and Act-Magnogel AcA 44, IBF Réactifs, France) are useful for batch extraction from large volumes of diluted and/or turbid solutions. Several attempts have been made to synthesize similar carriers under laboratory conditions (77), but limited availability of magnetite particles that have a chemically stable surface creates problems with ferric ion leakage.

F. Rarely Used Support Materials

Large amounts of a very cheap material may be needed when the sorbent is irreversibly degraded or contaminated during a few chromatographic runs. The list of appropriate materials includes agar gel (instead of agarose) (78), nylon (79), chitin (80), polystyrene (81,82), and perfluorocarbon polymers (83). Some unmodified materials can also bind proteins selectively: agar gel (84), coffee grounds (85), and tea leaves (86) were used successfully for the purification of trypsin.

G. Sorbents for HPLAC

The capacity of high-performance liquid affinity chromatography (HPLAC) sorbents is usually low, but subsequent separations may be less time consuming than one run on soft gel. Micropellicular sorbents are very promising for fast runs, and their capacity may be high enough even for preparative chromatography [protein A on 2-μm nonporous particles bound 4.5 mg/mL IgG (87)]. For the preparation of HPLC-grade silica, the method of Wójcik and Kwietniewski (88) seems to be the simplest. Some materials are still not used for affinity applications; for example, perfusion chromatography on open-pore polymeric column (POROS) is 10- to 100-fold faster than conventional HPLC (89), but has been proved only for ion exchange chromatography (IEC) and reverse phase chromatography (RPC) of proteins.

II. AFFINITY LIGAND: SELECTION, MODIFICATION, DESIGN, AND SYNTHESIS

A. Bioaffinity Ligands

Immobilization of a "natural" ligand requires addressing several possible problems:

Inadequate or unknown specificity
Lack of information as to the optimal immobilization site
Instability of ligand in the chromatography process
Inadequate binding strength
Price of the ligand and consistent availability

Affinity chromatography is generally used as an "on–off" method; that is, all of the substance of interest is adsorbed, contaminants are washed off, and the bound substances are eluted. The method requires strong binding (dissociation constants $K_d > 10^{-5}$ are barely acceptable) and harsh elution and column regeneration conditions are often needed. The advent of HPLC enabled the use of ligands with substantially lower affinity (K_d values of 10^{-2} to 5×10^{-3} and column capacity factors k' of 1 to 12) (90).

1. Proteins as Affinity Ligands

a. Immobilization

For site-directed immobilization various techniques are used (see Section III). In the case of nonspecific binding, *minimum support activation should be used for the necessary immobilized protein concentration, since the overactivation usually causes a decrease in the sorbent's capacity.* It is especially important in the case of soft gels, because the protein may be completely wrapped into the gel network, due to the multipoint immobilization. When "ready-to-use" activated gels are chosen, they may need limited alkaline hydrolysis for the removal of excess activity; about 5 μm/mL of active groups is usually sufficient. Numerous results indicate that addition of polyethylene glycol and high concentrations of water-structuring salts to the immobilization buffer increases coupling yield (e.g., Refs. 91 and 92).

b. Quantification of Immobilized Protein

The simplest method of quantifying immobilized protein is direct spectrophotometric measurement using a mixture of unsubstituted sorbent and protein solution as a standard. This method is only applicable for transparent (or at least translucent) gels. The refraction index of the buffer may be adjusted by the addition of ethylene glycol, glycerol, polyethylene glycol (PEG) and so on so that gels become more transparent. Moreover, the use of viscous additives such as PEG eliminates the need for continuous stirring during measurement, since the sedimentation of gels in such media is very slow. For higher sensitivity the standard Bradford (93) and bicinchonic acid methods (94) can be adapted successfully for use with immobilized proteins.

Antibodies

Any method of protein immobilization is applicable to antibodies, but periodate oxidation followed by reaction with a hydrazide-bearing gel seems to give the best results (95,96). The simplest way to immobilize antibodies is to adsorb them onto microparticulate macroporous polystyrene (97). When polyclonal antibodies are used, they are usually first purified using immobilized antigens followed by removing cross-reacting antibodies on another column, if necessary. For conservative mammalian proteins, sometimes only avian (i.e., egg yolk) antibodies can be developed (98). Using different eluants during elution from an antigen column makes possible control of the binding properties of antibody column. A new method consists of using *paralogs*, short peptides that mimic the binding site (paratope), or rather the binding properties of moderate affinity antibodies (see Ref. 99 and Chapter 12 of this book).

Bacterial IgG Fc Receptors

At least six types of IgG Fc receptors are produced by gram-positive cocci: type I, protein A by *Staphylococcus aureus* (100); type II, group A streptococci (101); type III, protein G by group C and G streptococci (102), type IV, by some group G streptococcal strains; and types V and VI, by some *Streptococcus zooepidemicus* strains. Only types I and III are widely used for affinity purification of immunoglobulins. The binding properties of a variety of bacterial Fc receptors are described in numerous papers (103–107).

Proteins A and G both also contain albumin-binding regions, which may be utilized for their purification under very mild conditions but may cause problems with IgG purification. Genetic engineering can remove both the albumin-binding domain and hydrophobic regions from protein G, and highly IgG-specific sorbent is commercially available (108). Another genetically engineered product is available from Pierce: Immobilized protein A/G is a gene fusion product that binds all IgG subclasses that proteins A and G bind, without binding to mouse or human albumin (109).

Method of Gene Fusion Products

There exists an excellent method for purification of genetically engineered proteins. The gene of a readily purified protein (e.g., protein A) is fused to the coding sequence of a second protein that is to be produced and purified. These linked proteins are expressed in an appropriate system and then purified with nearly absolute specificity using affinity chromatography on immobilized IgG (110) or albumin (in the case of protein A), or monoclonal antibodies. After enzymatic or chemical cleavage of the peptide bond between two proteins [the specificity of this reaction may be increased by using factor Xa as a specific tool for cleaving the sequence Ile–Glu–Gly–Arg inserted at the Arg (111)], another passage of the mixture through the same column removes contaminating partner protein. β-Galactosidase is often used instead of protein A (112), and various expression systems are commercially available.

The use of immobilized antigens for purification of antibodies with distinct specificity from polyclonal sera is usually effective. Immobilization is generally simple unless the most reactive site is the most active epitope. Haptenes are to be immobilized in exactly the same manner as they were bound to carrier macromolecules for immunization.

Lectins

If any glycoprotein is to be separated, appropriate immobilized lectins will serve as selective and powerful tools. Most common lectins are commercially available as ready-to-use sorbents, but some must be immobilized by investigators (Table 3). The sugar-binding site of the lectin should be protected during immobilization by saturation with the lectin's specific sugar (amino sugars must be used in N-acylated form to avoid reaction with activated matrix). Activated carriers, which can bind carbohydrates through their hydroxyl groups (e.g., divinylsulfone) also arouse suspicion. The approach to the immobilization of other protein ligands (receptors, effectors, etc.) should be similar—whenever possible, the active site should be protected.

Inhibitory Proteins

A wide variety of protein enzyme inhibitors can be used for the resolution of complex enzyme mixtures by pH and ionic strength manipulation in a single

Table 3 Commercially Available Immobilized Lectins[a]

Latin name	Specificity	Manufacturers[b]
Abrus precatorius	D-Gal	S
Arachis hypogaea	β-D-gal(1-3)-D-galNAc	S
Bandeirea simplicifolia	α-D-gal, α-D-galNAc	S
Canavalia ensiformis	α-D-man, α-D-glc	S, Se, Ph
Dolichos biflorus	α-D-galNAc	S
Glycine max	D-galNAc	S
Helix pomatia	D-galNAc	S
Lens culinaris	α-D-man	S, Se
Phaseolus vulgaris	oligosaccharide	S, Se
Phytolacca americana	$(D\text{-glcNAc})_3$	S
Pisum sativum	α-D-man	S
Ricinus communis	D-galNAc, β-D-gal	S
Tetragonolobus purpureas	α-L-fuc	S
Triticum vulgaris	$(D\text{-glcNAc})_2$, NeuNAc	S
Ulex europaeus		
I	α-L-fuc	S
II	$(D\text{-glcNAc})_2$	S
Vicia villosa	D-galNAc	S

[a]Jacalin from jack fruit seed (*Artocarpus integrifolia*) is of unique specificity: it binds only IgA1 (113).
[b]S, Sigma Chemical Co., United States; Se, Serva Feinbiochemica, Germany; Ph, Kabi Pharmacia, Sweden.
Source: Compiled from manufacturers' catalogs.

column. Some selected examples: Acrosin (114) and kallikrein (115) are purified using soybean trypsin inhibitor (SBTI), and alkaline protease from *Aspergillus oryzae* on ovoinhibitor (116). Immobilized hemoglobin in sometimes used instead of protease inhibitors (117,118).

Enzymes

Conversely, immobilized enzymes may offer the single possibility of purifying some inhibitors and high-molecular-weight substrates. The approach to the purification of reversible inhibitors is trivial, but the situation with other classes of compounds may be complicated:

1. The inhibitor or substrate gels covalently modified during the binding process and nothing remains to elute [protease/α_2-macroglobulin complex (119)].
2. Noncovalent binding is too strong to enable elution under nondestructive conditions.

A solution may be found through modification of the immobilized enzyme's structure. The most instructive example is the use of anhydrotrypsin* (120), anhydrochymotrypsin (121), and anhydrothrombin (122) as excellent affinity ligands for the purification of numerous inhibitors, including one as vulnerable as α_1-antitrypsin. The geometry and chemical constitution of anhydro and native enzymes are close to identity; only the nucleophilic serine is omitted. Anhydrotrypsin also binds tryptic peptides (123).

Structural and Denatured Proteins

The most commonly used denatured protein for immobilization is gelatin [for affinity chromatography of fibronectin (124)]. In this and similar cases there are no problems about the binding site, but new problems may arise. Some results indicate that the efficiency of the affinity sorbent is more dependent on the pH, salt concentration, and temperature regime of the immobilization procedure than on the final concentration of the ligand.

When an insoluble protein is used to be used for affinity chromatography, home-made beaded chromatography gels are useful as elastin for elastase purification (125) or collagen for collagenase. This method involves the entrapment of protein particles in the supporting gel network. Limited acid hydrolysis to soluble peptide fragments, which are then immobilized, circumvents this problem. Cross-linking need not damage the binding capacity of immobilized proteins, and regular treatment with a solution of glutaraldehyde, for example, can also serve as an efficient sterilization method (126).

2. *Nucleic Acids as Affinity Ligands*

Immobilized nucleic acids are used both as general ligands for all DNA-binding proteins and for sequence-recognizing molecules. The approach to the synthesis of these sorbents is completely different. Among the general-use sorbents, DNA–cellulose remains the most popular. The sorbent is prepared by drying the cellulose after impregnation with DNA solution, sometimes followed by ultraviolet irradiation to enhance binding. In recent years, covalent binding of DNA to activated agarose gels is prevalent.

High-specificity nucleic acid sorbents are usually synthesized by periodate oxidation followed by binding onto hydrazide-containing agarose gel. Where the nucleobases are sterically more available, short fragments of double-stranded DNA may be bound by diazotation via their single-stranded ends.

Although giving reasonable coupling yields, the above-mentioned protocols are not flexible enough. More promising in this context appears to be the preparation of a universal primed matrix where a specially synthesized partially palindromic DNA sequence containing single 4-triazolylthymidine is coupled to Sepharose (127). The reported yield of this procedure is about 35%, and that of subsequent ligation of the sequence-specific DNA via cohesive ends to primed Sepharose is 40% (128), giving an overall efficiency of about 15%.

Selected examples of the use of immobilized DNA:

RNA ligase (129)
mRNA guanlyltransferase and mRNA(guanine-7)methyltransferase (130)
DNA photoreactivating enzyme (131)
Endonuclease (specific for apurinic sites in DNA) (132)
ATP-dependent DNase (133)
DNase (134,135)
Protein kinase (135)
DNA polymerase (137,138)
RNA polymerase (139,140)
DNA-dependent RNA polymerase (141)

3. *Carbohydrates as Affinity Ligands*

A wide variety of proteins bind to carbohydrates. Usually, monosaccharides are immobilized via their distinct reactive group, whereas for polysaccharides only sterical availability is taken into account. The choice of immobilization method is generally determined by the presence or absence of functional groups other than hydroxyls in the sugar molecules.

a. Monosaccharides

Those sugars that contain hydroxyl groups only can be directly immobilized by epoxy-activated matrixes via primary hydroxyls (142) or on divinylsulfone (DVS)-activated gels. For the binding of sugar molecules by any distinct secondary hydroxyl, the others must be protected by easily cleavable protective groups.

The classical route for binding monosaccharides via their semiacetal hydroxyl starts with the synthesis of a glucoside bearing a terminal reactive group [usually aromatic amino (143)]. Often, thioglucosides are used to prevent any possible action of glucosidases (144,145).

Another possible way is to attach the sugar onto an amino-bearing carrier by reductive alkylation. 2-Amino-2-deoxysugars may be immobilized directly by their amino group, whose high reactivity practically excludes binding via hydroxyls. If spacer manipulation is needed, a variety of ways are available (146).

If the enzyme to be purified attacks any hydroxyl than the glycosidic hydroxyl, use of an appropriate deoxysugar may cause stronger binding of the enzyme to the ligand and avoid enzymatic modification of the sorbent. Otherwise, the enzyme may also be damaged (e.g., sugar oxidases can produce large amounts of H_2O_2).

It may be very difficult to direct the reaction of oligosaccharide immobilization to the right residue except when it is much more reactive than others. This problem may be overcome by using glycoproteins instead of pure sugars. Thus mucin, α_1-acid glycoprotein (147), ovalbumin (148), ovomucoid (149), thyroglobulin (150), fetuin (151,152), and lactalbumin (153) are used for purification of numerous lectins, sugar transferases, hydrolases, and so on.

b. Polysaccharides

Immobilized natural polysaccharides are used for the purification of a wide variety of proteins from lectins to restriction endonucleases, the most powerful tool being matrix-bound heparin. Selected examples of immobilized heparin used in affinity chromatography:

RNA polymerase (154–158)
Casein kinase (159)
Coagulation factor IX (160)
Collagenase (161,162)
Lipoprotein lipase (163)
Antithrombin III (164)
Trehalosephosphate synthetase (165)
Reverse transcriptase (166)
Tumor growth factor (167)
Phospholipase C (168)

The difference between the binding capacities of end-linked and multipoint-attached heparin is dramatic. (See Ref. 169 for an exhaustive review of heparin immobilization methods.) α- (170) or β-Cyclodextrins (171) are not degraded by enzymes and may be used instead of linear starch [the latter was first used in 1930 (172)].

A reversed order of immobilization (i.e., activation of the polysaccharide with, for example, cyanogen bromide followed by reaction with an amine-containing carrier) may be useful for labile polysaccharides (173). Sometimes polysaccharides need not be immobilized but may be used cross-linked or per se, if insoluble. Some following limited acid hydrolysis may significantly improve the protein binding. The use of a polyglucan–Sephadex (174,175) and a polygalactan–Sepharose (176) for the purification of glucose- and galactose-binding proteins (e.g., lectins), respectively, is essential. But chitin [poly(*N*-acetylglucosamin)] (177), chitosan (polyglucosamin), cross-linked guarana gum (galactomannan) (178), gum arabic (179), and starch (180), are among compounds also used. Soluble polysaccharides can serve as spacers to enhance the binding properties of many affinity sorbents (181,182).

4. Amino Acids and Peptides

a. Amino Acids

Amino acids and their derivatives are used primarily for the purification of numerous proteases and amino acid–metabolizing enzymes (Figure 3). They may also mimic an inhibitor (e.g., lysine), immobilized by means of its α-amino group, which acts like a plasmin(ogen) inhibitor—ε-aminocaproic acid. Sometimes derivatization of the carboxylate may be necessary. Sorption is often highly sensitive to pH, salt, temperature, and other conditions more easily modified than the ligand itself. [Even *p*-nitroanilide substrates can be used as ligands (183).]

D-Amino acid derivates are also used to avoid enzymatic hydrolysis (e.g., Ref. 184 and 185), and sometimes racemates are used. Following are specific examples of amino acid immobilization:

1. At moderate pH values lysine is bound through its α- (pK = 8.95) rather than its ε-amino group (pK = 10.53), but addition of Cu^{2+} ions blocks the first reaction and allows immobilization of the lysine through the ε group.
2. If an amino acid must be immobilized through its amino group by means of a spacer, lacking any oxygen atoms, it is treated with nitrous acid in a saturated

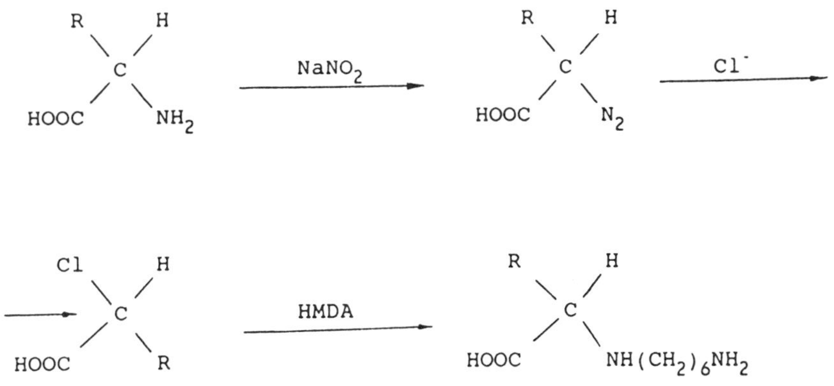

$$\mathrm{R(H)C(HOOC)(NH_2)} \xrightarrow{NaNO_2} \mathrm{R(H)C(HOOC)(N_2)} \xrightarrow{Cl^-} \mathrm{Cl(H)C(HOOC)(R)} \xrightarrow{HMDA} \mathrm{R(H)C(HOOC)(NH(CH_2)_6NH_2)}$$

Figure 3 Synthesis of N-alkyl-substituted amino acids.

solution of sodium chloride (an analog of the Sandmeyer reaction), and the α-D-chloro acid obtained is then reacted with hexamethylenediamine. Both reactions are of *s*$_N$2 type, and two subsequent Walden rearrangements return the starting anomer (186).

Selected examples of the use of immobilized amino acids in affinity chromatography are listed in Table 4.

b. Peptides

In addition to synthesis, two good sources of oligopeptides are proteolytic degradation of a protein (e.g., tryptic hydrolysate of protamine for trypsin purification (209) and microbially derived protease inhibitors). Cyclic octapeptides such as gramicidin (210) and bacitracin (211,212) offer a wide spectrum of inhibitory activity and are therefore used for the preparative purification of different (especially carboxylic) proteolytic enzymes; *Actinomyces*-produced pepstatin is used for purification of chymosin and pepsin (213). Synthetic peptide

Table 4 Examples of Immobilized Amino Acid Use in Affinity Chromatography

Protein	Amino acid	Ref.
Serum proteases	Arg-*O*-Hex	187
Threonine deaminase *E. coli*	Leu, Ile	188
Carboxypeptidase A and B	Arg	189
Carboxypeptidase B	D-Trp	190
Aspartase	Asp	191
Chymotrypsin	Trp-*O*-me	192
Asparaginase	Asn	193
α-Chymotrypsin	Cbz-Phe	194
Subtilisin	Cbz-Phe	194
Metalloendopeptidase	Cbz-Phe	194
Thrombin	Lys	195
Glutamine synthetase	Glu	196
Aromatic aminotransferase	Phe	197
Various proteases	Cbz-D-Phe	198
Skin esteroprotease	Arg-*O*-Me	199
Phenylalanine ammonia-lyase	Phe	200
α-Isopropylmalate synthase	Leu	201
Galactosyltransferase	Norleu	202
Asparaginase	Asp	203
Plasminogen	Lys	204,205
Anthranilate synthetase	Trp	206
Fibronectin	Arg	207
Prekallikrein	Arg	208

aldehydes have been used in the affinity chromatography of trypsin-like serine endopeptidases and cathepsin C (214–216) and for semicarbazone-blocked aldehydes for cysteine endopeptidases (217,218).

5. Nucleotides

The main body of available data concerns 5′-adenosine monophosphate (AMP), immobilized by various methods (Figure 4). The significance of the immobilization site should be stressed again (e.g., acetate kinase adsorbs onto N^6- and C^8- but not onto ribose-bound AMP (219).

The amino group *N*^{6}-aminohexyl-5′-AMP is not reactive enough; therefore, synthesis usually begins from the 6-mercaptopurine riboside 5′-phosphate (a metabolite of *Brevibacterium ammoniagenes*), which is converted into the desired ligand by heating with an excess of hexamethylenediamine (220). Similar but not identical results may be obtained by prior attack on the N^1 of adenine by any alkylating agent (iodoacetic acid, aziridine, epoxides, etc.), followed by the Dimroth rearrangement (Figure 5). When reacted with epoxy-activated Sepharose at pH 5 to 6, adenosine nucleotides are also bound in the N^1 position. This followed by alkaline treatment causes Dimroth rearrangement and yields N^6-substituted derivative (221). Direct coupling at alkaline pH is believed to lead to substitution at sugar hydroxyls (222).

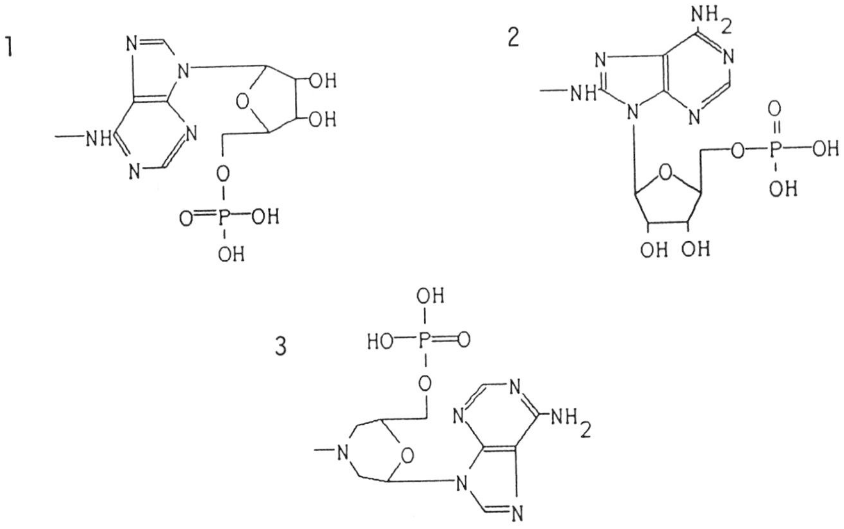

Figure 4 5′-AMP immobilized via 1-N^6, 2-C^8, and 3-ribose hydroxyls.

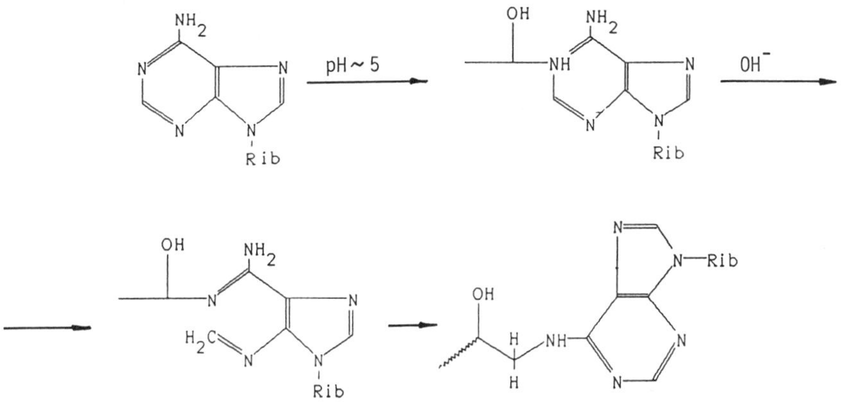

Figure 5 Immobilization of AMP onto epoxy-activated gel via Dimroth rearrangement.

The C^8 carbon atom of adenine residue can be brominated and then reacted with hexamethylenediamine, for example, for subsequent direct immobilization. The ribose residue may be oxidized to the corresponding dialdehyde using periodate followed by immobilization onto the matrix via reductive alkylation. Using adipic acid dihydrazide as a spacer avoids the appearance of a positively charged product (223).

It is possible to attach 5′-AMP to the carrier via its phosphate group, but this sorbent is not widely used. Binding of adenine nucleotides onto glutaraldehyde-activated Sepharose yields high coenzyme activity; the binding mechanism is unknown (224).

Some additional problems arise with nicotinamide-adenine dinucleotide phosphate (NADP). Since the corresponding 6-mercapto derivative is not available, the synthesis must be either started from condensation of N^6(6-aminohexyl)-5′-AMP with nicotinamide mononucleotide (225), or by using Dimroth rearrangement after carboxymethylation in the N^1 position (226). Other N^1-alkylating agents [aziridines (227), epoxides (228)] may also be used. Coupling via C^8 is possible by the same method as that used for 5′-AMP, with one difference: After bromination to the C^8 position, the nucleotide must be reduced, and after introduction of the arm, it must be oxidized again.

Binding NAD via ribose hydroxyls cannot give a predefined result since it is impossible to direct the reaction to one distinct sugar residue. Nevertheless, ribose-bound NAD is commercially available and widely used. NADP is treated in exactly the same manner as NAD in all immobilization procedures.

The oxidation status of dinucleotides may change during affinity chromatography, but there are some ways to overcome the problem. Pharmacia suggests using 5′-AMP instead of NAD and 2′,5′-adenosine diphosphate (ADP) instead of NADP, and NAD and 5′-AMP may usually be replaced by their sterical analog, Cibacron Blue F3GA dye, and NADP by Procion red HE3B dye. The immobilization technique of nucleotides other than cofactors has been less well investigated; they are usually immobilized via ribose hydroxyls.

If an enzyme is used to distinguish between DNA nad RNA, it can sometimes also distinguish between nucleotides and deoxyribonucleotides. RNase contamination is removed from DNase by 5′-uridylic acid (UMP) (229) or uridine 5′-diphosphate (UDP) (230), and RNase is purified using 5′-guanylic acid (GMP) (231) and cytidine monophosphate (CMP) (232).

6. Low-Molecular-Weight Specialty Ligands

Searching for a drug receptor or its separation is a common problem, but several drugs cannot be immobilized (e.g., among a distinct class of neuroleptics, only Haloperidol contains a reactive group). On the other hand, one can find a surplus of inhibitors for different enzymes among drugs, many of which are immobilizable and work in affinity chromatography (Table 5).

A general approach to specialty ligand design is that replacement of the aliphatic hydrocarbon chain in the ligand with an aromatic chain often enhances binding. Examples are:

Table 5 Examples of Drug Use in Affinity Chromatography

Drug	Protein	Ref.
Amethopterin	Acetyl-COA:*N*-acetyltransferase	233
Methotrexate	Tetrahydrofolate dehydrogenase	234
Chloramphenicol	Chloramphenicol acetyltransferase	235
Penicillin	D-Alanine carboxypeptidase	236
Kanamycin	Aminoglycoside 3′-phosphotransferase	237
Sulfanilamid	Carbonic anhydrase	238
Rifamycin	RNA polymerase	239
Edrophonium	Acetylcholinesterase	240
Ampicillin	β-Lactamase	241
Methicillin	β-Lactamase	242
Lumazine	Riboflavin synthase	243
Gentamicin	Gentamicin acetyltransferase I	244
Perphenasine	Calmodulin	245
Cephalosporin C	Penicillinase	246

1. *m*-Aminophenyltrimethylammonium on CH-Sepharose is used for binding acetylcholine esterase (247). (The distance between quaternary ammonium and carbonyl groups is the single similarity with the natural substrate molecule.)
2. Phosphatases are purified using aromatic but not aliphatic phosphonates or arsonates, and azo coupling of the ligand via another aromatic molecule (e.g., tyrosine or histidine) adds enzyme-binding capacity (248,249).
3. *p*-Aminophenyl spacer is widely used for the immobilization of carbohydrates (250,251).

Easily cleavable bonds in the ligands should be avoided whenever possible: Phosphonates but not phosphate esters are used for the purification of phosphatases, other esterase ligands contain an amide bond instead of an ester, thioglucosides are used instead of glucosides, and so on.

B. Nonbiospecific Ligands

1. Binding Molecules via SH Group

Many proteins contain more-or-less exposed free SH groups of cysteine which can form covalent bounds with heavy metal–containing groups or can form disulfides with another mercaptans. Organomercurial sorbents are synthesized by immobilization of p-acetoxymercurianiline (252), *p*-chloro- or hydroxymercybenzoic acid (253), Mersalyl ([(*O*-(e-hydroxymercury)-2-methoxypropyl)carbonyl] phenoxyacetic acid) (254), or by *p*-hydroxymercuryphenylsulfenyl chloride treatment of aminoethyl agarose (255). Covalently bound proteins are eluted either by a mercury salt (protein becomes mercurated and needs regeneration with, for example, β-mercaptoethanol) or by a competitive thiol compound (the column needs regeneration with a mercury salt). The high binding capacity of organomercurial sorbents often makes it possible to bind all proteins from complex mixtures except the desired non-SH protein (Figure 6). The sorbent can barely discriminate between different SH proteins. Following are examples of use:

Myoglobin (a non-SH protein) (255)
Papain (256)
Ficin (257)
S-Adenoxylmethionine decarboxylase (258)
Adenylate cyclase (259)
Creatine phosphokinase (260)

Three types of SH-containing affinity sorbents are commercially available and several different ones may be synthesized. The column must be activated for protein binding–i.e., converted into mixed disulfide by treatment with 2,2′-dipyridyldisulfide or 5,5′-dithiobisnitrobenzoic acids. Thiopropyl-Sepharose is prepared by reaction of epichlorohydrine-activated matrix with sodium thiosulfate, followed by reduction with dithiotreitol (261). Activated thiol-Sepharose is nothing

Figure 6 Selected organomercurial affinity ligands: (1) *p*-hydroxymercurybenzoic acid; (2) *p*-acetoxymercuryaniline; (3) Mersalyl.

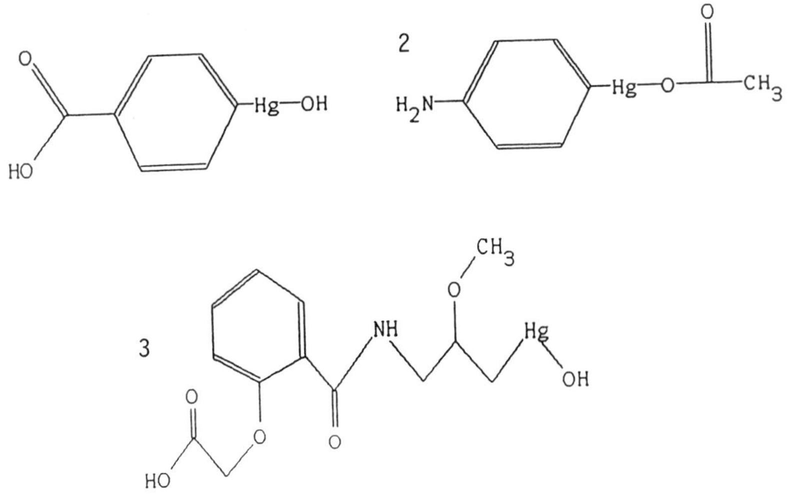

but glutathione immobilized onto a CNBr-activated matrix. *N*-Acetylhomocysteine is a derivative of aminoethylcellulose (Sigma). 5,5′-Dithiobisnitrobenzoic acid may be immobilized via its carboxylate end and does not require activation (263).

Commercially available SH sorbents are activated with 2,2′-dipyridyldisulfide, but in laboratory practice 5,5′-dithiobisnitrobenzoic acid (Ellmann's reagent) is more popular, because it forms a colored compound when dissociated from the ligand and permits visual control of the process (Figure 7).

If the selectivity of listed SH gels is insufficient, one can manipulate the spacer length and ligand structure–e.g., IgG κ chains activated with Ellmann's reagent were used for purification of α_1-antitrypsin (263). Using different leaving groups adds additional selectivity (264). Thiol sorbents are also used for separation of mercurated nucleic acids (265,266). Some immobilized arsenical compounds also bind thiols (267).

2. *Binding cis-Diols: Boronate Sorbents*

Carrier-bound *m*-aminophenylboronic acid forms borate esters with vicinal diols and makes possible separation of ribonucleotides from deoxyribonucleotides, glucosylated proteins from nonglycosylated proteins, and so on (Figure 8). It is also used for purification of serine proteases such as trypsin, chymotrypsin (268), subtilisins (269), and lipase (270).

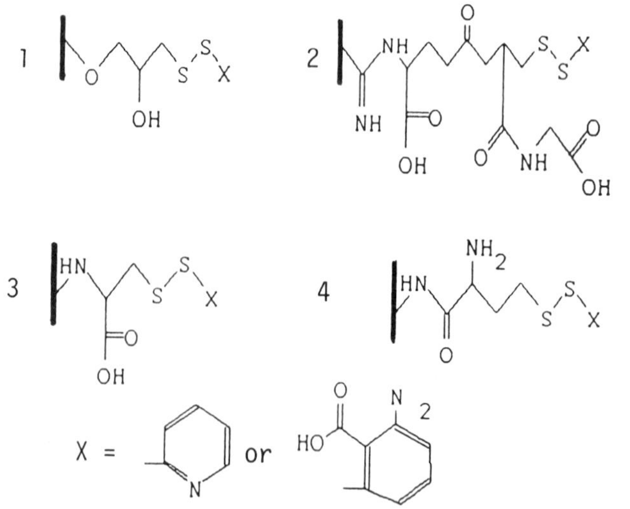

Figure 7 Sorbents for thiol-disulfide interchange: (1) thiopropyl; (2) glutathione; (3) cysteine (immobilized via amino group); and (4) sorbent obtained by treatment of amino-bearing carrier with homocysteine thiolactone.

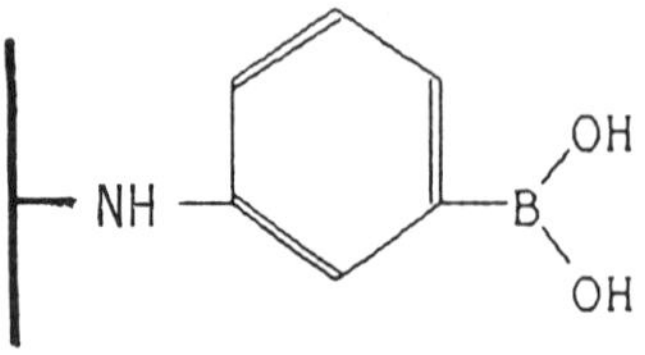

Figure 8 *m*-Aminophenylboronic acid affinity sorbent.

3. *Hydrophobic Sorbents*

If a protein is salted out in the presence of a somewhat hydrophobic material, the necessary salt concentration for precipitation is lower than in the free solution, and the selectivity is increased. Amphiphilic gels (e.g., Sepharose) can be used (271) [or Eupergit C 30N Diol for HPLC (272)], but if some hydrophobic groups are introduced into the matrix, the requisite salt concentration is diminished accordingly, even down to zero. In the latter case the proteins are eluted with the solution of ethylene glycol or isopropyl alcohol.

If a solely hydrophobic interaction is needed, sorbents are prepared using corresponding alkyl (or aryl)glycidyl ether (273) [synthesis: (274)], phenyl- and octyl-Sepharoses, alkyl (i.e., neopentyl)-Superose, butyl-Fractogel, and TSK-Gel Butyl (for HPLC).

Numerous mono- and diamines were immobilized onto CNBr-activated Sepharose for hydrophobic interaction chromatography, but positively charged isourea bonds and terminal amino groups interact with the sorption. Moreover, immobilized diamines bind specifically mono- (275,276) and diamine oxidases (277), aminotransferases (278), and so on; aminoethyl-Sepharose is used for purification of ceruloplasmin (279). Highly specific chymotrypsin ligand–phenylbutylamine (280) is also used as an hydrophobic group (281).

Noncharged hydrophobic agarose derivatives may also be prepared by immobilization of alkyl sulfides onto epoxy-activated gel (282) or aliphatic alcohols onto glycidoxypropyltrimethoxysilane-activated agarose (283). Immobilized polyethylene oxide was used for mild hydrophobic interaction chromatography of cells (284) and proteins (285).

4. *Chelating Sorbents*

Chelator-bound metal ions bind proteins via their exposed chelating groups. Usually, matrix-bound iminodiacetic acid (IDA) is utilized for this technique (286), but tris(carboxymethyl)ethylenediamine (TMED) (287), immobilized EDTA (288,289), and hydroxamic acids (290) have also been used. The strongest binding is achieved by immobilized Zn^{2+} and Cu^{2+}, but Ni^{2+}, Co^{3+}, and Fe^{3+} are also widely used.

Proteins are usually eluted by decreasing the pH, or by using a weak competitive ligand (e.g., ammonium salts), but ethylenediaminetetracetic acid (EDTA) solutions may be needed if the binding is very strong. In the latter case metal ions are coeluted and cause the product contamination. It was shown that the order of elution of serum proteins generally did not depend on either metal ion or buffer choice (291).

The properties of IDA-bound Fe^{3+} are unique: It binds primarily phosphoproteins and phosphopeptides (292). Selected examples of use include the following:

Cu^{2+}: Lactoferrin (293)
Interferons (294,295)
Superoxide dismutase (296–298)

Zn^{2+}: α_1-Proteinase inhibitor (299)
α_2-Macroglobulin (299)
α_2-SH glycoprotein (300)
Interferon (301)
Rabbit bone metalloproteinase (302)
Plasminogen activator (303)

Fe^{3+}: Lactate dehydrogenase (304)
Glycogen phosphorylase (304)

5. *Charge Transfer Chromatography*

The charge transfer branch of affinity chromatography is based on the interaction between immobilized and dissolved aromatic compounds, leading to the formation of a molecular complex (305). Aromatic π–π interactions are often accompanied by several side effects: electrostatic, hydrophobic, and others. The binding of molecules by charge transfer is rather weak (306). The most commonly used ligands are acriflavine, acridine yellow, trityl group, pentachlorophenol, and malachite green. Immobilized acriflavine enables us to separate nucleotides, oligonucleotides (order of elution: C < T < U < G < I < A) (307), single-chain nucleic acids (adsorbed) from double-chain nucleic acids (308), and a wide variety of aromatic compounds.

6. *Thiophilic Interaction Chromatography*

Thiophilic interaction chromatography is becoming one most powerful tools for the purification of antibodies, due to its high specificity and low process cost (especially when compared with protein A or G) (309). The presence of a sulfone group and an electron donor atom (not necessarily sulfur) is mandatory in these positions, just as they are in Structure 1, but the mechanism of the interaction is still not exactly known. The sorbent also binds α_2-macroglobulin, lentil lectin, carboxypeptidase A, SBTI (310), and peroxidase (311); desorption is achieved by omitting $(NH_4)_2SO_4$ from the buffer. The method was employed in the FPLC mode using Toyopearl (67) and in the HPLC mode using silica (312). In the last case three sulfur atom–containing groups were used, and elution conditions were different from those of the standard T-gel (Structure 2).

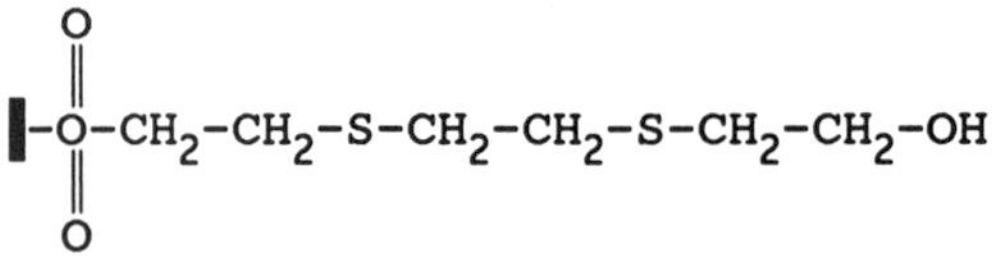

$$-O(=O)_2-CH_2-CH_2-S-CH_2-CH_2-S-CH_2-CH_2-OH$$

Structure 1 T-gel adsorbent.

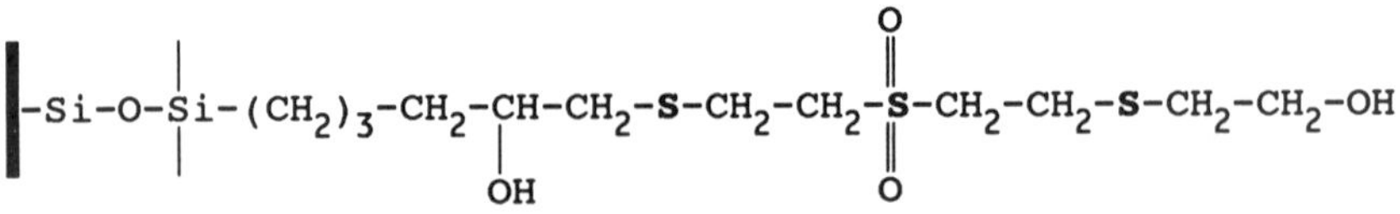

$$-Si-O-Si-(CH_2)_3-CH_2-CH(OH)-CH_2-\mathbf{S}-CH_2-CH_2-\mathbf{S}(=O)_2-CH_2-CH_2-\mathbf{S}-CH_2-CH_2-OH$$

Structure 2 Three-sulfur thiophilic adsorbent.

7. Immobilized Dyes

In one study pyruvate kinase was gel filtered in the presence of a void volume marker, blue dextran (dextran–Cibacron Blue F3GA dye conjugate), and they were eluted together in the void volume (313). The same dye was then immobilized onto Sephadex (314), and blue dextran onto Sepharose (315), and the Sepharose was stained blue (316). It was further shown that this compound mimics the structure of NAD (317), but several proteins bind very strongly to blue sorbents despite the fact that they have no relations to NAD. Strong binding of human serum albumin (but not that of other animals) is explained by the fact that Cibacron Blue dye may also mimic bilirubin (318). Later, different dyes were tested, but the experimental data obtained could not be systematized because generally the chemical structure of the dyes used was not known. Reactive Red 120 (Procion red HE3B) binds primarily NADP-requiring enzymes (319). Examples of the use of immobilized dyes are given in Table 6.

Cibacron Blue F3GA is still often used as immobilized blue dextran, and also in two combinations: coupled to DEAE (400) and CM ion exchangers (401) for use in IgG purification. A method for the preparation of "homemade" triazine dyes has also been published (402).

8. Molecular Imprinting

When gel monomers are polymerized in the presence of the molecules to be separated, hollows are formed according to the size of the molecules. The substance may be either bound to monomers by cleavable covalent bonds (403), or a couple of acrylic monomers, for example, bearing different ionic and hydrophobic groups are polymerized around the print molecule (404). The method is only applicable to the chromatography of low-molecular-weight compounds.

III. IMMOBILIZATION METHODS

A. Activation Methods

Some 10 to 20 methods for support activation are widely used and cover all possible needs, while usually the selectivity of the sorbent is slightly affected by the choice of activation method, but some ligands are very susceptible. Cyanogen bromide (Figure 9) is the first widespread activation method of agarose gels (405), but associated with several problems (according to personal communications from Kabi Pharmacia, the demands for cyanogen bromide–activated agarose are increasing). CNBr is highly toxic, cationic charges are formed on the gel (406,407), and ligand is slowly hydrolyzed under alkaline pH (408,409) or during elution with, for example, sodium thiocyanate (410). However, the method has been investigated thoroughly and is well documented. It is suitable for all hydroxyl-containing support materials. Reaction is carried out either in aqueous

Table 6 Examples of Immobilized Dye Use in Affinity Chromatography
A. Cibacron Blue F3GA

Protein	Ref.
α-Fetoprotein	320
α_2-Macroglobulin	321
15-Hydroxyprostaglandin dehydrogenase	322
2-Oxoaldehyde dehydrogenase	323
3(17)β-Hydroxysteroid dehydrogenase	324
6-Phosphogluconate dehydrogenase	325
Acyl-CoA:glycine *N*-acyltransferase	326
Acyl-acyl carrier protein synthetase	327
Adenylate cyclase	328
Adenylate kinase	325
Adenylosuccinate synthetase	329
Alcohol dehydrogenase	330
Aldehyde reductase	331
Aldehyde dehydrogenase	332
Aldose reductase	333,334
Aminoacyl tRNA synthetases	335
Aspartic proteinase	336
Biphosphoglycerate synthase	337
Butyrylcholinesterase	R. Toomik, personal communication
cAMP phosphodiesterase	338
Carbamyl phosphate synthetase I	339
Choline acetyltransferase	340
Chymosin	341
Citrate synthase	342
CMP kinase	343
CoA transferase	344
Creatine kinase	330
Cyclic nucleotide phosphodiesterase	345,346
D-3-phosphoglycerate dehydrogenase	347
Deoxy-D-arabinoheptulosonate-7-phosphate synthetase	348
Deoxycytidine kinase	349
Dihydrofolate reductase	350
Dihydropterine oxidase	351
DNA polymerase	325
Fructose diphosphatase	325
Fumarase	352
Glucose dehydrogenase	353
Glucose-6-phosophate dehydrogenase	330
Glutamate dehydrogenase	354
Glutamate synthase	355
Glutamine phosphoribosylpyrophosphate amidotransferase	356
Glyceraldehyde-3-phosphate dehydrogenase	330

Table 6 (Continued)

Protein	Ref.
Glycerol phosphate dehydrogenase	357
Glyoxalase II	358
GMP reductase	359
Guanylate cyclase	360
Guanylate kinase	361
Haemopexin	362
Hexokinase	330
HMG-CoA reductase	363
Interferon	364,365
Isocitrate dehydrogenase	366
Lactate dehydrogenase	367
Lecithin:cholesterol acyl transferase	368
Luciferase	369
Malate dehydrogenase	325
Methylenetetrahydrofolate reductase	370
myo-Inositol-1-phosphate synthase	371
NAD kinase	372
NAD(P)H:nitrate reductase	373,374
NAD-glycohydrolase	375
NADH: nitrate reductase	376
NADPH:nitrate reductase	377
Nucleoside diphosphate kinase	343
Orotate phosphoribosyltransferase	378
Orotidylate decarboxylase	378
Phenol sulfotransferase	379
Phosphodiesterase	380
Phosphofructokinase	325
Phosphoglycerate kinase	325
Phosphoglycerate mutase	325
Polynucleotide kinase	381
Polynucleotide phosphorylase	382
Prolyl hydrolase	383
Protein kinase	384
Pyruvate kinase	385
Pyrivate kinase	330
Serine hydroxymethyltransferase	386
Serine transhydroxymethylase	387
Squalene epoxidase	388
Terminal deoxynucleotidyl transferase	389
Thymidine kinase	361
Thymidylate synthetase	390
Tyrosine phenol-lyase	391
Uridine kinase	391

Table 6 (Continued)
B. Other dyes

Enzyme	Dye	Ref.
Carboxypeptidase G	Red H-8BN	392
Glycerokinase	Blue MX-3B	392
3-Hydroxybutyrate dehydrogenase	Red H-3B	392
	Blue MX-4GD	392
Malate dehydrogenase	Red H-3B	392
	Blue MX-4GD	392
Butyrylcholinesterase	Red HE-7B	392
Alkaline phosphatase	Reactive Yellow 13	393
Plasminogen	Red HE 3B	394
Porphobilinogen deaminase	Reactive Red 120	395
Malate dehydrogenase	Reactive Orange[a]	396
Lactate dehydrogenase	Reactive Orange[a]	396
Citrate synthetase	Reactive Orange[a]	396
Carnitine acetyltransferase	Reactive Orange[a]	396
Pyruvate kinase*[b]	Reactive Orange[a]	396
Adenylate kinase*	Reactive Orange[a]	396
Phosphoglycerate kinase*	Reactive Orange[a]	396
Lactate dehydrogenase	Scarlet MX-G[c]	397
Glyoxalate dehydrogenase	Red P4B	398
Alcohol dehydrogenase	Congo Red	399

[a]Matrex Gel Orange A.
[b]Binds in the presence of 10 mM $MgCl_2$.
[c]Bound via six-atom spacer arm.

media* or in an acetone–water mixture using triethylamine as catalyst (i.e., cyanotransfer)* (411). The reaction in aqueous alkali leads to the formation of cyanate ester, cyclic imidocarbonates, and nonreactive carbamates; cyano transfer yields only cyanate ester (Figure 10). Stable cyanylating reagents [e.g., *N*-cyano-triethylammonium tetrafluoroborate, 1-cyano-4-dimethylaminopyridinium tetra-fluoroborate (412), or *p*-nitrophenylcyanate (413)] make it possible to obtain active cyanate esters without the use of hazardous CNBr.

Bisepoxides (414), usually 1,4-butanediol diglycidyl ether*, are applicable to all alkali-resistant nonshrinking hydroxyl-containing polymers. This method introduces charge-free hydrophilic spacer arm and cross-linkages during the activation process. The linkage between matrix ligands is highly stable and never causes problems. All amino-, hydroxyl-, and thiol-containing compounds can be immobilized, so this is one of the most universal of all activation methods. Despite

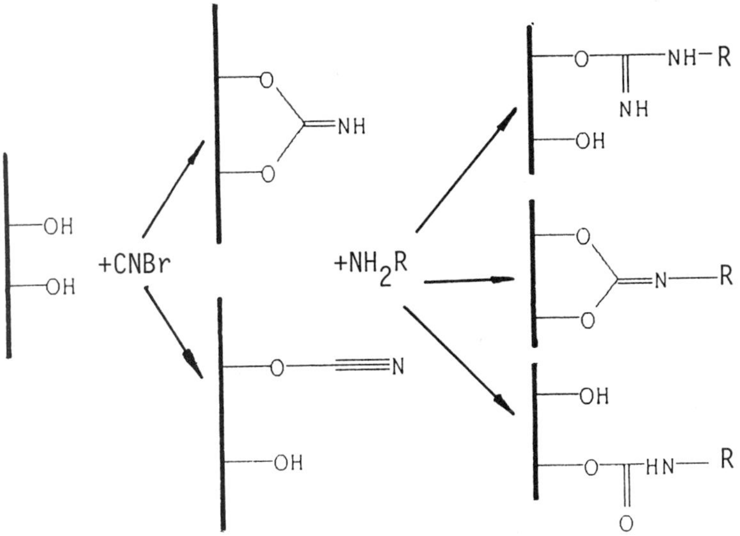

Figure 9 Matrix activation with cyanogen bromide and subsequent ligand immobilization.

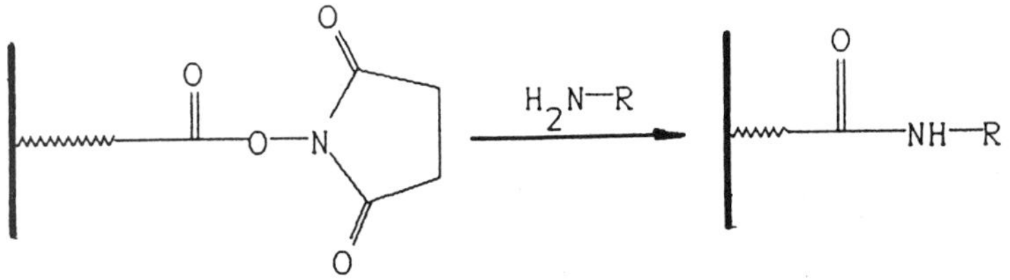

Figure 10 Ligand immobilization onto active ester gel.

widespread opinion to the contrary, epoxy-activated gels can bind proteins under mild conditions with high recovery (415).

Epichlorohydrine* (Figure 11) (416) is one of the best activation methods for industrial purposes, due to the moderate toxicity and low price, simultaneous cross-linking, and extreme stability of the sorbent. Some results indicate advantages from the use of epibromohydrine (417), but this reagent is not widely used.

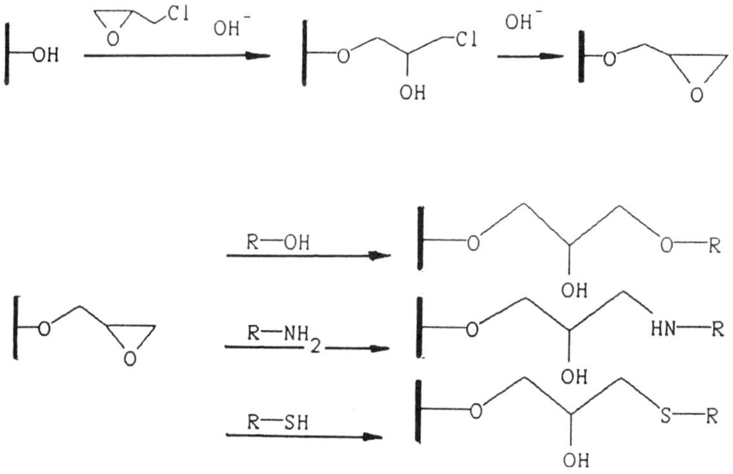

Figure 11 Matrix activation with epichlorohydrine and subsequent ligand immobilization.

Divinylsulfone (DVS) (418) is useful with alkali-resistant nonshrinking hydroxyl-containing polymers, and DVS-activated agarose provides extreme rigidity and can be used for HPLC. Some contradictive data about the influence of DVS on gel porosity and alkali stability have been published, but our experience has been that the porosity is not decreased and the ligand does not leach in the pH ranges commonly used. Ligands can be immobilized via their SH, NH_2, and OH groups. DVS is toxic and expensive.

Cyanuric chloride (Figure 12) is used in alkali-resistant hydroxyl-containing carriers. The method yields a highly stable bond between the matrix and the immobilized molecule. The main advantage is the possibility of immobilization of molecules at slightly acidic pH [important for proteins, which cannot tolerate even neutral media (e.g., pepsin)]. The triazine group exhibits both anionic and π–π interaction capabilities, and the reagent is toxic. The method is often used for cellulose activation* (419): for example, for the preparation of DNA–cellulose (420), 2,4,6-trifluoro-6-chloropyrimidine may be used instead of cyanuric chloride (421).

Sodium periodate (Figure 13) is useful with all *cis*-diol–containing polymers, including diol-coated silica. The immobilization of proteins onto cross-linked or soluble dextran gives very good results, but the situation with agarose is unclear. It need not react at all, because it contains no such groups, but use of enormous periodate quantities* yields a low degree of substitution (422). On the other hand, cross-linked Sepharose contains vicinal diol groups due to incomplete

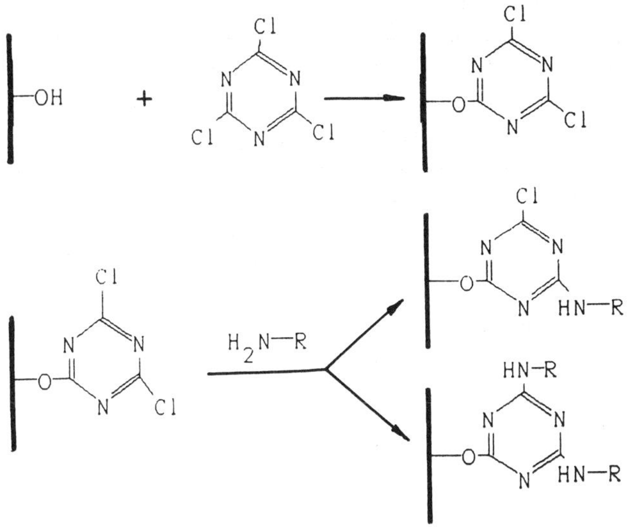

Figure 12 Matrix activation with cyanuric chloride and subsequent ligand immobilization.

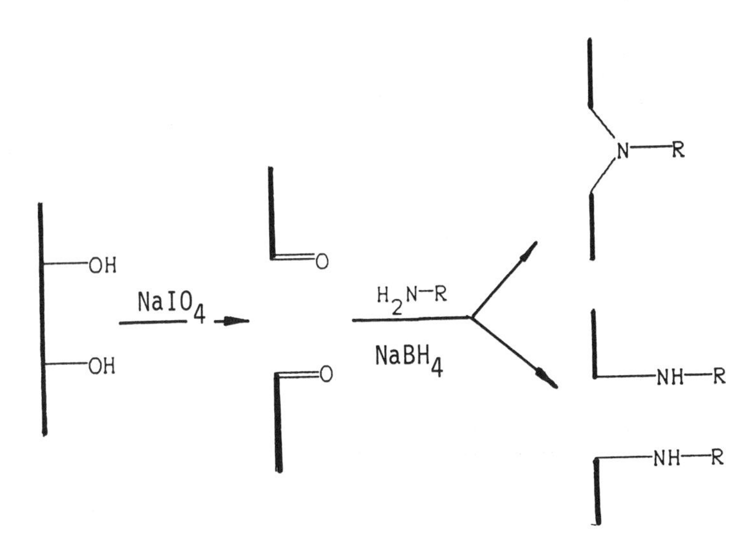

Figure 13 Matrix activation with sodium periodate and subsequent ligand immobilization.

cross-linking with dibromopropanol and therefore can be periodate-oxidized (423), and the number of reactive groups can be increased further by etherification with glycidol (424). Mild acid hydrolysis also deliberates some terminal aldehyde groups of agarose and enables us to immobilize amino compounds (425).

The binding of amines to periodate-activated matrices by Schiff base formation is reversible, and to obtain a stable product this intermediate should be reduced with $NaBH_4$ or preferably with $NaCNBH_3$ (the latter does not react with aldehydes at or above pH 5). In some cases ascorbate (426) and borane–amine complexes are used for reduction (427,428). Attempts were made to introduce aldehyde groups into agarose molecules through galactose oxidase treatment, but low activity levels were obtained (429).

Carbonylating reagents [carbonyldiimidazole (Figure 14), *p*-nitrophenylchloroformate, *N*-hydroxysuccinimide chloroformate, and trichlorophenylchloroformate) (430)] belong to synthetic analogs of phosgene; that is, the final product has the structure

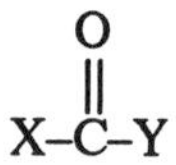

$$X{-}\overset{\overset{\displaystyle O}{\|}}{C}{-}Y$$

Phosgene cannot be used, due to its extreme toxicity as well as the instability of the activated matrix, 1,1′-Carbonyldiimidazole is the most common choice, but if

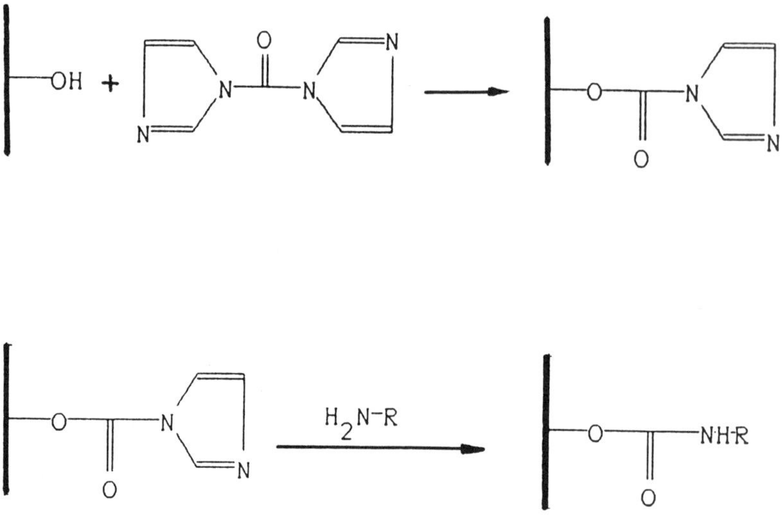

Figure 14 Matrix activation with carbonyldiimidazole and subsequent ligand immobilization.

higher reactivity of the matrix is needed, 1,1′-carbonyldi-1,2,4-triasole may be used (431). *p*-Nitrophenylchloroformate enables to follow the immobilization reaction visually, since the leaving *p*-nitrophenolate anion is yellow in color.

This method is applicable to all hydroxyl-containing matrices that can be dehydrated with organic solvents (432). The activated matrix is stable at slightly alkaline pH, but reacts rapidly with primary amines to form a stable noncharged bond.

Benzoquinone (Figure 15) is used in hydroxyl-containing supports; it yields strongly colored, very reactive gels that exhibit some nonspecific sorption due to the π–π interactions (433).

Sulfonyl halides [Figure 16; e.g., tosyl (434) and tresyl (trifluoroethylsulfonyl) (435) chlorides] make it possible to bind molecules under very mild conditions. The method is applicable to dehydratable hydroxyl-containing carriers. Stable bonds are formed, and no significant disadvantages of the method are known, except the absence of a spacer arm. A tresyl-activated matrix is more reactive, but tosyl chloride is often used, due to its better availability, lower price, and nontoxicity.

Glutaraldehyde activation is used with amine- or amide-containing support materials (436). The actual structure of the reagent is unclear: it may react as a monomer (437), as a polymeric product of aldol condensation, or as a trimeric product (claimed to be the most probable). Therefore, different binding

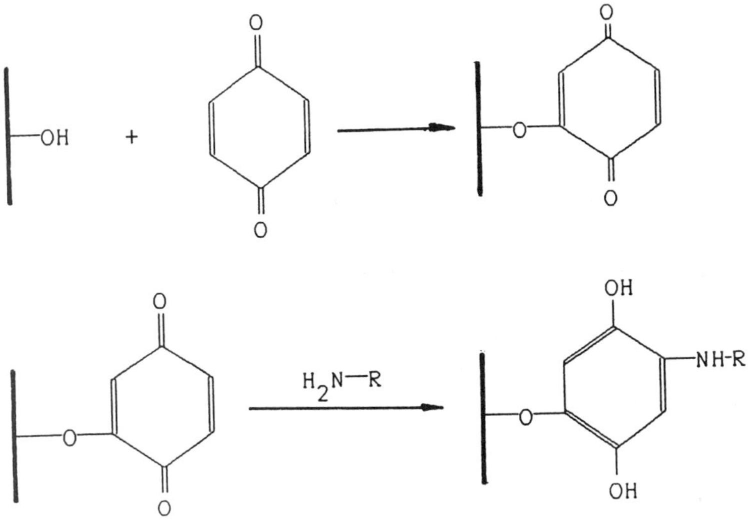

Figure 15 Matrix activation with benzoquinone and subsequent ligand immobilization.

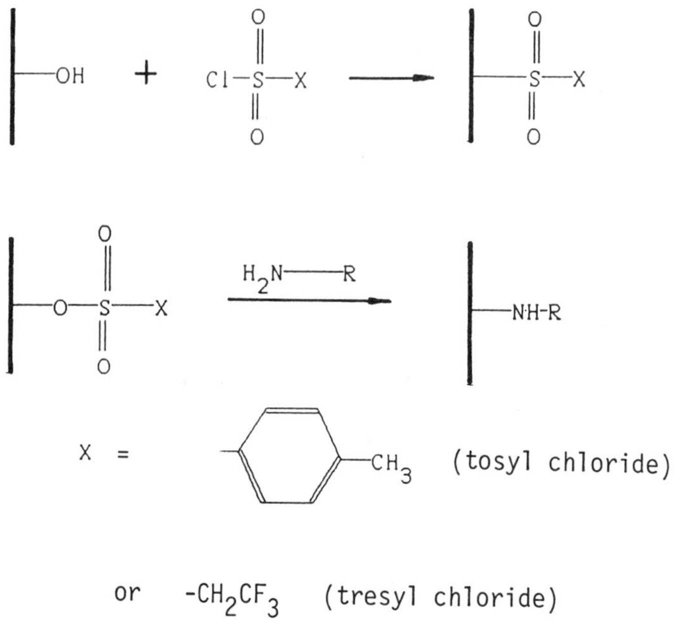

—OH + Cl—SO_2—X ⟶ —SO_2—X

—O—SO_2—X $\xrightarrow{H_2N—R}$ —NH-R

X = —C_6H_4—CH_3 (tosyl chloride)

or $-CH_2CF_3$ (tresyl chloride)

Figure 16 Matrix activation with sulfonyl halides and subsequent ligand immobilization.

mechanisms are also possible: a primary amine may react with an activated double bond (438) or by Schiff base formation (Figure 17). The high stability of glutaraldehyde-immobilized proteins obtained without a reduction step necessitates the first reaction route. Aldehyde-containing gels were also prepared by immobilization of 4-butyraldehyde diethyl acetal onto CNBr-activated Sepharose followed by acid hydrolysis of protective groups (439).

Direct substitution at the amide group is used for polyacrylamide activation. If it is impossible to treat the gel with pure amine, ethylene glycol is recommended as a solvent (440). Hydrazine reacts with amide groups through the formation of hydrazides, which are then converted into very reactive acyl azides by treatment with nitrous acid (441). The method is applicable preferably for immobilization of small ligands.

Carboxymethylation is not activation in the conventional sense, but is used as a starting step for spacer arm extension and immobilization by means of carboiimides or *N*-ethoxycarbonyl-2-ethoxy-1,2-dihydroquinoline (EEDQ). This

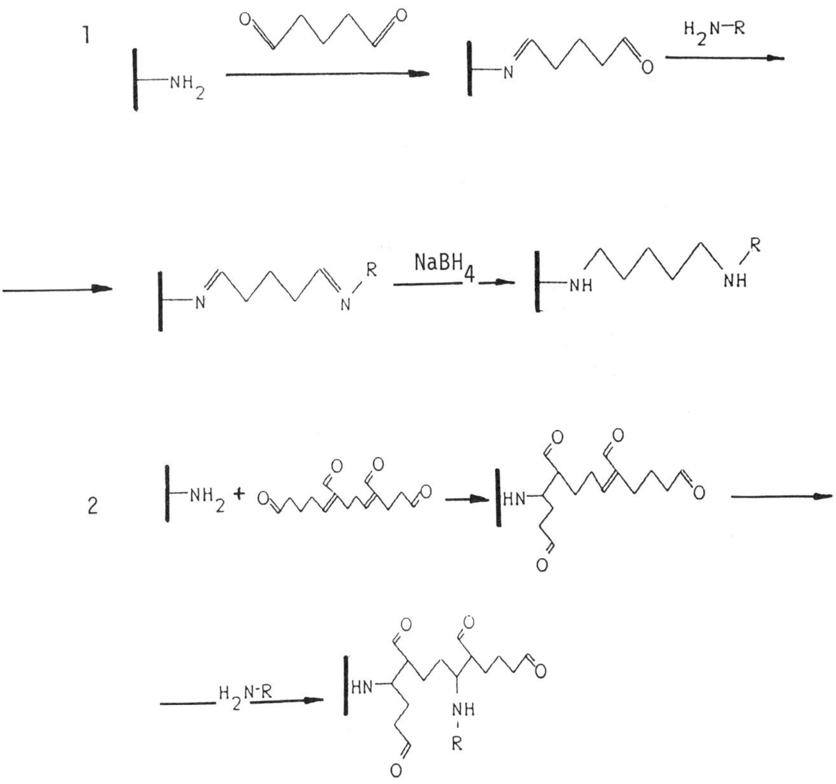

Figure 17 Two possible methods of matrix activation and ligand immobilization with glutaraldehyde: (1) via Schiff base formation; (2) via addition to double bond.

method yields stable, noncharged amide bonds and is used for the preparation of various affinity sorbents. Proper 4% agarose gel is commercially available (CM-Bio-Gel-A), but the charge density of CM-Sepharose (120 μ*M*/mL) is probably too high for affinity applications.

Other available activation methods include 4-(oxyacetyl)phenoxyacetic acid (OAPA), which is immobilized by its carboxyl end onto amino-bearing carriers and used for protein immobilization via the arginine residues (442), and bromoacetyl bromide activation of cellulose (443), which has lost its significance, although similar gels with an iodoacetylated spacer arm may be used to immobilize proteins via their sulfhydryl groups (444).

B. Indirect Immobilization of Proteins

Indirect immobilization of proteins enables binding them by a predetermined site: for example, immobilized IgG Fc receptor binds IgG molecules through the Fc region so that the antigen-binding site remains free. Another possible method is biotinylation of periodate-oxidized antibodies with biotinyl hydrazide followed by immobilization on avidin-coated support (445).

Similar results are achieved when monoclonal antibodies against a distinct site of a molecule to be immobilized are preliminarily bound to solid matrix. Some technical problems are associated with this approach: it may be difficult to achieve selective elution from this "sandwich" without cross-linking of the primarily immobilized compound, but the destructivity of the procedure is unpredictable. The list of reagents used for cross-linking of adsorbed proteins includes glutaraldehyde, dimethylsuberimidate, and tea phenols (446). When separate elution is omitted and the entire complex coeluted, column lifetime is increased manyfold (447).

C. Spacers

If steric limitations for ligand–solute interactions are to be avoided, some ligands must not be located close to the surface of the solid support. Usually, four to six atoms between the matrix and the ligand will be sufficient (hexamethylenediamine and ε-aminocaproic acid are generally used), but sometimes up to 16 atoms are needed (448). A hydrophilic spacer of any length can be prepared by subsequent reaction of activated gel with 1,3-diamino-2-propanol and bromoacetyl bromide (449). Carboxy-terminal groups are often used activated with *N*-hydroxysuccinimide (NHS)* (450). A rarely used method for the preparation of hydrazide-bearing gels consists of the reaction of hydrazide hydrate with active ester gel (451).

Activation of agarose with butanediol–diglycidyl ether or amine (amide)-containing gels with glutaraldehyde introduces spacer groups by themselves. *p*-Phenylenediamine and benzidine are used as aromatic spacers. They are usually diazotized and used for azo coupling of proteins via aromatic amino acids. This method enables selection of the predominant binding site: Trp is most reactive at about pH 5, His at pH 5 to 6, and at pH > 7 His and Tyr react with equal speed (452). A reversed order of coupling can be used: Aromatic amino ligand is diazotized and then reacted with, for example, immobilized Tyr or His (453).

D. Coupling with Carbodiimides or EEDQ

Carbodiimides form an active ester with carboxylic acid and the later is turned into an amide or ester by a substitution reaction with the substance to be immobilized (or vice versa: carboxyl-bearing ligand is turned into an active ester and reacted with an amino-containing sorbent). The reactions may be performed in a single step or separately (which makes it possible to omit protective groups when the ligand contains both amino and carboxyl groups):

$$\text{R–COOH} + \text{R}^1\text{–N=C=N–R}^2 \longrightarrow \text{R–}\overset{\overset{\text{O}}{\|}}{\text{C}}\text{–O–}\overset{\overset{\text{NR}^1}{\|}}{\text{C}}\text{–NHR}^2 \xrightarrow{\text{H}_2\text{N–R}^3}$$

$$\text{R–CO–NHR}^3 + \text{R}^1\text{HN–CO–NHR}^2$$

Three different carbodimides are routinely used: (1) dicyclohexylcarbodiimide (DCC), (2) *N*-ethyl-*N*′-(3-dimethylaminopropyl)carbodiimide (EDC), and (3) *N*-cyclohexyl-*N*′-3(2-morpholinoethyl)metho-*p*-toluenesulfonate (CMC).

DCC and its reaction product, dicyclohexylurea, are not soluble in water, and organic solvents (usually dioxane) are used. It is generally used to link active ester groups to carboxyl-bearing carriers and is the most efficient reagent for NHS–silica preparation (454). Water-soluble carbodiimides are used for direct ligand coupling in aqueous solutions; EDC is usually more efficient than CMC. The pH optimum is 4.5 to 6; it usually decreases at the beginning of the reaction and should be adjusted during the first 1 to 1.5 hours [appropriate concentrated buffer (e.g., morpholinoethane-sulfonic acid) is a good alternative]. Some suppliers believe DDC to be extremely toxic. Esters other than carboxylic acids can also be prepared using carbodiimides*.

N-ethoxycarbonyl-2-ethoxy-1,2-dihydroquinoline (EEDQ; Figure 18) forms mixed anhydrides with carboxyl groups, which then react with primary amine to form an amide bond (455). It can be used to couple proteins to CM-cellulose and CM-Sephadex (456) and for low-molecular-weight ligand immobilization (457). The reaction is usually performed in nonbuffered 50% aqueous ethanol (extreme pH values should be avoided). The results are as good as with CMC, but EEDQ is less expensive.

E. Reversible Ligand Coupling

Ligand may be bound onto a matrix via a cleavable covalent bond (e.g., S–S or S–Hg) or by hydrophobic interaction (458,459), which may be useful for separation of whole enzyme inhibitor or another type of complex for further investigation, for reuse of an expensive matrice, or for mild elution of complex followed by dissociation under conditions not tolerated by the matrix.

F. Activation Methods for Silica

Silica is activated with reactive silanes containing (potentially) reactive groups on their hydrocarbon chain. The route most commonly used starts with treatment

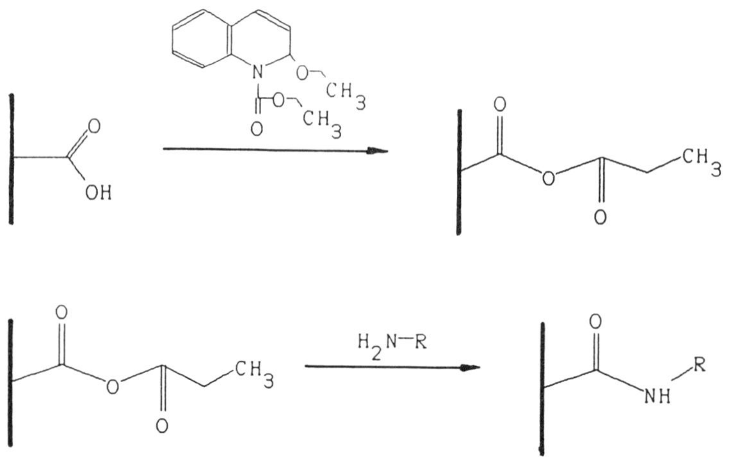

Figure 18 Ligand immobilization with EEDQ.

using 3-glycidoxypropyltrimethoxysilane, which can be used directly for immobilization or converted into "diol-silica" by aqueous acid treatment. Both reactions may be performed in aqueous solution in a single step* (460). Diol-silica is then activated at neutral or acidic pH by any diol-touching method.

Another route starts with 3-aminopropyltrimethoxysilane treatment and leads to aminopropylsilica, which can be activated further by glutaraldehyde or chloroformates, used for carbodiimide coupling, succinylated and then converted into *N*-hydroxysuccinimidyl ester* (461), and so on.

Primary hydroxylsilica can be prepared by one of the following methods (462):

1. Diol-silica is periodate-oxidized to aldehyde and reduced with sodium borohydride.
2. Methacryloxypropylsilica is hydrolyzed with 1 *M* HCl at 60°C for several hours.
3. Aminopropylsilica is suspended in 6 *M* HCl and reacted with sodium nitrite.

Thiol-bearing silica can be prepared using corresponding thiol-terminal silanes (463).

IV. SOLUBILIZATION OF BIOMOLECULES FOR AFFINITY CHROMATOGRAPHY

If a protein to be purified is associated with lipids, it must be solubilized prior to chromatography. If lipids are not needed for protein stability, extraction with cold

(> –20°C) acetone is often used. Grounded plant seeds are generally extracted with hexane or petroleum ether. Peripheral membrane solutions are held in place by electrostatic forces and can be solubilized by changes in pH or increased salt concentrations.

Sometimes it is possible to extract proteins from a fatty starting material with a mixture of butanol and water. Three layers are formed: insoluble particles on the bottom, an intermediate water containing solubilized proteins, and an upper organic layer containing lipids.

Solubilization of integral membrane proteins is more difficult because the presence of lipids is necessary for retaining the native status of the protein (for a more detailed review, see Ref. 464). A portion of the lipids may be replaced with an appropriate detergent, but the choice of the detergent may be very difficult [e.g., the muscarinic acetylcholine receptor remains active only if solubilized with digitonin (A. Rinken, personal communication)].

Triton X-100 is poorly compatible with ultraviolet-absorbance detection, but hydrogenation depletes this drawback without alteration of the detergent properties (465). Detergent removal and exchange can be performed by chromatography on Toyopearl HW 40 (466), phenyl-Superose (467), Bio-Beads SM-2, SM-4, ion exchanger AG11-A8 (Bio-Rad Laboratories), and Extracti-Gel D (Pierce). Sodium dodecyl sulfate (SDS) can be removed by the precipitation with K^+ ions followed by centrifugation (468).

V. ELUTION

Adsorbed compounds can be washed off the column in two ways: by specific (concurrent) or by nonspecific elution. The first (often expensive) way is based on direct interruption by analogs of either ligand or adsorbate into the complex formed on the affinity resin. A possible disadvantage is the formation of new stable complexes.

If the affinity interaction is mainly hydrophobic, elution with detergent or mixed solvents (up to 50% ethylene glycol or isopropanol) is promising; but if the interaction is mainly ionic, high salt concentrations or pH alteration are used. The harshest eluants are 6 *M* urea, 2 to 4 *M* thiocyanate, and guanidinium hydrochloride. Contact time with denaturing solutions must be minimized (e.g., a single hollow fiber may be connected to the column outlet or a concentrated neutralizing buffer prepoured into the fraction collector tubes). Depletion of metal ions (469), or vice versa, restoration of depleted ions (470) responsible for the nativity of the protein, provides a mild elution method. All denaturation problems can be avoided by electrophoretic desorption (471). Distilled water (but not *very* diluted buffer) may elute something even after chaotropic salts are used.

Elution from antigen–antibody complexes can be modeled on polystyrene microplates (472): Immune complexes are dissociated without eluting the adsorbed antigen, while all parameters are controlled by a conventional ELISA technique. In the majority of cases gradient elution gives higher-purity products than does stepwise elution and should be used at least during method development.

VI. MONITORING

A. Spectrophotometry

Chromatography processes are usually monitored by an ultraviolet detector so that there will be no problems in detecting proteins or nucleic acids. Proteins that contain chromogenic groups can be detected separately on another wavelength, so a dual-wavelength detector may be useful.

B. Chemical Detection

The column outflow is divided into various parts by a multichannel peristaltic pump. The main part is directed into a fraction collector and the minor one is mixed with a detecting reagent from another pump channel. The reaction mixture is allowed to react in a capillary tube and then monitored by an optical nerve. Alternatively, eluate samples are automatically directed to a stopped-flow instrument (473). Usually, enzyme activities are analyzed in collected fractions, but if an inhibitor was used for elution, dialysis of samples is cumbersome.

C. Electrophoresis

Electrophoresis is used to check the purity of the product obtained. Possible problems are:

1. Inhibitors may alter the electrophoretic behavior of proteins and results obtained by isoelectric focusing, for example, may be changed beyond recognition.
2. Isoelectric focusing (IEF) only weakly is compatible with high salt concentrations, a problem that may be overcome by drop dialysis (474). A membrane filter is floated on the required solution, then the sample drops are pipetted onto the filter and recovered with tolerable losses after $^1/_2$ h. If the starting salt concentration is moderate, it is possible to perform the IEF; the ampholyte concentration should be increased to 3% and prefocusing is necessary. The process should be started at low current densities, and further replacement(s) of electrode strips will add safety. IEF with PhastSystem can tolerate up to 0.5 *M* of nonbuffering salt (P. Kårsnäs, personal communication).
3. SDS electrophoresis is compatible with high neutral salt concentrations (475) but is difficult to perform in the presence of K^+ or NH_4^+ ions, so Na^+ salts should be preferred for elution. Alternatively, volatile ammonium buffers are

removed by freeze-drying [the applicability of $(NH_4)_2CO_3$ for hydrophobic interaction chromatography has been demonstrated (476)].

VII. METHODS

1. Preparation of Beaded Agarose Gel

Mix equal volumes of agarose solution and non-water-miscible organic solvent(s) (kerosene, mineral oil, toluene, toluene–carbon tetrachloride 3:1, or similar) containing 0.1 to 1% of some very hydrophobic oil-soluble detergent at 60°C in a round-bottomed flask positioned in an empty water bath. The flask must be filled completely. After about 1 min of stirring, start to cool by filling the water bath with an ice–water mixture; when cooled to room temperature, wash the agarose with water–dioxane, isopropyl alcohol, acetone, or a similar solvent (the choice depends on both the organic phase and the detergent used; dioxane–water is a universal washing solution but is the most toxic). Adjust the detergent concentration, which is strongly dependent on the starting agarose quality and concentration.

2. Preparation of Cross-Linked and Desulfated Agarose Gel

Mix equal volumes of agarose gel and 1 *M* NaOH 5 g/L NBH_4. Add 0.02 vol of epichlorohydrine. Heat to 60°C under continuous stirring and mix for 1 h. Wash with water until neutral conditions, mix with an equal volume of 2 *M* NaOH and 5 g/*L* $NaBH_4$, autoclave at 120°C for 1 h, then wash with water, diluted acid, and water (477).

3. Preparation of Trisacryl-like Gels

Dissolve in 1 L of water 330 g of *N*-acryloyl-2-amino-1,3-propane and 40 g of *N,N'*-diallyltartardiamide; warm to 55°C, add 0.12 g of ammonium persulfate and 1.6 mL of TEMED, mix immediately with 2 L of kerosene, and stir intensively until polymerized. Wash with 1% Triton X-100 (478).

4. Cyanogen Bromide Activation of Agarose

Suspend the gel in an equal volume of water, add 50 to 300 mg of CNBr (as a solution in dioxane or acetonitrile) per milliliter of gel under vigorous stirring. Keep the pH at 11 by the addition of 4 to 8 *M* alkali using an autotitrator. Keep the temperature below 20°C (with ice). The reaction takes 8 to 12 min (titration stops). Wash immediately with binding buffer; add ligand solution and stir overnight.

Warning: CNBr bromide is extremely toxic and volatile! All laboratory glassware must be detoxified with ferric chloride solution. CNBr as an efficient lacrimator can be easily recognized in air at sublethal concentrations, but avoid situations, when you even do not reach to recognize it!

5. Cyanogen Bromide Activation of Agarose (Simplified)

Mix equal volumes of the gel and a 2 *M* Na_2CO_3 solution, both cooled on ice. Add 0.2 g of CNBr per milliliter of agarose (as a solution in dioxane or acetonitrile). Stir for some minutes, filtrate, wash immediately with binding buffer, and use for immobilization (479).

6. Cyanogen Bromide Activation of Agarose (Cyano Transfer)

Wash the gel with 30% and 60% aqueous acetone, suspend in an equal volume of 60% acetone, and cool to –15°C; add 1 *M* CNBr in acetone (each milligram of cyanogen bromide per milliliter of gel will yield activity of about 2 μ*M*/mL), then an equal volume of 1.5 *M* triethylamine in 60% acetone under intensive stirring and cooling. Pour the mixture into a tenfold volume of an ice-cooled mixture of acetone and 0.1 *m* HCl (1:1). The active groups remain stable for about 1 h.

For coupling, wash with 60% and 30% acetone, cold water, and coupling buffer. For storage, wash with acetone–dioxane–water (60:35:5) and store at –20°C. The loss of activity is about 10% per month. This relationship is linear up to 80 μ*M*/mL (480).

7. Bisoxirane Activation of Agarose

Mix equal volumes of gel, 0.6 *M* NaOH, 2.5 g/L $NaBH_4$, and bisoxirane. Stir for 6 to 10 h; wash with water. This recipe yields activity of about 30 and 45 μ*M*/mL for Sepharose 4B and 6B, respectively. If 10 to 15 μ*M*/mL is needed, decrease the bisoxirane concentration three- to fourfold. Never perform this reaction at temperatures below 20°C (481). For an activity determination, suspend the gel in an excess of $Na_2S_2O_3$, and titrate at pH 7 the alkali formed.

8. Epichlorohydrin Activation of Agarose

Mix with 2 volumes of 1 *M* NaOH, 5 g/L $NaBH_4$, and 0.2 volume of ECH. Warm to 60°C and stir for 2 h (or 4 h at 40°C); wash with water. Usually, the active group concentration will be about 25 to 30 μ*M*/mL for 4% gel. This mixture will be stable for some weeks at 4°C (482,483).

9. Epichlorohydrin Activation of Toyopearl

For HW-55, mix 1.3 g of gel with 15 *M* NaOH and 2.5 mL of ECH, and shake for 2 h at 50°C. Achievable activity is about 330 μ*M*/g wet weight.

For HW-65 (diol type), mix 1.3 g of gel with 1 mL of water, 1 mL of 10 *M* NaOH, and 1.5 mL of ECH; shake for 1 h at 50°C. Achievable activity is about 150 μ*M*/g wet weight (484).

10. Divinylsulfone Activation of Agarose

Mix equal volumes of agarose gel and 1 *M* Na_2CO_3, add 10% gel volume of DVS, and stir for 0.5 to 1.5 h; wash with water. This mixture will be stable for several weeks in water at 4°C (420). For an activity determination, mix the gel with an equal volume of 3 *M* sodium thiosulfate and titrate the alkali formed. The reaction is very slow and may last 10 to 15 h.

11. Trichlorotriazine Activation of Agarose

Mix the gel with an equal volume of acetone and stir for 0.5 h. Then add cyanuric chloride (0.5 to 1% of gel weight dissolved in a minimal quantity of acetone, then diluted with water 1:1); stir for 5 min; add 2 *M* Na_2CO_3 (10% gel volume), stir for another 5 min; wash with 50% aqueous acetone.

12. Trichlorotriazine Activation of Cellulose

Suspend cellulose in 3 *M* NaOH, allow to swell for 15 min; filter off excess liquid; add an equal weight of cyanuric chloride as 5% solution in dioxane–xylene (1:1, w/w), stir for 30 min, filter off; wash with dioxane (twice), acetic acid–dioxane–water (1:2:1 w/w/w), and acetone (twice). Dry under vacuum (485).

13. 2,4,6-Trifluoro-5-Chloropyrimidine Activation of Agarose

Suspend Sepharose 4B in two volumes of 3 *M* NaOH, rotate for 15 min, then filter off. Suspend the gel in about 2 volumes of xylene–dioxane (1:1 w/w) and cool to 0°C. Slowly add FCP solution (about 25% of agarose volume, dissolved in an equal volume of dioxane–xylene mixture) so that temperature does not exceed 5°C (about 1 h). Stop the reaction with concentrated AcOH (10% of gel volume). Filter, wash with xylene–dioxane (10 volumes), dioxane (10 volumes), and water (40 volumes). Store at 4°C (486).

14. Periodate Activation of Agarose

Mix agarose gel with an equal volume of 0.5 *M* sodium periodate, agitate for 18 h; wash with plenty of water followed by an immobilization buffer (422).

15. Carbonyldiimidazole Activation of Agarose

Transfer the gel to dry dioxane and suspend in an equal volume of the same solvent. Add 3 to 4% of carbonyldiimidazole and stir for 15 min; wash with dry dioxane. The activity of 6% agarose gel will be approximately 45 μ*M*/mL. It is stable in dioxane (487).

16. Benzoquinone Activation of Agarose

Wash Sepharose with 0.1 *M* carbonate or phosphate buffer (pH 8) (may contain 20% EtOH); add benzoquinone (1% of the gel weight) dissolved in EtOH. Stir for 1 h; wash with 20% EtOH, then with 1 *M* NaCl (488).

17. Tosyl (tresyl) Activation of Agarose

Transfer the gel to dry acetone; suspend in an 0.5 volume of dry acetone, add 2 mL of dry pyridine per milligram of tosyl (tresyl) chloride to be added, add tosyl (tresyl) chloride (0.5 to 2% of gel volume); stir for 10 to 15 min. Wash with 20 volumes of acetone. Store in dry acetone. For binding, transfer to water by washings with acetone–1 m*M* HCl (aqueous) mixtures and finally, with 1 m*M* HCl (489,490).

18. Glutaraldehyde Activation of Ultrogel AcA

Wash with 0.5 *M* potassium phosphate (pH 7.6); filter off; mix with an equal volume of 25% aqueous glutaraldehyde, adjust to pH 7.4, stir for 18 h at 37°C, and wash with water. Activated gel is stable (422).

19. Glutaraldehyde Activation of Polyacrylamide

Leave 1 g of Biogel P-300 to swell in water, wash with 0.1 *M* phosphate (pH 6.9); adjust the volume to 45 mL with the same buffer; add 4.8 mL of 25% glutaraldehyde; incubate at 37°C for 17 h, wash with 0.1 *M* phosphate (pH 6.9) (4 × 4 volumes) and pH 7.7 (three times) (491).

20. Hydrazine Activation of Ultrogel AcA

Suspend the gel in 0.5 volume of water; warm to 50°C; add hydrazine (75% of the gel volume); stir for 2 h, cool down (by adding of ice); wash with water; suspend in 0.5 volume of ice-cold water; place mixture in an ice bath and add 0.5 *M* HCl (0.5 gel volume) at 4°C under continuous agitation; after 5 min add a solution of $NaNO_2$ (8.4 mg/mL of gel). Stir for 2 min and use immediately (422).

21. Preparation of Carboxymethyl-agarose

Mix cross-linked agarose with 0.84 volume of ice-cold 10 *M* NaOH at 0°C; leave for 0.5 h; add chloracetic acid under continuous stirring (75 g per liter of agarose as a saturated solution in water), warm to 70°C, and stir for 1 h. Wash with water. The gel will contain about 20 μ*M*/mL carboxymethyl groups. If higher activity is needed, increase the amount of chloroacetic acid and add an equivalent amount of alkali (procedure used in the author's lab).

22. Binding of myoInositol-2-phosphate to AH-Sepharose

Suspend 20 mL of AH-Sepharose in 36 mL of water and 36 mL of *tert*-butanol, add a 1.5-fold molar excess of *myo*-inositol-2-phosphate, adjust pH with NaOH to 6.1. Warm to 70°C with continuous stirring (use a reflux cooler) for 1 h and add a 6.7-fold molar excess of dicyclohexylcarbodiimide (DCC) in 54 mL of *tert*-butanol; stir at 70°C for 1 h. Wash with 500 mL of *n*-butanol, ethanol, water, butanol, and ethanol, then with 3 L of water (492).

23. Immobilization of IgG by Periodate Oxidation

Dissolve 50 to 75 mg of IgG in 2 mL of 0.05 *M* NaOAc, add 0.2 mL of 100 m*M* $NaIO_4$, and mix for 1 h in dark. Desalt by gel filtration. Wash the hydrazide-containing gel (e.g., AvidGel Ax, BioProbe International) with 0.05 *M* NaOAc (pH 5), and 8 to 11 mg of oxidized IgG in 1 mL of the same buffer per milliliter of gel. Tumble overnight at 4°C. Wash with 2 gel volumes of water, 1 *M* NaCl, PBS (493).

24. Binding of Antibodies to Sephacryl S-1000

Wash Sephacryl gel with 0.1 *M* sodium acetate pH 4.7, and add 5 gel volumes of 3 m*M* sodium periodate in the same buffer. Allow to react for 15 min in the dark; wash with water. Transfer the gel to freshly prepared coupling buffer consisting of 0.1 *M* trisodium citrate, 50 m*M* $NaCO_3$, and 5 m*M* ascorbic acid (pH 10). Add antibody solution in the same buffer and leave for 20 h. Wash with pH 10 buffer, block unreacted groups with 1 *M* ethanolamine, 5 m*M* ascorbic acid (pH 9) for 1 h. Wash with water, 0.1 *M* glycine (pH 3), 1 *M* ethanolamine (pH 9), 5 *M* urea (pH 6), 3 *M* potassium thiocyanate, and finally with 15 m*M* citrate and 150 m*M* NaCl (pH 7) (494).

25. Preparation of N-Hydroxysuccinimidylsilica

Mix 1 g of succinylated aminopropyl-silica with 2 mL of 2-propanol and 58 mg of NHS. Add 96 mg of ethyl-3-(3-dimethylaminopropyl)-carbodiimide (EDAC) in 1 mL of 2-propanol; sonicate under vacuum for 1 min. Shake for 1 h and wash with 2-propanol (495). Or use dioxane instead of 2-propanol and dicyclohexylcarbodiimide instead of EDAC. [The latter method is claimed to be preferable (496).]

26. Preparation of Diol-silica

In aqueous solution, reflux silica in 10% (v/v) glycidoxypropyltrimethoxysilane in 0.01 *M* Na_2HPO_4 (pH 8.8) for 16 to 18 h (497).

In nonaqueous solution, dry 10 g of silica at 150°C in vacuo. Suspend in 150 mL of dry toluene, add 10 mL of 3-glycidoxypropyltrimethoxysilane and 0.25 mL of toluene, and reflux for 16 h. Convert to the diol by 1 h of treatment with 10 m*M* H_2SO_4 at 90°C.

In mixed solvent, reflux 20 g of zirconium-stabilized silica in 10 mL of 3-glycidoxypropyltrimethoxysilane + 30 mL of tetrahydrofuran and 1 mL of 0.1 *M* NaOH. Remove liquid by rotatory evaporator, heat for 2 h at 150°C, then wash with tetrahydrofuran, toluene, methanol, and acetone. Dry in vacuo (498).

27. Preparation of Zirconium-Stabilized Silica

Suspend 20 g of silica in 200 mL of 5% (w/v) zirconium oxychloride in water, dearate under vacuo, and reflux for 1 h; filter off, dry overnight at 150°C, and fire at 800°C for 5 h. Or reflux silica in aqueous 1% (w/v) zirconium oxychloride, filter off, dry at 100°C in vacuo, and fire above 800°C for 16 to 18 h (497,498).

28. Preparation of Anhydrotrypsin

Dissolve 1 g of trypsin in 1 L of 0.075 *M* Tris–HCl (pH 7.2), 3 m*M* $CaCl_2$; add 250 mg of phenylmethylsulfonyl fluoride in 250 mL of 2-propanol + 750 mL of H_2O.

Allow to react for 1/2 h; dialyze against water at 0 to 4°C; freeze-dry. Treat with 0.05 *M* KOH (2.5 mL per milligram) at 0°C, adjust to pH 7, dialyze, and freeze-dry. The product contains 5 to 10% of active trypsin, but four subsequent treatments as above reduce its quantity below 1% with about 60% recovery (499).

29. Preparation of Anhydrothrombin

Dissolve α-thrombin in 50 m*M* Tris–HCl (pH 7.5), makes 0.03 m*M* in PMSF, allow to stand for 2 h; adjust pH at 9.0 with 0.1 *M* NaOH, leave for 24 h at 4°C; adjust pH at 7.5 with 0.1 *M* HCl. Apply to benzamidine–Sepharose rB [50 m*M* Tris–HCl (pH 7.5)], and wash with starting buffer; elute anhydrothrombin with 0.2 *M* benzamidine (pH 7.5). Transfer to 50 m*M* Tris–HCl (pH 7.5), 1 *M* NaCl by gel filtration of Sephadex G-25. The product contains 0.017% of starting clotting activity and 0.05% of S-2160 hydrolysis activity (122).

REFERENCES

1. P. D. G. Dean and D. H. Watson, *J. Chromatogr., 165*:301 (1979).
2. R. Arshady, *J. Chromatogr., 586*:181 (1991).
3. R. Arshady, *J. Chromatogr., 586*:199 (1991).
4. C. Araki, *Bull. Chem. Soc.* (Jpn.), *29*:543 (1956).
5. M. Samec and V. Isajevic, *Colloidchem. Beih., 16*:285 (1922).
6. C. Araki, *The polysaccharides of agarophytes, Proc. 5th International Seaweed Symposium*, Halifax, 1956, pp. 3–17.
7. A. I. Usov, Ya. G. Ivanova, and V. F. Makienko, *Bioorg. Khim.* (Rus.), *5*:1647 (1979).
8. S. Hjertén, *Biochim. Biophys. Acta, 779*:393 (1964).
9. S. Bengtsson and L. Philipson, *Biochim. Biophys. Acta, 79*:399 (1964).
10. E. Presecan, H. Porumb, and I. Lascu, *J. Chromatogr., 469*:396 (1989).
11. K. B. Guiseley, U.S. Patent 3,956,273, May 1976.
12. J. Porath, J.-C. Jansson, and T. Låås, *J. Chromatogr., 60*:167 (1971).
13. J. Porath, T. Låås, and J.-C. Jansson, *J. Chromatogr., 103*:49 (1975).
14. A. I. Usov and L. I. Miroshnikova, *Carbohydr. Res., 43*:204 (1975).
15. M. W. C. Hatton and E. Regoeczi, *Biochim. Biophys. Acta, 438*:339 (1976).
16. V. C. Maino, M. J. Hayman, and M. J. Crumpton, *Biochem. J., 146*:247 (1975).
17. N. Gilboa-Garber, L. Mizrahi, and N. Garber, *FEBS Lett., 28*:93 (1972).
18. C. J. Bacchi, S. L. Marcus, C. Lambros, B. Goldberg, L. Messina, and S. H. Hutner, *Biochem. Biophys. Res. Commun., 58*:778 (1974).
19. J. M. Fernandes-Sousa, R. Perez-Castello, and R. Rodrigues, *Biochim. Biophys. Acta, 523*:430 (1978).
20. S. Hjertén, *Acta Chem. Scand., B36*:203 (1982).
21. S. Hjertén, Jia-li Liao, *J. Chromatogr., 457*:165 (1988).
22. Jina-ping Li, K.-O. Eriksson, and S. Hjerten, *Prep. Biochem., 20*(2):107 (1990).
23. T. Andersson, M. Carlsson, L. Hagel, P.-A. Pernemalm, and J.-C. Jansson, *J. Chromatogr., 326*:33 (1985).

24. M. G. Bite, in *Process Engineering Today*, Proc. Eurochem '86, Online Publications, Londo, 1986, p. 137.
25. B. Ericsson, K. Aspberg, and J. Porath, *Biochim. Biophys. Acta, 310*:446 (1973).
26. G. Getland, P. Garred, H. B. Pettersen, T. E. Mollnes, and E. Johnson, *Clin. Exp. Immunol., 79*:459 (1990).
27. D. H. Campbell, E. Luescher, and L. S. Lerman, *Proc. Natl. Acad. Sci. USA, 37*:575 (1951).
28. H. V. Aposhian and A. Kornberg, *J. Biol. Chem., 237*:519 (1962).
29. C. Rüttimann, M. Cotorás, J. Zaldivar, and R. Vicuna, *Eur. J. Biochem., 149*:41 (1985).
30. G. Rhodes, K. D. Jentsch, and T. M. Jovin, *J. Biol. Chem., 254*:7465 (1979).
31. L. L. Nolan, J. H. Rivera, and N. N. Khan, *Biochem. Biophys. Acta, 1120*:322 (1992).
32. R. A. Rubin and P. Modrich, *Methods Enzymol., 65*:96 (1980).
33. A. Venegas, R. Vicuña, F. Valdes, and A. Yudelevich, *FEBS Lett., 109*:156 (1980).
34. P. R. Whitehead and N. L. Brown, *Arch. Microbiol., 141*:70 (1985).
35. J. D. Hendrix and N. E. Welker, *J. Bacteriol., 162*:682 (1985).
36. H. G. Stunnenber, L. M. J., Wennekes, T. Spierings, and H. W. J. van der Broek, *Eur. J. Biochem., 117*:121 (1981).
37. T. J. Guifoyle, *Biochemistry, 19*:5966 (1980).
38. M. Golomb and D. P. Grandgenett, *J. Biol. Chem., 254*:1606 (1979).
39. P. D. G. Dean, S. R. Willetts, and J. E. Blanch, *Anal. Biochem., 41*:344 (1971).
40. M. S. Le, J. P. Robertson, and K. L. Gollan, *Int. Labmate, 12*:23 (1987).
41. V. S. Horsney, C. V. Prowse, and D. S. Pepper, *J. Immunol. Methods, 93*:83 (1986).
42. M. Belew, J. Porath, J. Fohlman, and J.-C. Janson, *J. Chromatogr., 147*:205 (1978).
43. H. Dellweg, M. John, and J. Schmidt, *Eur. J. Appl. Microbiol., 1*:191 (1975).
44. S. Sivakami and A. Radhakrishnan, *Int. J. Biochem. Biophys., 10*:283 (1973).
45. I. Ishiyama and G. Uhlenbruck, *Immunoforschung, 143*:147 (1972).
46. B. B. L. Agrawal and I. Goldstein, *Biochim. Biophys. Acta, 147*:262 (1967).
47. B. Karlstam, *Biochim. Biophys. Acta, 329*:295 (1973).
48. H. O. J. Olson and I. A. Liener, *Biochemistry, 6*:105 (1967).
49. N. Fornstedt and J. Porath, *FEBS Lett., 57*:187 (1975).
50. N. L. Stults, L. M. Asta, and Y. C. Lee, *Anal. Biochem., 180*:114 (1989).
51. *Biotech Prod. Int., 2*:34 (1990).
52. S. Hjertén and R. Mosbach, *Anal. Biochem., 3*:109 (1962).
53. J. Lead and A. H. Sehon, *Can. J. Chem., 40*:159 (1962).
54. L. Genaud, J. Guillot, M. Damez, and M. Coulet, *J. Immunol. Methods, 22*:339 (1978).
55. V. Horesji and J. Kocourek, *Biochim. Biophys. Acta, 297*:346 (1973).
56. V. Pavliak, V. Kéry, K. Tichlárik, and J. Sandula, *J. Chromatogr., 490*:418 (1989).
57. T. A. Valueva, E. A. Kostanova, and V. V. Mosolov, in *New Methods in Practical Biochemistry*, Nauka, Moscow, 1988, p. 13.
58. R. L. Schnaar and Yuan Chuan Lee, *Biochemistry, 14*:1535 (1975).
59. J. Coupek, M. Krivakova, and S. Pokorny, *J. Polym. Sci. Symp., 42*:182 (1973).
60. J. Coupek, in *Affinity Chromatography and Related Techniques* (T. C. J. Gribnau, J. Visser, and R. J. F. Nivard, eds.), Elsevier, Amsterdam, 1982.
61. E. Boschetti, in *Affinity Chromatography* (P. D. J. Dean, W. S. Johnson, and F. A. Middle, eds.), IRL Press, Oxford, 1985.

62. M. Kozulic, B. Kozulic, and K. Mosbach, *Anal. Biochem., 163*:506 (1987).
63. J. Saint-Blanchard, J. M. Kirzin, P. Riberon, F. Petit, J. Fourcart, P. Girot, and E. Boschetti, *Affinity Chromatography and Related Techniques* (T. C. J. Gribbnau, J. Visser, and R. J. F. Nivard, eds.), Elsevier, Amsterdam, 1982, p. 305.
64. E. Boschetti, *J. Biochem. Biophys. Methods, 19*:21 (1989).
65. I. Matsumoto, Y. Ito, and N. Seno, *J. Chromatogr., 239*:747 (1982).
66. A. Siimar, J. Parik, and P. Toomik, unpublished data.
67. E. Juronen, J. Parik, and P. Toomik, *J. Immunol. Methods, 136*:103 (1991).
68. K. Shimura, M. Kazama, and K.-I. Kasai, *J. Chromatogr., 292*:369 (1984).
69. E. Boshetti, R. Tixier, and J. Uriel, *Biochimie, 54*:439 (1972).
70. M. J. Harvey and P. D. G. Dean, *Sci. Tools, 23*:36 (1976).
71. S. G. Doley, M. J. Harvey, and P. D. G. Dean, *FEBS Lett., 65*:87 (1976).
72. E. Boschetti and M. Delay, *Sci. Tools, 25*:18 (1978).
73. P. J. Robinson, P. Dunnill, and M. D. Lilly, *Biochim. Biophys. Acta, 242*:659 (1971).
74. G. Szabó, K. Offenmüller, and E. Csató, *Anal. Chem., 60*:213 (1988).
75. R. W. Stout and J. J. DeStefano, *J. Chromatogr., 326*:63 (1985).
76. A. E. Ivanov, L. V. Kozlov, B. B. Shoibonov, and V. K. Antonov, *Biomed. Chromatogr., 5*:90 (1991).
77. P. Dunnill and M. D. Lilly, *Biotechnol. Bioeng., 16*:987 (1974).
78. T. Porumb, H. Porumb, I. Lascu, and I. Proinov, *J. Chromatogr., 319*:218 (1985).
79. D. O. Sweeney, *Experientia, 33*:1405 (1977).
80. S. Y. Huang and C. H. Lin, *Chromatographia, 27*:449 (1989).
81. A. M. Fischer, X. J. Yu, J. Tapon-Bretaudiere, D. Muller, A. Bros, and J. Jozefonvicz, *J. Chromatogr., 363*:95 (1986).
82. M. A. Jacquot-Dourges, D. Muller, D. Barritault, and J. Jozefonvicz, *J. Chromatogr., 510*:141 (1990).
83. D. J. Stewart, D. R. Purvis, and C. R. Lowe, *J. Chromatogr., 510*:177 (1990).
84. I. Safarik, Z. Laudová, and B. Králová, *J. Chromatogr., 303*:283 (1984).
85. I. Safarik, *Biomed. Biochim. Acta, 46*:293 (1987).
86. I. Safarik, *Chem. Listy, 79*:766 (1985).
87. L. Várady, K. Kalgathi, and C. Horváth, *J. Chromatogr., 458*:207 (1988).
88. A. Wójcik and L. Kwietniewski, *J. Chromatogr., 435*:55 (1988).
89. N. B. Afeyan, S. P. Fulton, N. F. Gordon, I. Mazsaroff, and F. E. Regnier, *Biotechnology, 8*:203 (1990).
90. D. Zopf and S. Ohlson, *Nature, 346*:87 (1990).
91. K. Nakamura, T. Hashimoto, Y. Kato, K. Shimura, and K.-I. Kasai, *J. Chromatogr., 510*:101 (1990).
92. G. S. Murthy and N. R. Mougdal, *J. Biosci., 10*:351 (1986).
93. R. A. Asryants, I. V. Duszenkova, and N. K. Nagradova, *Anal. Biochem., 151*:571 (1985).
94. T. M. Stich, *Anal. Biochem., 191*:343 (1990).
95. M. C. Cress and That T. Ngo, *Am. Biotechnol. Lab., 7*:16 (1989).
96. P. L. Domen, J. R. Nevens, A. Krishna-Mallia, G. T. Hermannson, and D. C. Klenk, *J. Chromatogr., 510*:293 (1990).
97. L. J. Janis and F. E. Regnier, *J. Chromatogr., 444*:1 (1988).

98. M. Gassmann, P. Thömmes, T. Weiser, and U. Hübsher, *FASEB J.*, *4*:2528 (1990).
99. L. M. Kauvar, P. Y. K. Cheung, R. H. Gomer, and A. A. Fleischer, *BioChromatogr.*, *5*:22 (1990).
100. A. Forsgren and J. Sjöquist, *J. Immunol.*, *97*:822 (1966).
101. A. Grubb, R. Grubb, P. Christensen, and C. Shalén, *Int. Arch. Allergy Appl. Immunol.*, *67*:369 (1982).
102. L. Björck and G. Kronvall, *J. Immunol.*, *133*:969 (1984).
103. E. B. Myhre and G. Kronvall, *Mol. Immunol.*, *17*:1563 (1980).
104. E. B. Myhre and G. Kronvall, *Infect. Immun.*, *27*:808 (1980).
105. K. J. Reis, E. J. Siden, and M. D. P. Boyle, *Biotechniques*, *6*:130 (1988).
106. M. Yarnell and M. D. P. Boyle, *J. Immunol.*, *136*:2670 (1986).
107. M. Yarnell and M. D. P. Boyle, *Scand. J. Immunol.*, *24*:549 (1986).
108. J. McGuire, *Int. Biotechnol. Lab.*, June 1990, p. 14.
109. *Biotech Prod. Int.*, *2*:15 (1990).
110. B. Lövenadler, B. Nilsson, L. Abrahamsén, T. Moks, L. Ljungqvist, E. Holmgren, S. Paleus, S. Josephson, L. Philipson, and M. Uhlén, *EMBO J.*, *5*:2393 (1986).
111. K. Nagai and H. C. Thøgersen, *Nature*, *309*:810 (1984).
112. K. K. Stanley, in *Recombinant Systems in Protein Expression* (K. K. Alitalo et al., eds.), Elsevier, Amsterdam, 1990, pp. 3–12.
113. H. Kondoh, K. Kobayashi, K. Hagiwara, and T. Kajii, *J. Immunol. Methods*, *88*:171 (1986).
114. E. Fink, H. Schiessler, M. Arnhold, and H. Fritz, *Hoppe-Seyler's Z. Physiol. Chem.*, *353*:1633 (1972).
115. H. Fritz, G. Wunderer, and B. Dittmann, *Hoppe-Seiler's Z. Physiol. Chem.*, *353*:893 (1972).
116. G. Feinstein and A. Gertler, *Biochim. Biophys. Acta*, *309*:196 (1973).
117. G. K. Chua and W. Bushuk, *Biochem. Biophys. Res. Commun.*, *37*:545 (1969).
118. R. Smith and V. Turk, *Eur. J. Biochem.*, *48*:245 (1974).
119. L. Sottrup-Jensen, T. E. Petersen, and S. Magnusson, *FEBS Lett.*, *121*:275 (1980).
120. H. Ako, R. J. Foster, and C. A. Ryan, *Biochem. Biophys. Res. Commun.*, *47*:1402 (1972).
121. H. Ako, C. A. Ryan, and R. J. Foster, *Biochem. Biophys. Res. Commun.*, *46*:1637 (1972).
122. T. Tomono, E. Sawada, and E. Tokunaga, in *Separation of Plasma Proteins* (J. M. Curling, ed.), Pharmacia Fine Chemicals, Uppsala, Sweden, 1983, pp. 157–167.
123. S. I. Ishii, H. Yokosawa, T. Kumuzaki, and I. Nakamura, *Methods Enzymol.*, *91*:378 (1983).
124. M. Vuento and A. Vaheri, *Biochem. J.*, *183*:331 (1979).
125. Y. Legrand, G. Pignaud, J. P. Caen, and B. Robert, *Biochem. Biophys. Res. Commun.*, *63*:224 (1975).
126. M. F. Roulleau and E. Boschetti, in *Affinity Chromatography and Related Techniques* (T. C. J. Gribbnau, J. Visser, and R. J. F. Nivard, eds.), Elsevier, Amsterdam, 1982, p. 323.
127. S. Pochet, T. Huynh-Dinh, and J. Igolen, *Tetrahedron*, *43*:3481 (1987).
128. B. Arcangioli, S. Pochet, R. Sousa, and T. Huynh-Dinh, *Eur. J. Biochem.*, *79*:359 (1989).

129. J. W. Cranston, R. Silber, V. G. Malathi, and J. Hurwitz, *J. Biol. Chem., 249*:7447 (1974).
130. S. A. Martin, E. Paoletti, and B. Moss, *J. Biol. Chem., 250*:9322 (1975).
131. A. P. M. Eker and A. M. J. Fichtinger-Schepman, *Biochim. Biophys. Acta, 378*:54 (1975).
132. J. P. Kuebler and D. A. Goldthwait, *Biochemistry, 16*:1370 (1977).
133. B. Van Dorp, R. Benne, and F. Palitti, *Biochim. Biophys. Acta, 395*:446 (1975).
134. J. C. Schabort, *J. Chromatogr., 73*:253 (1972).
135. M.-L. Greth and M.-R. Chevallier, *Biochim. Biophys. Acta, 390*:168 (1975).
136. J. Schlepper and R. Knippers, *Eur. J. Biochem., 60*:209 (1975).
137. C. A. Ross and W. J. Harris, *Biochem. J., 171*:231 (1978).
138. L. L. Nolan, J. H. Rivera, and N. N. Khan, *Biophys. Acta, 1120*:322 (1992).
139. G. L. Clark, K. W. Peden, and R. H. Symons, *Virology, 62*:434 (1974).
140. T. M. Wandzilak and R. W. Benson, *Biochem. Biophys. Res. Commun., 76*:247 (1977).
141. P. A. Lowe, D. A. Hager, and R. R. Burgess, *Biochemistry, 18*:1344 (1979).
142. R. Uy and F. Wold, *J. Chromatogr., 81*:98 (1977).
143. E. Junowicz and J. E. Paris, *Biochim. Biophys. Acta, 321*:234 (1973).
144. E. E. Grebner and I. Parikh, *Biochim. Biophys. Acta, 350*:437 (1974).
145. A. L. Miller, R. G. Frost, and J. S. O'Brien, *Anal. Biochem., 74*:537 (1976).
146. A. Koshy, D. Robinson, and J. L. Stirling, *Biochem. Soc. Trans., 3*:244 (1975).
147. D. Ziegler, G. Keilich, and R. Brossmer, *Fresenius' Z. Anal. Chem., 301*:99 (1980).
148. R. T. Dean, *J. Chromatogr., 150*:279 (1978).
149. P. Cuatrecasas, I. Parikh, and M. D. Hollenberg, *Biochemistry, 12*:4253 (1973).
150. I. Matsumoto and T. Osawa, *Biochem. Biophys. Res. Commun., 46*:1810 (1972).
151. B. A. Sela, J. V. Wang, and G. M. Edelman, *J. Biol. Chem., 250*:7535 (1975).
152. J. P. Kabayo and D. W. Hutchinson, *FEBS Lett., 78*:221 (1977).
153. R. Mawal, J. F. Morrison, and K. E. Ebner, *J. Biol. Chem., 246*:7106 (1971).
154. H. Sternbach, R. Engelhardt, and A. G. Lezius, *Eur. J. Biochem., 60*:51 (1975).
155. J. A. Jaehning, P. S. Woods, and R. G. Roeder, *J. Biol. Chem., 252*:8762 (1977).
156. G. Pflugfelder and J. Sonnenbichler, *FEBS Lett., 93*:361 (1978).
157. B. L. Davidson, T. Leighton, and J. C. Rabinowitz, *J. Biol. Chem., 254*:9220 (1979).
158. P. Ballario, E. Mauro, C. Giuliani, and F. Pedone, *Eur. J. Biochem., 105*:225 (1980).
159. G. Muszynska, E. Ber, and G. Dobrowolska, in *Affinity Chromatography* (P. D. G. Dean, W. S. Johnson, and F. A. Middle, eds.), IRL Press, Oxford, 1985, p. 125.
160. J. S. Rosenberg, P. W. McKenna, and R. D. Rosenberg, *J. Biol. Chem., 250*:8883 (1975).
161. S. Sakamoto, M. Sakamoto, P. Goldhaber, and M. J. Glimcher, *Biochim. Biophys. Acta, 384*:41 (1975).
162. G. Murphy, J. J. Reynolds, U. Bertz, and M. Baggiolini, *Biochem. J., 203*:209 (1982).
163. T. Olivercrons, T. Egelrud, P. H. Iverius, and U. Lindahl, *Biochem. Biophys. Res. Commun., 43*:524 (1971).
164. G. Mitra, E. Hall, and I. Mitra, *Biotechnol. Bioeng., 28*:217 (1986).
165. A. D. Elbein and M. Mitchell, *Arch. Biochem. Biophys., 168*:369 (1975).
166. M. Golomb, *J. Virol. Methods, 1*:157 (1980).

167. R. Lupu, A. Wellstein, J. Sheridan, B. W. Ennis, G. Zugmaier, D. Katz, M. E. Lippmann, and R. B. Dickson, *Biochemistry, 31*:7330 (1992).
168. F. W. Perrella, R. Jankewicz, and E. D. Dandrow, *Biochem. Biophys. Acta, 1076*:209 (1991).
169. H. Sasaki, A. Hayashi, H. Kitagaki-Ogawa, I. Matsumoto, and N. Seno, *J. Chromatogr., 400*:123 (1987).
170. P. Vretblad, *FEBS Lett., 47*:86 (1974).
171. M. P. Silvanovich and R. D. Hill, *Anal. Biochem., 73*:430 (1976).
172. O. Holmberg, *Biochem. Z., 258*:134 (1930).
173. G. Mitra, E. Hall, and I. Mitra, *Biotechnol. Bioeng., 28*:217 (1986).
174. B. B. L. Agrawal and I. J. Goldstein, *Biochem. Biophys. Res. Commun., 147*:262 (1967).
175. H. O. J. Olson and I. A. Liener, *Biochemistry, 6*:105 (1967).
176. N. Gilboa-Garber, L. Mizrahi, and N. Garber, *FEBS Lett., 28*:93 (1972).
177. R. Bloch and M. M. Burger, *Biochem. Biophys. Res. Commun., 58*:13 (1974).
178. J. Lönngren, I. J. Goldstein, and R. Bywater, *FEBS Lett., 68*:31 (1976).
179. Y. Fujita, K. Oishi, K. Suzuki, and H. Imahori, *Biochemistry, 14*:4465 (1975).
180. S. K. Dube and P. Nordin, *Arch. Biochem. Biophys., 94*:121 (1961).
181. B. A. Klyaschchitsky and V. Kh. Mitina, *J. Chromatogr., 210*:55 (1981).
182. B. A. Klyaschchitsky, V. Kh. Mitina, G. E. Morozevich, and R. I. Yakubovskaya, *J. Chromatogr., 210*:67 (1981).
183. K. K. Andersson, C. Balny, P. Douzou, and J. G. Bieth, *J. Chromatogr., 192*:236 (1980).
184. P. Cuatrecasas, M. Wilchek, and C. B. Anfinsen, *Proc. Natl. Acad. Sci. USA, 77*:636 (1968).
185. M. Pacaud, *Eur. J. Biochem., 64*:199 (1976).
186. T. Tosa, T. Sato, R. Sano, K. Yamamoto, Y. Matuo, and I. Chibata, *Biochim. Biophys. Acta, 334*:1 (1974).
187. S. Fujii and M. Muramatu, *13th Annual Meeting, Tanabe Amino Acid Research Foundation*, Osaka, 1972.
188. K. Koerner, I. Rahimi-Laridjani, and H. Grimminger, *Biochim. Biophys. Acta, 397*: 220 (1975).
189. B. J. Gates and J. Travis, *Biochemistry, 12*:1867 (1973).
190. M. Sokolowsky, *Methods Enzymol., 34B*:411 (1974).
191. T. Tosa, T. Sato, R. Sano, K. Yamamoto, Y. Matuo, and I. Chibata, *Biochim. Biophys. Acta, 334*:1 (1974).
192. P. Cuatrecasas, M. Wilchek, and C. B. Anfinsen, *Proc. Natl. Acad. Sci. USA, 61*:636 (1968).
193. S. R. Mardashev, A. Ya. Nikolayev, E. A. Kozlov, and O. N. Petri, *Biokhimiya, 40*:78 (1975).
194. K. Fujiwara, K. Osue, and D. Tsuru, *J. Biochem., 77*:739 (1975).
195. M. W. C. Hatton and E. Regoeczi, *Biochim. Biophys. Acta, 427*:575 (1976).
196. R. H. McParland, J. G. Guevara, R. R. Becker, and H. J. Evans, *Biochem. J., 153*:597 (1976).
197. D. A. Weigant and E. W. Nester, *J. Biol. Chem., 251*:6974 (1976).

198. K. Fujiwara and D. Tsuru, *Int. J. Peptide Protein Res., 9*:18 (1977).
199. H. E. J. Seppä and M. Järvinen, *J. Invest. Dermatol., 66*:165 (1976).
200. Y. Tanaka and I. Uritani, *J. Biochem., 81*:963 (1977).
201. J. W. Tracy and G. B. Kohlhaw, *J. Biol. Chem., 12*:4085 (1977).
202. D. K. Podolsky and M. M. Weisel, *J. Biol. Chem., 254*:3983 (1979).
203. I. Karube, S. Suzuki, and T. Sato, *Biotechnol. Bioeng., 20*:1775 (1978).
204. D. G. Deutsch and E. T. Mertz, *Science, 170*:1095 (1970).
205. T. H. Finlay, V. Troll, M. Levy, A. J. Johnson, and L. T. Hodgins, *Anal. Biochem., 87*:77 (1978).
206. S. Marcus, *Methods Enzymol., 34*:337 (1974).
207. M. Vuento and A. Vaheri, *Biochem. J., 183*:331 (1979).
208. H. Takahashi, *J. Biochem., 71*:471 (1972).
209. K. Kasai and S. Ishii, *J. Biochem., 78*:653 (1975).
210. V. M. Stepanov, G. I. Lavrenova, K. Adli, and M. V. Gonchar, *Biokhimiya, 41*:294 (1976).
211. V. M. Stepanov, A. Y. Strongin, L. S. Isotova, Z. T. Abramov, L. A. Lyublinskaya, L. M. Ermakova, L. A. Baralova, and L. P. Belyanova, *Biochem. Biophys. Res. Commun., 77*:298 (1977).
212. V. M. Stepanov, G. N. Rudenskaya, L. P. Revina, Y. B. Gryaznova, E. N. Lysogorskaya, I. Yu. Fillippova, and I. I. Ivanova, *Biochem. J., 285*:281 (1992).
213. K. Murakami, T. Inagami, A. M. Michelakis, and S. Cohen, *Biochem. Biophys. Res. Commun., 54*:482 (1973).
214. A. H. Patel and R. M. Schultz, *Biochem. Biophys. Res. Commun., 104*:181 (1982).
215. A. H. Patel, A. Ahsan, B. P. Suthar, and R. M. Schultz, *Biochem. Biophys. Acta, 748*:321 (1983).
216. P. H. Wang, Y. S. Do, L. Maculay, T. Shinagawa, P. W. Anderson, J. D. Baxter, and W. A. Hsueh, *J. Biol. Chem. 266*:12663 (1991).
217. H. D. Rich, M. A. Brown, and A. J. Barrett, *Biochem. J., 235*:731 (1986).
218. P. M. Dando and A. J. Barrett, *Anal. Biochem., 204*:328 (1992).
219. J. R. Schwartz, G. W. Pace, B. A. Solomon, C. K. Colton, and M. C. Archer, *Prep. Biochem., 8*:479 (1978).
220. D. B. Craven, M. J. Harvey, C. R. Lowe, and P. D. G. Dean, *Eur. J. Biochem., 41*:329 (1974).
221. H.-L. Schmidt and G. Grenner, *Eur. J. Biochem., 67*:295 (1976).
222. W. P. Schrader, A. S. Stacy, and B. Pollara, *J. Biol. Chem., 251*:4026 (1976).
223. M. Wilchek and R. Lamed, *Methods Enzymol., 34*:475 (1974).
224. H. Ukeda, K. Ono, M. Imabayashi, K. Matsumoto, and Y. Osayama, *Agric. Biol. Chem., 53*:235 (1989).
225. D. B. Craven, M. J. Harvey, and P. D. G. Dean, *FEBS Lett., 38*:320 (1974).
226. M. Lindberg, P. O. Larsson, and K. Mosbach, *Eur. J. Biochem., 40*:187 (1973).
227. H. L. Schmidt and G. Grenner, *Eur. J. Biochem., 67*:295 (1976).
228. P. Zappelli, A. Rossodivita, and L. Re, *Eur. J. Biochem., 54*:355 (1975).
229. O. Brison and P. Chambon, *Anal. Biochem., 75*:402 (1976).
230. I. H. Maxwell and W. E. Hahn, *Nucleic Acids Res., 4*:241 (1977).
231. L. Jervis, *Phytochemistry, 13*:723 (1974).

232. J. W. Baynes and F. Wold, *J. Biol. Chem., 251*:6016 (1976).
233. H. H. Andres, H. J. Kolb, and L. Weiss, *Biochim. Biophys. Acta, 746*:182 (1983).
234. J. Gaudlie and B. L. Hillcoat, *Biochim. Biophys. Acta, 268*:35 (1972).
235. Y. Zaidenzaig and W. V. Shaw, *FEBS Lett., 62*:266 (1976).
236. M. Gorecki, A. Bar-Eli, Y. Burstein, and A. Patchornik, *Biochem. J., 147*:131 (1975).
237. Y. Matsuhashi, T. Sawa, T. Takeuchi, and H. Umezawa, *J. Antibiot., 29*:204 (1976).
238. P. J. Wistrand and T. Wahlstrand, *Biochim. Biophys. Acta, 481*:712 (1977).
239. K.-H. Scheit and A. Stutz, *FEBS Lett., 50*:25 (1975).
240. A. J. Hogdson, I. W. Chubb, and A. D. Smith, *Biochem. Soc. Trans., 6*:648 (1978).
241. K. Okamura, M. Sakamoto, and T. Ishikura, *J. Ferment. Technol., 57*:300 (1979).
242. R. G. Coombe and A. M. George, *Anal. Biochem., 75*:652 (1976).
243. T. Lang, C. J. Suckling, and H. C. S. Wood, *J. Chem. Soc. Perkin Trans., 1*(19):2189 (1977).
244. J. W. Williams and D. B. Northrop, *Biochemistry, 15*:125 (1976).
245. M. A. Rocher, M. P. Martin, M. J. Toro, and E. Montoya, *J. Chromatogr., 368*:462 (1986).
246. L. J. Crane, G. E. Bettinger, and J. O. Lampen, *Biochem. Biophys. Res. Commun., 50*:220 (1973).
247. J. D. Berman and M. Young, *Proc. Natl. Acad. Sci. USA, 68*:395 (1971).
248. M. Landt, S. Boltz, and L. Butler, *Biochemistry, 17*:915 (1978).
249. E. Mössner, M. Boll, and G. Pfleiderer, *Hoppe-Seyler's Z. Physiol. Chem., 361*:543 (1980).
250. E. Steers, P. Cuatrecasas, and H. B. Pollard, *J. Biol. Chem., 246*:196 (1971).
251. M. E. Rafestin, O. Oblin, and M. Monsigny, *FEBS Lett., 40*:62 (1974).
252. L. A. E. Sluyterman and J. Wijdenes, *Biochim. Biophys. Acta, 200*:593 (1970).
253. P. Cuatrecasas, *J. Biol. Chem., 245*:3059 (1970).
254. J. C. Bonnafous, J. Dornand, J. Favero, M. Sizes, E. Boschetti, and J. C. Mani, *J. Immunol. Methods, 58*:93 (1983).
255. D. J. Goss and L. J. Parkhurst, *J. Biochem. Biophys. Methods, 3*:315 (1980).
256. L. A. E. Sluyterman and J. Wijdenes, *Biochim. Biophys. Acta, 200*:593 (1970).
257. C. D. Anderson and P. L. Hall, *Anal. Biochem., 60*:417 (1974).
258. T. Maciag, M. Weibel, and E. K. Pye, *Methods Enzymol., 34*:451 (1974).
259. P. Cuatrecasas, S. Jacobs, and V. Bennett, *Proc. Natl. Acad. Sci. USA, 72*:1739 (1975).
260. V. Madelian and W. A. Warren, *Anal. Biochem., 64*:517 (1975).
261. K. Brocklehurst, J. Carlsson, M. P. J. Kierstan, and E. M. Crook, *Biochem. J., 133*:573 (1973).
262. J. Lin and J. F. Foster, *Anal. Biochem., 63*:485 (1978).
263. C.-B. Laurell, E. Thulin, and B. P. Bywater, *Anal. Biochem., 81*:336 (1977).
264. C.-B. Laurell and E. Thulin, *J. Chromatogr., 159*:25 (1978).
265. D. J. Gross and L. S. Parkhurst, *J. Biol. Chem., 253*:7804 (1978).
266. T. J. Beebee and D. S. Carty, *Anal. Biochem., 101*:7 (1980).
267. U. Hannestad and P. Lindqvist, *Anal. Biochem., 126*:200 (1982).
268. V. Kh. Akrapov and V. M. Stepanov, *J. Chromatogr., 155*:329 (1978).
269. E. Millquist, H. Petersson, and M. Ranby, *Anal. Biochem., 170*:289 (1988).

270. V. K. Akparov, N. N. Nutsibidze, and T. V. Romanova, *Bioorg. Khim.*, *6*:609 (1980).
271. F. von der Haar, *Biochem. Biophys. Res. Commun.*, *70*:1009 (1976).
272. Dj. Josic, W. Hofmann, R. Habermann, J.-D. Schulzke, and W. Reutter, *J. Clin. Chem. Clin. Biochem.*, *26*:559 (1988).
273. S. Hjertén, J. Rosengren, and S. Påhlman, *J. Chromatogr.*, *101*:281 (1974).
274. V. Ulbrich, J. Makes, and M. Jurecek, *Coll. Czech. Chem. Commun.*, *29*:1466 (1964).
275. T. Toraya, M. Fujimura, S. Ikeda, S. Fukui, H. Yamada, and H. Kumagai, *Biochim. Biophys. Acta*, *420*:316 (1976).
276. A. Svenson and P.-A. Hynning, *Prep. Biochem.*, *11*:99 (1981).
277. G. Floris, M. B. Fadda, M. Pellegrini, M. Corda, and A. Finazzi Agro, *FEBS Lett.*, *72*:179 (1976).
278. G. B. Henderson, S. Shaltiel, and E. E. Snell, *Biochemistry*, *13*:4335 (1974).
279. M. A. Mateescu, M. Carbonaro, L. Calabrese, and B. Mondovi, *Proc. 4th European Congress on Biotechnology, 1987*, Vol. 2 (O. M. Neissel, R. R. van der Meer, and K. Ch. A. M. Luyben, eds.), Elsevier, Amsterdam, 1987, p. 310.
280. K. Slavik, W. Rode, and V. Slaviková, *Biochemistry*, *15*:4222 (1976).
281. V. Kasche, F. Löffler, T. Scholzen, D. M. Krämer, and Th. Boller, *J. Chromatogr.*, *510*:149 (1990).
282. F. Maisano, M. Belew, and J. Porath, *J. Chromatogr.*, *321*:305 (1985).
283. S. Hjertén, Kunquan Yao, Zhao-quian Liu, Duan Yang, and Bo-liang Wu, *J. Chromatogr.*, *354*:203 (1986).
284. U. Matsumoto and Y. Shibusawa, *J. Chromatogr.*, *187*:351 (1980).
285. R. Mathis, P. Hubert, and E. Dellacherie, *J. Chromatogr.*, *347*:291 (1985).
286. J. Porath, J. Carlsson, I. Olsson, and G. Belfrage, *Nature (London)*, *258*:598 (1975).
287. J. Porath and B. Olin, *Biochemistry*, *22*:1621 (1983).
288. M. Haner, M. T. Hezl, B. Rassount, and E. R. Birnbaum, *Anal. Biochem.*, *138*:229 (1984).
289. Y. Moroux, E. Boschetti, and J. M. Egly, *Sci. Tools*, *32*:1 (1985).
290. N. Ramadan and J. Porath, *J. Chromatogr.*, *321*:81 (1985).
291. S. A. Margolis, A. J. Fatiadi, L. Alexander, and J. J. Edwards, *Anal. Biochem.*, *183*:108 (1989).
292. J. Porath, G. Dobrowolska, A. Medin, P. Ekman, and G. Muszynska, *J. Chromatogr.*, *604*:19 (1992).
293. B. Lönnerdal, J. Carlsson, and J. Porath, *FEBS Lett.*, *75*:89 (1977).
294. E. Bollin, Jr., and E. Sulkowski, *Arch. Virol.*, *58*:149 (1978).
295. K. C. Chadha, P. M. Grob, A. J. Mikolski, L. R. Davis, Jr., and E. Sulkowski, *J. Gen. Virol.*, *43*:701 (1979).
296. B. Lönnerdal and C. L. Keen, *J. Appl. Biochem.*, *4*:203 (1982).
297. M. Miyata-Asano, K. Ito, H. Ikeda, S. Sekiguchi, K. Arai, and N. Taniguchi, *J. Chromatogr.*, *370*:501 (1986).
298. R. J. Weselake, S. L. Chesney, A. Petkau, and A. D. Friesen, *Anal. Biochem.*, *155*:193 (1986).
299. T. Kurecki, L. F. Kress, and M. Laskowski, *Anal. Biochem.*, *99*:415 (1979).
300. J. P. Lebreton, *FEBS Lett.*, *80*:351 (1977).
301. V. G. Edy, A. Billiau, and P. DeSomer, *J. Biol. Chem.*, *252*:5934 (1977).

302. W. A. Galloway, G. Murphy, J. D. Sandy, J. Gavrilovic, T. E. Cawston, and J. J. Reynolds, *Biochem. J., 209*:741 (1983).
303. D. C. Rijken, G. Wijngaards, M. Zaal-De Jong, and J. Welbergen, *Biochem. Biophys. Acta, 580*:140 (1979).
304. G. Chaga, L. Andersson, B. Ersson, and J. Porath, *Biotechnol. Appl. Biochem., 11*, 424 (1989).
305. J. M. Egly and E. Boschetti, in *Affinity Chromatography and Related Techniques* (T. C. J. Gribnau, J. Visser, and R. J. F. Nivard, eds.), Elsevier, Amsterdam, 1982, p. 445.
306. D. A. P. Small, *J. Chromatogr., 248*:271 (1982).
307. E. Boschetti, P. Girot, A. Straub, and J. M. Egly, *FEBS Lett., 139*:193 (1982).
308. J. M. Egly, J. L. Plassat, and E. Boschetti, *J. Chromatogr., 243*:301 (1982).
309. J. Porath, F. Maisano, and M. Belew, *FEBS Lett., 185*:306 (1985).
310. J. Porath and T. W. Hutchens, *Int. J. Quantum Chem. Quantum Biol. Symp., 14*:297 (1987).
311. G. Chage, L. Andersson, B. Ersson, and M. Berg, *Biomed. Chromatogr., 6*:172 (1992).
312. B. Nopper, F. Kohen, and M. Wilchek, *Anal. Biochem., 180*:66 (1989).
313. R. Haeckel, B. Hess, W. Lauterborn, and K. Wuster, *Hoppe-Seyler's Z. Physiol. Chem., 249*:699 (1968).
314. P. Roschlau and B. Hess, *Hoppe-Seyler's Z. Physiol. Chem., 353*:441 (1971).
315. L. Ryan and C. Vestling, *Arch. Biochem. Biophys., 160*:279 (1974).
316. W. Heyns and P. DeMoor, *Biochim. Biophys. Acta, 358*:1 (1974).
317. S. Thompson, K. Cass, and E. Stellwagen, *Proc. Natl. Acad. Sci. USA, 72*:669 (1975).
318. P. Dean, M. Morgan, E. George, F. Quadri, V. Bouriotis, P. Leatherbarrow, and J. Angal, *Proc. 3rd International Symposium on Affinity Chromatography and Molecular Interaction* (J. Egly, ed.), Strasbourg, France, June 26–29, Inserm, Paris, 1979, p. L8.
319. D. Watson, M. Harvey, and P. Dean, *Biochem. J., 173*:591 (1978).
320. J. Young and B. Webb, *Anal. Biochem., 88*:619 (1978).
321. P. Arnaud and E. Gianazza, *FEBS Lett., 137*:157 (1982).
322. C. Westbrook, Y. Lin, and J. Jarabak, *Biochem. Biophys. Res. Commun., 76*:943 (1977).
323. D. van der Jagt and L. Davidson, *Biochim. Biophys. Acta, 484*:260 (1977).
324. W. Heyns and P. DeMoore, *Biochim. Biophys. Acta, 370*:102 (1974).
325. S. Thompson, K. Cass, and E. Stellwagen, *Proc. Natl. Acad. Sci. USA, 72*:669 (1975).
326. E. P. Lau, B. E. Haley, and R. E. Berden, *Biochemistry, 16*:2581 (1977).
327. Z. Shen, D. Fice, and D. M. Byers, *Anal. Biochem., 204*:34 (1992).
328. E. Stellwagen and B. Baker, *Nature, 261*:719 (1976).
329. G. D. Markham and G. H. Reed, *Arch. Biochem. Biophys., 184*:24 (1977).
330. R. Easterday and I. Easterday, *Adv. Exp. Med. Biol., 42*:123 (1974).
331. A. J. Turner and J. Hryszko, *Biochim. Biophys. Acta, 613*:256 (1980).
332. N. Tamaki, M. Nakamura, K. Kimura, and T. Hama, *J. Biochem. (Jpn.), 82*:73 (1977).
333. R. A. Boghosian and E. T. McGuinness, *Biochim. Biophys. Acta, 567*:278 (1979).

334. B. Wermuth, H. Bürgisser, K. Bohren, and J. P. von Wartburg, *Eur. J. Biochem., 127*:443 (1982).
335. V. Nikodem, R. Johnson, and J. Fresco, *Fed. Proc., 36*:822 (1977).
336. J. O. Capobianco, C. G. Lerner, and R. C. Goldman, *Anal. Biochem., 204*:96 (1992).
337. Z. Rose and S. Dube, *Arch. Biochem. Biophys., 177*:284 (1976).
338. E. Dicou and P. Brachet, *Biochem. Biophys. Res. Commun., 102*:1172 (1981).
339. M. Mori and P. Cohen, *Fed. Proc., 37*:1341 (1978).
340. R. Roskoski, C. Lim, nad L. Roskoski, *Biochemistry, 14*:5105 (1975).
341. S. Subramanian, *Prep. Biochem., 17*:297 (1987).
342. A. H.-J. Wang, M. L. Sherman, and A. Rich, *Biochem. Biophys. Res. Commun., 82*:150 (1978).
343. Y. Cheng and B. Domin, *Anal. Biochem., 85*:425 (1978).
344. H. White and W. Jencks, *J. Biol. Chem., 251*:1708 (1976).
345. M. Morrill, S. Thompson, and E. Stellwagen, *J. Biol. Chem., 254*:43271 (1979).
346. M. Gorgi, D. Piscitelli, P. Rossi, and R. Geremia, *Biochem. Biophys. Acta, 1121*:178 (1992).
347. G. A. Grant, L. M. Keeper, and R. A. Bradshaw, *J. Biol. Chem., 253*:2724 (1978).
348. R. J. McCandliss, M. D. Poling, and K. M. Herrmann, *J. Biol. Chem., 253*:4259 (1978).
349. M. Deibel and D. Ives, *J. Biol. Chem., 252*:8235 (1977).
350. B. Chambers and R. Dunlap, *J. Biol. Chem., 254*:6515 (1979).
351. T. R. Unnasch and G. M. Brown, *J. Biol. Chem., 257*:14211 (1982).
352. J. S. Keruchenko, I. D. Keruchenko, K. L. Gladilin, V. N. Zaitsev, and N. Y. Chirgadze, *Biochem. Biophys. Acta, 1122*:85 (1992).
353. W. Hilt, G. Pfleiderer, and P. Fortnagel, *Biochem. Biophys. Acta, 1076*:298 (1991).
354. B. A. Hemmings, *J. Biol. Chem., 253*:5255 (1978).
355. D. Masters and B. Rowe, *Fed. Proc., 38*:724 (1979).
356. L. Messenger and H. Zalkin, *J. Biol. Chem., 254*:3382 (1979).
357. J. Edgar and R. Bell, *Fed. Proc., 36*:857 (1977).
358. J. C. Ball and D. L. van der Jagt, *Anal. Biochem., 98*:472 (1979).
359. T. Spector, T. E. Jones, and R. L. Miller, *J. Biol. Chem., 254*:2308 (1979).
360. J. Zwiller and P. Mandel, *C. R. Acad. Sci. Paris, 286D*:423 (1978).
361. Yung-Chi Cheng and B. Domin, *Anal. Biochem., 85*:425 (1978).
362. Z. Hrkal, P. Cabart, and I. Kalousek, *Biomed. Chromatogr., 6*:212 (1992).
363. Z. Beg, J. Stonik, and H. Brewer, *FEBS Lett., 80*:123 (1977).
364. W. Jankowski, W. von Münchhausen, E. Sulkowski, and W. Carter, *Biochemistry, 15*:5182 (1976).
365. S. Yonehara, Y. Yanase, T. Sano, M. Imai, S. Nakasawa, and H. Mori, *J. Biol. Chem., 256*:3770 (1981).
366. G. Seelig and R. Colman, *J. Biol. Chem., 252*:3671 (1977).
367. L. Ryan and C. Vestling, *Arch. Biochem. Biophys., 160*:279 (1974).
368. J. Chung, D. Abano, G. Fless, and A. Scanu, *J. Biol. Chem., 254*:7456 (1979).
369. S. Rajgopal and M. Vijayalakshmi, *J. Chromatogr., 243*:164 (1982).
370. R. G. Mathews and B. J. Haywood, *Biochemistry, 18*:4845 (1979).
371. J. Aradi, A. Zsindely, A. Kiss, and M. Schablik, *Prep. Biochem., 12*:137 (1982).

372. D. Apps and C. Gleed, *Biochem. J., 159*:6145 (1976).
373. M. Guerrero and M. Gutierrez, *Biochim. Biophys. Acta, 482*:272 (1977).
374. P. Greenbaum, K. Prodouz, and R. Garrett, *Biochim. Biophys. Acta, 526*:52 (1978).
375. F. Schuber and M. Pascal, *Biochim., 59*:735 (1977).
376. L. Solomonson, *Plant Physiol., 56*:853 (1975).
377. N. Amy, R. Garrett, and B. Anderson, *Biochim. Biophys. Acta, 480*:83 (1977).
378. P. Reyes and R. Sandquist, *Anal. Biochem., 88*:522 (1978).
379. R. Sekura and W. Jakoby, *J. Biol. Chem., 254*:5658 (1979).
380. J. Oka, K. Ueda, and O. Havaishi, *Biochim. Biophys. Res. Commun., 80*:841 (1978).
381. B. Nichols, T. Lindell, E. Stellwagen, and J. Donelson, *Biochim. Biophys. Acta, 526*:410 (1978).
382. J. Drocourt, D. Thang, and M. Thang, *Eur. J. Biochem., 82*:355 (1978).
383. R. Pannell and R. Newman, *Fed. Proc., 37*:1528 (1978).
384. J. Demaille, K. Peters, and E. Fischer, *Biochemistry, 16*:3080 (1977).
385. K. Abbe and T. Yamada, *J. Bacteriol., 149*:299 (1982).
386. D. N. Rao and N. A. Rao, *Plant Physiol., 69*:11 (1982).
387. J. C. Braman, M. J. Black, and J. H. Mangum, *Prep. Biochem., 11*:23 (1981).
388. T. Ono, K. Nakazono, and H. Kosaka, *Biochim. Biophys. Acta, 709*:84 (1982).
389. M. Deibel and M. Coleman, *Fed. Proc., 38*:484 (1979).
390. L. F. Bisson and J. Thorner, *J. Biol. Chem., 256*:12456 (1981).
391. G. G. Meadows and G. S. Cantwell, *Res. Commun. Chem. Pathol. Pharmacol., 30*:535 (1980).
392. T. Atkinson, P. M. Hammond, R. D. Hartwell, P. Hughes, M. D. Scawen, R. F. Sherwood, D. A. P. Small, C. J. Bruton, M. J. Harvey, and C. R. Lowe, *Biochem. Soc. Trans., 9*:290 (1981).
393. D. G. Williams, B. G. H. Byfield, and D. W. Moss, *Enzyme, 28*:28 (1982).
394. N. Harris and P. Byfield, *FEBS Lett., 103*:162 (1979).
395. A. J. Spano and M. Timko, *Biochem. Biophys. Acta, 1076*:29 (1991).
396. S. Fulton, *Dye-Ligand Chromatography*, Amicon Corporation, Lexington, Mass., 1980.
397. M. Naumann, R. Reuter, P. Metz, and G. Kopperschläger, *J. Chromatogr., 466*:319 (1989).
398. A. J. Balmforth and A. Thomson, *Biochem. J., 218*:113 (1984).
399. A. H. Schmidt and H. Wombacher, *J. Chromatogr., 519*, 379 (1990).
400. C. Bruck, D. Portetelle, C. Glineur, and A. Bollen, *J. Immunol. Methods, 53*:313 (1982).
401. J. G. Izant, J. A. Weatherbee, and J. R. McIntosh, *Nature, 295*:248 (1982).
402. V. E. Barskii, V. B. Ivanov, G. I. Mikhailov, and Yu. E. Sklyar, *Cytologia (Rus.), 17*:987 (1975).
403. G. Wulff, J. Vietmeier, and H. J. Poll, *Makromol. Chem., 188*:731 (1987).
404. R. Arshady and K. Mosbach, *Makromol. Chem., 182*:687 (1981).
405. R. Axen, J. Porath, and S. Ernback, *Nature, 214*:1302 (1967).
406. M. Wilchek and T. Miron, *Methods Enzymol., 34*:74 (1974).
407. J. F. Kennedy, J. A. Barnes, and J. B. Matthews, *J. Chromatogr., 196*:379 (1980).
408. I. Parikh, S. March, and P. Cuatrecasas, *Methods Enzymol., 34*:96 (1974).

409. T. H. Hseu, S. L. Lan, and M. D. Yang, *Anal. Biochem., 116*:181 (1981).
410. Lin Peng, G. J. Calton, and J. W. Burnett, *Enzyme Microb. Technol., 8*:681 (1986).
411. J. Kohn and M. Wilchek, *Biochem. Biophys. Res. Commun., 107*:878 (1982).
412. J. Kohn and M. Wilchek, *FEBS Lett., 154*:209 (1983).
413. J. Kohn, R. Lenger, and M. Wilchek, *Appl. Biochem. Biotechnol., 8*:227 (1983).
414. L. Sundberg and J. Porath, *J. Chromatogr., 90*:87 (1974).
415. G. S. Murthy and N. R. Mougdal, *J. Biosci., 10*:351 (1986).
416. J. Porath and N. Fornstedt, *J. Chromatogr., 51*:479 (1970).
417. A. Nishikawa and P. Bailon, *J. Solid-Phase Biochem., 1*:33 (1976).
418. L. Sundberg and J. Porath, *Protides and Biological Fluids, Proc. 23rd Colloquium*, Brugge, Belgium, 1975, Oxford University Press, Oxford, 1976, p. 517.
419. N. L. Smith and H. M. Lenhoff, *Anal. Biochem., 61*:392 (1974).
420. S. Biagioni, R. Sisto, A. Ferraro, P. Caiafa, and C. Turano, *Anal. Biochem., 89*:616 (1978).
421. T. C. Gribnau, G. I. Tessler, and R. J. F. Nivard, *J. Solid-Phase Biochem., 3*:1 (1978).
422. J. M. Egly, E. Boschetti, and M. Monsigny, *Ultrogel, Magnogel and Trisacryl. Practical Guide for Use in Affinity Chromatography and Related Techniques*, Reactifs IBF, Société Chimique Pointet-Girard, 1983.
423. N. L. Stults, L. M. Asta, and Y. C. Lee, *Anal. Biochem., 180*:114 (1989).
424. J. M. Guisán, *Enzyme Microb. Technol., 10*:375 (1988).
425. N. L. Stults, P. Lin, M. Hardy, Y. C. Lee, Y. Uchida, Y. Tsukada, and T. Sugimori, *Anal. Biochem., 135*:392 (1983).
426. D. J. Tuma, T. M. Donohue, V. A. Medina, and M. F. Sorrell, *Arch. Biochem. Biophys., 234*:377 (1984).
427. J. C. Cabacungan, A. I. Ahmed, and R. E. Freenay, *Anal. Biochem., 124*:272 (1982).
428. W. S. D. Wong, D. T. Osuga, and R. E. Feeney, *Anal. Biochem., 139*:58 (1984).
429. M. Wilchek, T. Miron, and J. Kohn, *Ann. N.Y. Acad. Sci., 434*:254 (1986).
430. M. Wilchek and T. Miron, *Biochem. Int., 4*:629 (1982).
431. G. S. Bethell, S. J. Ayers, M. T. W. Hearn, and W. S. Hancock, *J. Chromatogr., 219*:353 (1981).
432. G. S. Bethell, J. Ayers, S. Hancock, and M. T. W. Hearn, *J. Biol. Chem., 254*:2572 (1979).
433. J. Brandt, L. O. Andersson, and J. Porath, *Biochim. Biophys. Acta, 386*:196 (1975).
434. K. Nilsson and K. Mosbach, *Eur. J. Biochem., 112*:397 (1980).
435. K. Nilsson and K. Mosbach, *Biochem. Biophys. Res. Commun., 102*:449 (1981).
436. P. D. Weston and S. Avrameas, *Biochem. Biophys. Res. Commun., 45*:1574 (1971).
437. J. Turkova, *Affinity Chromatography*, Elsevier, Amsterdam, 1978.
438. L. A. Osterman, *Methods of Protein and Nucleic Acid Research*, Vol. 3, *Chromatography*, Springer-Verlag, New York, 1986.
439. T. Korpela and A. Hinkkanen, *Anal. Biochem., 71*:322 (1976).
440. S. Reuveny, A. Mizrahi, M. Kotler, and A. Freeman, *5th General Meeting of ESACT*, Copenhagen, Denmark, *Dev. Biol. Stand., 55*:11 (1982).
441. J. K. Inman and H. M. Dintzis, *Biochemistry, 8*:4074 (1969).
442. P. J. Duerksen and K. D. Wilkinson, *Anal. Biochem., 160*:444 (1987).

443. A. T. Jagendorf, A. Patchornik, and M. P. Sela, *Biochim. Biophys. Acta, 78*:515 (1963).
444. P. L. Domen, J. R. Nevens, A. Krishna Mallia, G. T. Hermanson, and D. C. Klenk, *J. Chromatogr., 510*:293 (1990).
445. J. V. Babashak and T. M. Phillips, *J. Chromatogr., 476*:187 (1989).
446. J. F. Roland, R. J. Wargel, W. L. Alm, S. P. Kiang, and F. M. Bliss, *Appl. Biochem. Biotechnol., 9*:15 (1984).
447. Dj. Josic, W. Hoffmann, R. Habermann, J.-D. Schulzke, and W. Reutter, *J. Clin. Chem. Clin. Biochem., 26*:559 (1988).
448. S. Minoba, T. Watanabe, T. Sato, T. Tosa, and I. Chibata, *J. Chromatogr., 248*:401 (1982).
449. P. O'Carra, S. Barry, and T. Griffin, *FEBS Lett., 43*:169 (1974).
450. P. Cuatrecasas and I. Parikh, *Biochemistry, 11*:2291 (1972).
451. D. J. O'Shannessy and W. L. Hoffmann, *Biotechnol. Appl. Biochem., 9*:488 (1987).
452. L. A. Cohen, *Methods Enzymol., 34*:102 (1974).
453. M. Landt, S. Boltz, and L. Butler, *Biochemistry, 17*:915 (1978).
454. D. Wu and R. R. Walters, *J. Chromatogr., 458*:169 (1988).
455. B. Belleau and G. Malek, *J. Am. Chem. Soc., 90*:1651 (1968).
456. P. V. Sundaram, *Biochem. Biophys. Res. Commun., 61*:717 (1974).
457. E. Boschetti, M. Gorgier, and R. Garelle, *Biochimie, 60*:425 (1978).
458. J. L. Torres, R. Guzman, R. G. Carbonell, and P. K. Kilpatrick, *Anal. Biochem., 171*:411 (1988).
459. R. Kaul, U. Olsson, and B. Mattiasson, *J. Chromatogr., 438*:339 (1988).
460. R. W. Stout and J. J. DeStefano, *J. Chromatogr., 326*:63 (1985).
461. H. W. Jarrett, *J. Chromatogr., 405*:179 (1987).
462. K. Ernst-Cabrera and M. Wilchek, *J. Chromatogr., 397*:187 (1987).
463. S. K. Bhatia, L. C. Shriver-Lake, K. L. Prior, J. H. Georger, J. M. Calvert, R. Bredehorst, and F. S. Ligler, *Anal. Biochem., 178*:408 (1989).
464. L. Limbird, *Cell Surface Receptors: A Short Course of Theory and Methods*, Martinus Nijhoff, Hingham, Mass., pp. 133–158.
465. G. E. Tiller, T. J. Muller, M. E. Dockter, and W. G. Struve, *Anal. Biochem., 141*:262 (1984).
466. T. Horigone and H. Sugano, *J. Chromatogr., 283*:315 (1984).
467. P. C. Adams, F. D. Roberts, L. W. Powell, and J. W. Halliday, *J. Chromatogr., 427*:341 (1988).
468. H. Suzuki and T. Terada, *Anal. Biochem., 172*:259 (1988).
469. W. H. Velander, C. L. Orthner, J. P. Tarakhan, R. D. Madurawe, A. H. Ralston, D. K. Strickland, and W. N. Drohan, *Biotechnol. Prog., 5*:119 (1989).
470. P. Haezebrouck, W. Noppe, H. van Dael, and I. Hanssens, *Biochem. Biophys. Acta, 1122*:305 (1992).
471. P. D. G. Dean, P. Brown, M. J. Leyland, D. H. Watson, S. Angal, and M. J. Harvey, *Biochem. Soc. Trans., 5*:1111 (1977).
472. M. Weiss and Z. Eisenstein, *J. Liquid Chromatogr., 10*:2815 (1987).
473. G. R. Rashbaum and J. Everse, *Anal. Biochem., 90*:146 (1978).
474. H. Görisch, *Anal. Biochem., 173*:393 (1988).

475. Yew Phew See, P. M. Olley, and G. Jackowski, *Electrophoresis, 6*:283 (1985).
476. H. Nakamura, T. Konishi, and M. Kamada, *Anal. Sci., 6*:137 (1990).
477. J. Porath, J.-C. Janson, and T. Låås, *J. Chromatogr., 60*:167 (1971).
478. French Patent 7,702,391, February, 1979.
479. R. Axen, J. Porath, and S. Ernback, *Nature, 214*:1302 (1967).
480. J. Kohn and M. Wilchek, *Biochem. Biophys. Res. Commun., 107*:878 (1982).
481. L. Sundberg and J. Porath, *J. Chromatogr., 90*:87 (1974).
482. J. Porath and N. Fornstedt, *J. Chromatogr., 51*:479 (1970).
483. I. Matsumoto, N. Seno, A. M. Golovtchenko-Matsumoto, and T. Osava, *J. Biochem., 87*:535 (1980).
484. I. Matsumoto, Y. Ito, and N. Seno, *J. Chromatogr., 239*:747 (1982).
485. N. L. Smith and H. M. Lenhoff, *Anal. Biochem., 61*:392 (1974).
486. T. C. Gribnau, G. I. Tessler, and R. J. F. Nivard, *J. Solid-Phase Biochem., 3*:1 (1978).
487. M. Wilchek and T. Miron, *Biochem. Int., 4*:629 (1982).
488. J. Brandt, L. O. Andersson, and J. Porath, *Biochim. Biophys. Acta, 386*:196 (1975).
489. K. Nilsson and K. Mosbach, *Eur. J. Biochem., 112*:397 (1980).
490. K. Nilsson and K. Mosbach, *Biochem. Biophys. Res. Commun., 102*:449 (1981).
491. P. D. Weston and S. Avrameas, *Biochem. Biophys. Res. Commun., 45*:1574 (1971).
492. F. Koller and O. Hoffmann-Ostenhof, *Hoppe-Seyler's Z. Physiol. Chem., 360*:507 (1979).
493. M. C. Cress and That N. Ngo, *Am. Biotechnol. Lab., 7*:16 (1989).
494. V. S. Horsney, C. V. Prowse, and D. S. Pepper, *J. Immunol. Methods, 93*:83 (1986).
495. H. W. Jarrett, *J. Chromatogr., 405*:179 (1987).
496. Danlin Wu and R. R. Walters, *J. Chromatogr., 458*:169 (1988).
497. R. W. Stout and J. J. DeStefano, *J. Chromatogr., 326*:63 (1985).
498. G. Szabó, K. Offenmüller, and E. Csató, *Anal. Chem., 60*:213 (1988).
499. A. Pusztai, G. Grant, J. C. Stewart, and W. B. Watt, *Anal. Biochem., 172*:108 (1988).

2

Support Materials for Affinity Chromatography

Per-Olof Larsson

University of Lund, Lund, Sweden

I. INTRODUCTION

Ideally, an affinity chromatography support should allow the biomolecules rapid and unhindered access to the immobilized affinity ligand. Otherwise, it should adopt a completely passive role. These are not trivial requirements. The unhindered and speedy access by large molecules requires, for example, small particles with large pores. A completely passive role requires, for example, chemical inertness toward the chromatographic solvents and the solutes, absence of ionic or hydrophobic groups, and inertness toward the physical–mechanical strain exerted on the particles during the chromatographic process. To make such an ideal support useful, it must obviously be available at a reasonable price. Understandably, no such ideal support exists, since many of the requirements are in open conflict with each other. Available support materials are therefore compromises, although they may be geared toward one type of applications. Extreme examples are micrometer-sized nonporous silica particles which are optimized for extremely rapid analytical or micropreparative separations and soft 0.1-mm agarose particles, which are optimized for standard, preparative applications.

It may be rewarding to consider in some detail what properties an ideal support should have, thereby making it easier to select a real support material, well adapted for the separation task at hand.

II. PROPERTIES OF SUPPORT MATERIALS

A. Inertness Toward Proteins: Chemical Character

A successful affinity adsorbent binds the desired biomolecule firmly and specifically while leaving all other molecules in the process stream untouched. The requirement for such an ideal situation is first, that the affinity ligand is capable of forming a suitably firm complex with the biomolecule of interest. Second, the support material must be inert to proteins in general to avoid the simultaneous binding of nondesired proteins. In other words, the support material should have a chemical character very similar to that of the medium in which it is operating. Almost without exception, affinity chromatography separations occur in aqueous solutions. The support should thus be as hydrophilic as possible. As a rule, the medium has low ionic strength. The support should therefore contain as few charges as possible. Many support materials available today fulfill these requirements very satisfactorily, either because their basic structure has the desired properties or because they are provided with a protein-inert coating.

A well-known example is the polysaccharide agarose, which is by far the most used matrix in affinity chromatography applications. Agarose particles (e.g., Sepharose from Pharmacia or Bio-Gel A from Bio-Rad) consist of polymeric chains of the disaccharide agarobiose (D-galactose and 3,6-anhydro-1-galactose). The individual polymeric chains are clustered together in bundles and form a porous and hydrophilic network, where the groups facing the solvent have a minimum tendency to attract proteins (see Chapter 1).

Another example is silica. The silica-based material is certainly hydrophilic but is totally unsuitable for affinity chromatography unless it has been surface modified. The reason is that the native surface is covered primarily by silanol groups. The silanol groups are weak acids giving the silica surface a strong negative charge. These charges in combination with other binding forces often irreversibly adsorb proteins to native silica. However, several schemes have been worked out to render the surface inert toward proteins. One example is coating with γ-glycidoxypropyltriethoxysilane followed by acid hydrolysis of the epoxy groups, leaving the silica covered with a hydrophilic, noncharged surface layer (1) (see Chapter 1).

A support material should be inert toward proteins, but to make it suitable for affinity chromatography it should also be easily provided with ligands. Since most support materials are rich in hydroxyl groups, the chemistries developed for the attachment of ligands have been focused on these groups as anchoring points. A wealth of methods are available (2–5) (see Chapter 1).

A support material with a tendency to adsorb proteins in a nonspecific manner may still be useful, provided that the medium is modified accordingly. By changing the buffer medium in such a way that its chemical character (hydrophilicity, ion strength) coincides with that of the support, there will be a minimum tendency toward nonspecific adsorption. One simple remedy is often to carry out the affinity separations in a well-buffered system with an ionic strength of about 0.1 to suppress the effect of charged groups also present in "inert" support materials.

Even if the basic support material is essentially without nonspecific interactions with proteins, the final affinity adsorbent could have obtained a dramatically changed character. The coupling procedures may introduce undesired groups, or the ligand itself may confer nonspecific interactions. A well-known example is activation with cyanogen bromide of agarose and other polysaccharide supports. In its standard protocol a very large excess of BrCN reagent is used (Chapter 1). The excess is necessary because the alkaline conditions lead to a substantial breakdown of the active cyanate ester groups. Several breakdown products, including charged ones, may be formed, causing unspecific interactions with chromatographed molecules. A modified protocol includes activation at low temperature in the presence of a cyanogen transfer agent, resulting in a less altered support (Chapter 1).

Also, the ligand itself may be responsible for unwanted interactions. Assume that agarose beads are provided with an adenosine triphosphate derivative (ATP) and that the final support has a ligand concentration of 10 to 15 μmol per milliliter of packed bed. This is a quite normal degree of substitution and should be suitable for retaining ATP-dependent proteins. However, the substitution has also made the agarose a cation exchanger. At pH 7 the charge concentration on the gel will be about 50 μmol per milliliter of packed bed. Taking into account the multivalency of the charged triphosphate groups, the medium should probably be buffered with a 0.1 to 0.2 *M* buffer to cancel the ion-exchange properties of the affinity gel. This effect should be considered when an adsorbent is designed. To avoid unspecific effects and thereby improve the overall efficiency, it may be wise to avoid unnecessarily high ligand concentrations.

A related phenomenon is the introduction of unwanted binding via the ligand spacer. The spacer molecule serves the function of making the ligand more accessible to the chromatographed macromolecule. In many reported cases hexamethylene diamine has been used as a spacer. Each such spacer introduces both charges and a patch of hydrophobicity on the support surface. Sometimes these extra functionalities may act cooperatively with the ligand and give an even stronger binding of the desired protein. However, in most cases they will lower the performance of the adsorbent, due to retention of undesired proteins. This phenomenon was observed and investigated early in the development of affinity chromatography and led to the development of a special kind of affinity chromatography, hydrophobic interaction chromatography.

B. Chemical Stability

Obviously, the affinity adsorbent should be chemically stable under operating conditions. This includes stability toward enzymes and microbes in the process stream, elution buffers, and regenerating and cleaning agents. The commonly used agarose-based supports are in this respect almost ideal, especially in their cross-linked form (e.g., Sepharose CL-4B from Pharmacia). They are not attacked by enzymes, they can be used continuously between pH 3 and 12, and they withstand all commonly used water-based eluants without shrinking and swelling. An additional attractive feature is that cross-linked agarose easily withstands sanitation with 0.5 *M* sodium hydroxide. Especially in industry, regular sanitation is important to achieve desirable product quality and reliable performance. A preferred way of doing this is cleaning in place (CIP) with a strong sodium hydroxide solution. The strongly alkaline solution is an active bactericide and efficiently removes otherwise essentially irreversibly deposited material (particles, denatured proteins, lipids, and other compounds). Such deposits may contaminate the product in later runs and lead to clogging of the column and its ultimate collapse. Also of importance is the stability of agarose at high temperatures. Cross-linked agarose may thus be sterilized by autoclaving at 121°C.

Few other support materials can match the chemical stability of agarose. Organic polymeric supports made from, for example, polystyrene are inherently very stable but are usually provided with surface layers to make them protein compatible, layers that are less stable. Especially vulnerable to hydrolytic conditions are inorganic materials such as porous glass and silica. These supports should not be used above pH 8, preferably not even above pH 7, for any prolonged period of time, due to the hydrolysis of the silica structure. However, silica materials are rarely used without a coating to make them more inert toward proteins, as discussed above. To a certain extent such a coating will also shield and protect the silica from hydrolysis. Furthermore, certain brands of silica supports have been treated by a special process that incorporates zirconium in the surface structures, a procedure that considerably improves stability in an alkaline environment.

The discussion above may suggest that provided agarose is chosen, the affinity material will be satisfactorily stable in all situations encountered. This is not the case, since very often the weak point is not the matrix itself but the ligand or the anchoring chemistry. For example, derivatives of adenosine monophosphate (AMP) or nicotinamide-adenosine dinucleotide (NAD) are very efficient ligands for purifying NAD-dependent dehydrogenases but are delicate molecules prone to breakdown, either spontaneously or aided by hydrolytic enzymes. Less efficient but vastly more stable ligands such as reactive dyes (e.g., Cibacron Blue) may therefore be preferred, especially in large-scale applications, where the cost of the separation material may otherwise be prohibiting.

The anchoring of the ligands to the matrix is another weak point. The often used BrCN method, for example, is very convenient but leads to an isourea bridge between support and ligand which is slowly hydrolyzed at alkaline pH. Alternative methods give a more stable product: for example, coupling via ether linkages, using bisepoxides (see Chapter 1).

C. Mechanical Stability

A chromatographic support material should have mechanical stability sufficient to withstand the pressure drop generated over the column when the column is run at optimum speed for the separation process. Most packing materials meet this requirement satisfactorily in well-behaved systems. However, often the source from which the desired protein is purified contains many substances that may foul the separation bed. Deposits of lipids, denatured proteins, and particulate contaminants may restrict the flow and quickly raise the back pressure to unacceptable levels. In the case of soft gels such as standard agarose beads, high back pressure will compress the bed, thereby increasing the pressure even further and ultimately causing collapse of the bed. Mechanically strong supports such as silica or heavily cross-linked polymers will not collapse, but the high back pressure is unwanted. These problems are accentuated in large-scale purifications where the adsorbent is used repeatedly, allowing the gradual buildup of deposits.

Fine-grade support materials are more sensitive to particulate impurities in the process stream than are larger particles, due to their filtering effect. Also, column fouling due to the deposition of (denatured) proteins, lipids, and other substances is more accentuated with small-particle beds. High-performance liquid chromatography (HPLC)-grade materials (0.05 to 0.010 mm) thus require carefully filtered solutions and more attention to details to avoid situations that might lead to clogging of the column. Clogging may also occur with standard columns but is often easier to deal with from a practical point of view. For example, it is usually the topmost layer that becomes contaminated. It is easy to discard this contaminated layer without noticeable deterioration of separation performance in standard (open) columns. This is contrary to the situation in HPLC columns, which may require repacking.

An advantage with HPLC-grade materials, on the other hand, is their mechanical ruggedness, as they are constructed for high flow rates that will give high back pressure. Consequently, deposition of foreign materials may raise the back pressure considerably, but it may still be acceptable for the support.

D. Pore Size Requirements

The ideal affinity support material should allow unhindered access to the immobilized ligands. This requires large pores, of at least five times the diameter of the chromatographed protein, to avoid severely restricted diffusion rates. For a protein of normal size this means pores larger than 300 Å. Several common support materials are available in such pore sizes. For example, agarose supports based on 4% agarose have a pore size around 300 Å. Support materials based on silica, polystyrene, and polymethacrylates are available in many porosities, including very wide ones. Silica particles are, for example, available with 4000-Å pores. Support materials with very large pores give unhindered diffusion but will also have a correspondingly smaller surface area per milliliter of bed volume. The reduced surface area will lead to diminished binding capacity. As a rule of thumb, a pore size of 300 to 500 Å is usually a good compromise in most situations, giving fairly unrestricted diffusion and a large surface area. As discussed separately below, special benefits may be gained if the support materials contain both normal pores and very wide "through-pores."

E. Choice of Particle Size: Chromatographic Performance

When affinity chromatography was developed during the 1970s, the support materials that were readily available were used, in most cases 0.1-mm agarose particles. Later, affinity chromatography was combined with HPLC techniques (1) which rely on particles with a diameter of 0.005 to 0.010 mm. The development was prompted by the phenomenal success of HPLC, with its ability to separate complex mixtures of small molecules rapidly and later also, mixtures of proteins. The affinity chromatography variety of HPLC (HPAC or HPLAC) is steadily increasing in importance but has not experienced the dramatic improvement in performance that accompanied the transition of LC into HPLC. There are some good reasons for this that it could be illuminating to consider in some detail. To make the differences more clearcut, the discussion below is focused on a comparison between standard-grade particles (0.1 mm) and HPLC-grade particles (0.01 mm). However, it must be emphasized that several support materials are available in intermediate particle sizes, which may well be the optimum choice in many applications.

A general measure of chromatographic efficiency is the plate height [HETP (height equivalent to one theoretical plate)]: the smaller the plate height, the more efficient the separation process. The following expression gives the various contributions to the overall plate height (6):

$$H_{tot} = H_{e.disp.} + H_{l.diff.} + H_{e.diff.} + H_{i.diff.} + H_{kin} \quad (1)$$

$H_{e.disp.}$ (eddy dispersion) relates to the broadening of peaks due to irregularities in the packed bed. Even a well-packed bed will contain microscopic regions where the flow is faster or slower than the average flow, a fact that leads to spreading of an applied sample (Figure 1A). $H_{e.disp.}$ is not influenced by the flow rate. $H_{l.diff.}$ (longitudinal diffusion) relates to the broadening of peaks due to diffusion along the axis of the column (Figure 1B). It becomes important only at very low flow rates, flow rates seldom used in practice. $H_{e.diff.}$ (external diffusion) and $H_{i.diff.}$ (internal diffusion) relate to the broadening of peaks due to the limiting diffusion rate outside and inside the particles, respectively (Figure 1B). Finally, H_{kin} (kinetics) relates to the broadening of peaks due to slow kinetics (i.e., slow binding and slow release of molecules) (Figure 1B and C).

Let us first study the purification of a small peptide with reversed-phase chromatography (hydrophobic interaction chromatography) using either a standard-grade material (0.1 mm) or a HPLC-grade material (0.01 mm). Both packing materials are assumed to be identical with the exception of the particle size.

Table 1 shows that an impressively higher throughput is achieved with the HPLC adsorbent, theoretically a factor of 100 different. However, this improved performance is obtained at a cost—the necessity of using expensive support materials and expensive high-pressure equipment (at the same linear flow velocity (cm/min), with back pressure inversely proportional to the square of the particle diameter). Still, in many applications the use of HPLC technique for preparative separations is considered very cost-effective, especially if the substance to be purified is expensive and unstable.

The question is now to what extent the foregoing benefits of increased capacity will be gained if HPLC materials are used in affinity chromatography separations. The answer is dependent on the affinity system used. Two situations may be distinguished: dynamic affinity chromatography and "normal" affinity chromatography (adsorption–wash–desorption).

1. Dynamic Affinity Chromatography

Dynamic affinity chromatography involves the separation of several substances as a result of their differing migration rate through the affinity column (differing K_{diss} values). It is mainly used analytically and for basic studies of affinity phenomena but seldom for preparative purposes, due to its low capacity. Dynamic affinity separations are chromatographically very similar to the reversed-phase separation described in Table 1. However, it is hardly probable that the transfer to HPLC support will improve the separation capacity of the affinity column by a large factor as was the case for the reversed phase column described in Table 1. The reason can be found in Eq. (1). When discussing the reversed-phase isolation of peptides, it was assumed that the kinetic contribution to the plate height (H_{kin})

A

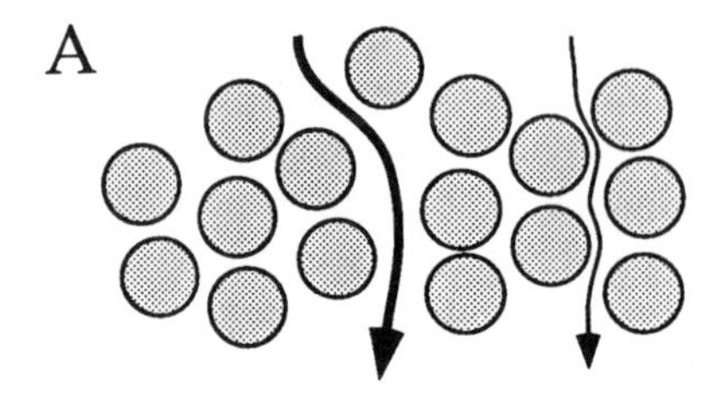

B

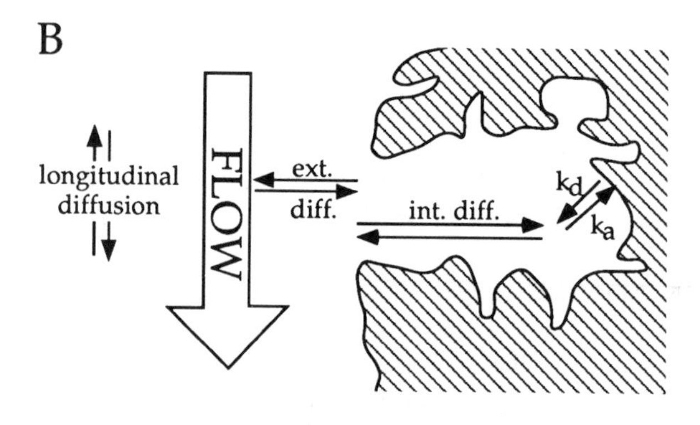

C

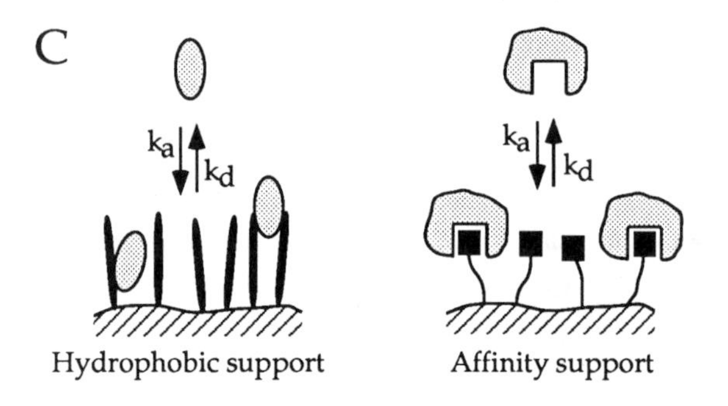

Figure 1 Factors influencing the chromatographic efficiency: (A) uneven flow through the bed: (B) diffusion inside and outside the porous chromatographic particle; (C) formation and dissociation of the complexes.

Table 1 Preparative Purification of a Peptide by Normal- and HPLC-Grade Support Material[a]

Parameter	LC technique	HPLC technique
Column volume (mL)	20	20
Particle size (mm)	0.1	0.01
Plate number	250	250
Flow rate (mL/min)	1	100
HETP (mm)	1	0.1
Column dimensions (cm)	25 × 1	2.5 × 3.2
Operational capacity (mg/h)	6	600
Pressure drop (bar)	0.1	10

[a]The basis for the calculations is the LC column, which is assumed to give a sufficient plate number (250) to resolve the peptide from its impurities when run at 1 mL/min. Based on these data, the HPLC column (same volume) is designed to give the same resolution (i.e., the same plate number). It is assumed that the terms $H_{e.diff.}$ and $H_{i.diff.}$ in Eq. (1) govern the plate height under the conditions used (the plate height will be approximately proportional to the linear flow rate and to the square of the particle diameter). In each separation run 10 mg of substance is loaded. The separation is carried out with a single solvent (10 column volumes of solvent is needed in each separation cycle).

could be neglected. This is not possible with affinity interactions, which may sometimes be very slow and make a substantial contribution to the plate height.

The formation of the affinity complex (k_a in Figure 1C) is usually rapid. The k_a constant may approach the limit set by diffusion, around $10^8\ M^{-1}\ s^{-1}$ or at least it should be within one to two orders of magnitude from this limit. On the other hand, the dissociation step (k_d in Figure 1C) is usually very slow, at least in strong-binding complexes. Intuitively, it is easy to understand that the eluted peaks will be very broad if the dissociation rate constants are of the magnitude $10^{-3}\ s^{-1}$ (i.e., the affinity complexes will need on an average 1000 s to dissociate).

A closer inspection of the H_{kin} term reveals that it is proportional to the flow rate and inversely proportional to the dissociation rate constant (6). Significantly, the H_{kin} term does not contain the particle diameter, meaning that its value will not decrease with diminishing particle size. If the dissociation rate constant is small, the H_{kin} term will dominate. Relatively little will be gained by reducing the particle size since it will only diminish the $H_{e.diff.}$ and $H_{i.diff.}$ terms. Consequently, the theoretically hundredfold improvement in capacity shown in Table 1 cannot be expected. A more realistic improvement is a factor of 1 to 10. The exact factor is clearly dependent on the dissociation rate constant, which usually is not readily available.

In quantitative applications of affinity chromatography this kinetic effect may be used to determine rate constants. If HPLC-type support materials are used, the broadening of peaks due to diffusion can be neglected and instead be attributed to the kinetics of the protein–ligand interaction. However, such studies usually result in considerably lower rate constants than other methods (one to two orders of magnitude), indicating that support-bound affinity complexes have a sluggish kinetics compared to free affinity complexes (7,8).

Dynamic affinity chromatography may be used preparatively, provided that the systems used have reasonably fast kinetics. The "weak affinity" concept (see Chapter 11) thus take advantage of the properties of weak affinity complexes. The moderate equilibrium constant (K_{eq}) for these systems ensures that the effective dissociation rate constant k_d will be fairly high ($k_d = k_a/K_{eq}$).

2. *Normal Affinity Chromatography (Adsorption–Wash–Desorption)*

Normal affinity chromatography purification is a very efficient purification procedure, due to the very high selectivity of the adsorbent for the desired substance (high α value in chromatographic terms). The protein to be purified binds very tightly to the support-bound ligand. The extract containing the protein can therefore be pumped through the column until finally the entire column is saturated with the protein and a more-or-less sharp breakthrough occurs. After washing off impurities, the desired protein is eluted. This is normally carried out by abruptly changing the mobile-phase composition in such a way that the protein immediately loses all its affinity for the immobilized ligand. This may be effected by a high concentration of salt or a substantial change of pH.

To what extent will this process benefit from an HPLC-type support? Clearly, the advantages must be somewhat different from the situation described for reversed-phase chromatography in Table 1, where a sufficiently high plate number had to be reached. The present affinity separation process does not involve dynamic chromatography. It is rather an adsorption–desorption process, and a plate number of 1 would therefore suffice to give a pure product. However, the adsorption and desorption steps may be carried out much faster with HPLC materials, since they will allow faster equilibration with the mobile phase. The desorbed substance may also be obtained in a less diluted form. Furthermore, less washing is needed as the HPLC particles would promote a faster equilibration with the washing medium.

It is difficult to quantify how many times faster the isolation will be with small particles. One uncertain factor is the magnitude of the association rate constant for the formation of the affinity complex. In a typical affinity interaction the association rate constant could be expected to be high, on the order of 10^6 M^{-1} s^{-1} for a protein, although there are exceptions. This would mean that the H_{kin} contribution to band broadening would be small compared to the diffusion effects, at least for large particles. Thus the binding process would greatly benefit from smaller particles, where diffusion problems are less pronounced. Admittedly, Eq. (1) is not intended to cover the precise behavior of affinity chromatography separations according to the adsorption–desorption principle described here. Still, the equation can be used for a qualitative discussion. Another uncertain factor is the dissociation rate constant during release of the adsorbed material. Clearly, it should be geared toward a very high value by the proper choice of eluant. However, this is not always possible.

The bottom line is that it is difficult to predict the improvement in separation capacity (mg/h) when going from normal grade (0.1 mm) particles to HPLC-grade material (10 μm), due to many uncertain factors. The adsorption–washing–deposition processes should in most cases be diffusion restricted with large support particles. This means that a reduction of particle size by a factor of 2 should theoretically improve the operational capacity by a factor of 4. With smaller support particles, on the other hand, these processes will probably also be kinetically controlled, making further reduction in size less rewarding. As a rule of thumb one could therefore assume that an improvement in throughput would be limited to a factor of about 10 when going from standard to HPLC-grade material. Settling for an intermediate particle size may be a profitable alternative, giving almost the same performance as HPLC-grade particles but with much less back pressure.

Large particles can be operated more efficiently than is first anticipated from their size in diffusion-controlled affinity adsorption processes. The trick is to use only a fraction of their total binding (9). Binding to the outer 20% of a large particle is thus very rapid—it takes only 10% of the time needed to saturate the entire particle. Still this outer layer has almost 50% of the total binding capacity of the particle. Admittedly, special arrangements have to be made to make full use of the principle (e.g., fast recirculation through the bed to ensure that all parts of the bed experience the same loading concentration). Also, the washing and elution steps are not substantially improved by this principle, as the substances released (impurities or desired compound) have diffusional access to the entire particle. Still, in a relative sense the principle diminishes the superiority of HPLC-grade particles.

Whether normal- or fine-grade support materials are selected, comparatively short, wide columns should be used when the affinity chromatography is operated according to the adsorption–wash–desorption principle (9,10). As discussed above, there is no need for a certain number of theoretical plates in this kind of separation process. With short, wide columns, much faster throughput is possible. Even a high flow rate (mL/min) would give only moderate back pressure. The large cross section of the column would ensure a moderate linear flow rate (cm/min) through the column, allowing diffusional processes and the binding process enough time for effective adsorption.

III. SUPPORTS WITH SPECIAL PROPERTIES

As discussed above, agarose has many attractive features as a base material for affinity chromatography supports, but in its native form it cannot withstand high back pressures. A satisfactory remedy is chemical cross-linking. An alternative way of making the agarose cope with high flow rates is to prepare composite materials of agarose and mechanically stable structures such as kieselguhr (e.g., Macrosorb K from Sterling Organics). Here the very wide pores of kieselguhr are filled with agarose gel. As an adsorbent the composite will have the properties of non-cross-linked agarose but with much higher mechanical strength (11).

Small particles improve the chromatographic process by diminishing diffusion effects as discussed above. Unfortunately, small particles also lead to high back pressure, as the pressure is inversely proportional to the square of the particle diameter. However, special supports are available that in a sense circumvent this unfortunate relation between performance and pressure.

Very thin (0.5 μm) nonporous silica fibers were coated with a molecular layer of dextran (Figure 2) to which an affinity ligand was coupled (12). The fibers were packed in short columns (e.g., 1 cm height, 10 cm diameter) and used to purify lactate dehydrogenase from crude extracts. Despite the thin fiber diameter, the back pressure was very low, which is explained by the fact that silica fibers form

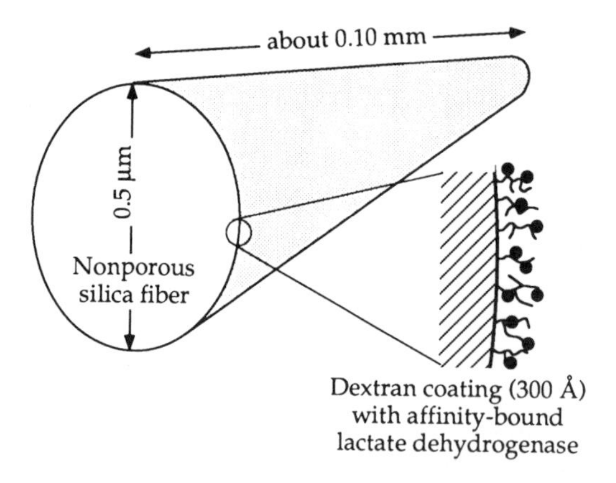

Figure 2 Affinity fibers. Nonporous silica fibers were coated with a molecular layer of dextran, containing a covalently bound NAD ligand. The filled circles represent affinity-bound lactate dehyrogenase molecules. The figure is drawn approximately to scale.

a comparatively loose bed structure. The adsorption desorption kinetics was, as expected, very fast for the essentially nonporous affinity fibers, giving them very high operational capacity (about 1 g of pure enzyme per 100 g of fiber material per hour). However, a distinct drawback with such fiber beds is the low binding capacity per unit bed volume, due to the limited surface area. The use of very thin porous fibers may be a remedy.

Nonporous spherical particles may also be used for affinity chromatography separations (13). Such particles (2 μm in diameter) will have excellent mass transfer characteristics, making them ideal for analytical purposes and for quantitative aspects of affinity chromatography. However, small, spherical particles form very dense beds and therefore give high back pressure. Also, the use of essentially nonporous agarose particles has been reported, suitable for rapid micropreparative affinity separations (14).

Recently, an unusual type of support material with exciting properties has been described (15), based on surface-modified polystyrene (10- to 20-μm particles). Called perfusion gels, these contain at least two sets of pores, very large "through-pores" and normal macropores (Figure 3). The through-pores are wide enough to permit part of the chromatographic flow to be channeled through each individual particle, a phenomenon that never occurs with normal chromatographic supports. A chromatographed substance will thus be transported by flow to the inner parts of a particle, leaving only short distances to be covered by the slow diffusion process. These particles thus combine the attractive features of small and large particles for chromatography: high efficiency and low back pressure. The new supports have been used very successfully in ion-exchange chromatography

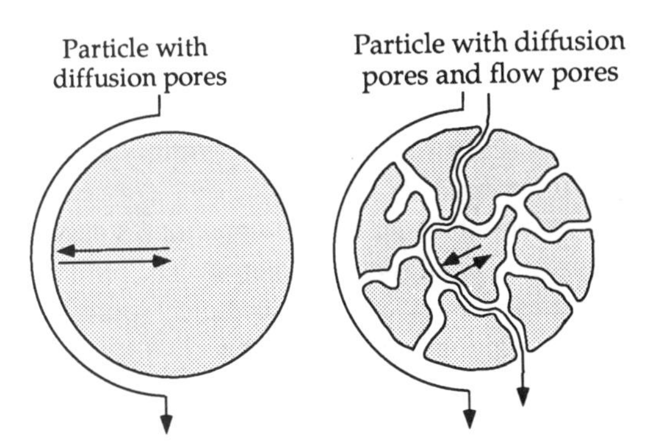

Figure 3 Comparison of chromatographic particles. Normal particles have long diffusion distances, particles with through-pores have short diffusion distances.

applications and should also be of interest in affinity chromatography separations, especially in diffusion-limited affinity systems.

At our laboratory we are presently studying large perfusion particles (0.1 to 0.2 mm) prepared from agarose, which in many aspects is an ideal base material. Such large particles give negligible flow resistance and have about fivefold higher purification capacity than that of corresponding particles without through-pores (16).

IV. OVERALL PERFORMANCE AND COSTS

HPLC-grade particles are much more expensive than standard-size particles, due to their more complicated preparation and due to smaller sales volume. In lab-scale quantities the cost for both 0.1-mm silica and 0.1-mm agarose particles is approximately $30 per 100 g. The cost for the same amount of 0.01-mm material is almost two orders of magnitude as expensive, about $2000. The cost of the base material for small-scale separations (a few milligrams of proteins) may not be important since only a few grams will be needed. In large-scale separations the cost will be substantial if high performance grades are chosen, even if bulk quantities are priced much lower than lab quantities. A careful analysis of costs, chromatographic performance, and a number of other factors must therefore be made before choosing support material for large-scale separations. These factors include the composition of the mixture to be separated, the properties and stability of the product desired, choice of affinity ligand, choice of binding chemistry, design of separation scheme, purity requirements of the product, including demands on absence of pyrogens, released column constituents, and so on, and the availability of equipment.

The spectacular improvement in operational capacity observed in standard liquid chromatography when turning to HPLC-grade support materials cannot be expected in affinity chromatography, the main reason being the slower kinetics of many affinity systems. However, the improvement possible may still be very significant and merit their use, especially if fast separation is of premium importance due to the lability of the protein to be isolated.

In lab-scale separations the cost of support material is less important (as only a few grams are needed) as is its capacity (mg/h/g of support material). The primary concern in laboratory separations is, rather, an overall convenient purification procedure and an easy way of preparing the adsorbent. This clearly favors standard-grade agarose materials, with a reputation for easy handling, easy derivatization, and a generally satisfactory performance. However, if suitable affinity materials are already commercially available in several grades and forms, including prepacked HPLC columns, the choice may be different. The choice of a prepacked column combined with appropriate HPLC equipment should permit rapid and facile development of a separation scheme. The convenient and precise delivery of gradients, washing liquids, and so on, that is possible with HPLC equipment is definitely a bonus. Obviously, it is also possible to combine standard-grade affinity materials with HPLC equipment, although with fewer advantages.

REFERENCES

1. P.-O. Larsson, M. Glad, L. Hansson, S. Ohlson, and K. Mosbach, in *Advances in Chromatography*, Vol. 21 (J. C. Giddings, E. Grushka, J. Cazes, and P. R. Brown, eds.), Marcel Dekker, New York, 1983, p. 19.
2. P. D. G. Dean, W. S. Johnson, and F. A. Middle, eds., *Affinity Chromatography: A Practical Approach*, IRL Press, Oxford, 1985.
3. W. B. Jacoby and M. Wilchek, eds., *Methods in Enzymology*, Vol. 34, *Affinity Techniques*, Academic Press, New York, 1974.
4. W. B. Jacoby, ed., *Methods in Enzymology*, Vol. 104, *Enzyme Purification and Related Techniques*, Academic Press, New York, 1974.
5. C. R. Lowe, *An Introduction to Affinity Chromatography*, North-Holland, Amsterdam, 1979.
6. C. Horváth and H.-J. Lin, *J. Chromatogr., 149*:43 (1970).
7. K. Nilsson and P.-O. Larsson, *Anal. Biochem., 134*:60 (1983).
8. A. J. Muller and P. W. Carr, *J. Chromatogr., 357*:11 (1986).
9. J.-C. Jansson, Large-scale affinity chromatography, *Proc. Bioseparations '90—New Technologies in Upstream and Downstream Processing*, Cambridge, 1990.
10. P. C. Wankat and Y. M. Koo, *A.I.Ch.E. J., 34*:1006 (1988).
11. M. G. Bitte, S. Berezenko, F. J. S. Reed, and L. Derry, *Appl. Biochem. Biotechnol., 18*:275 (1988).
12. P. Wikström and P.-O. Larsson, *J. Chromatogr., 388*:123 (1987).
13. B. Anspach, K. K. Unger, J. Davies, and M. T. V. Hearn, *J. Chromatogr., 457*:195 (1988).
14. J. Li, K.-O. Eriksson, and S. Hjertén, *Prep. Biochem., 20*:107 (1990).
15. N. B. Afeyan, N. F. Gordon, I. Mazsaroff, I. Varady, S. P. Fulton, Y. B. Yang, and F. E. Regnier, *J. Chromatogr., 519*:1 (1990).
16. P.-O. Larsson, Swedish patent application 9200827 (1992).

II

PREPARATIVE APPLICATIONS OF AFFINITY CHROMATOGRAPHY

3

Affinity Chromatography of Enzymes

Felix Friedberg and Allen R. Rhoads

Howard University, Washington, D.C.

I. INTRODUCTION

Purification of enzymes can exploit their generic protein characteristics (a relatively nonspecific interaction) or their binding specificities. In practice, these two recognition modes are often used sequentially, since attempts at bioselective separations at an early stage of purification can encounter nonspecific binding, while physical separation is usually inadequate to distinguish between proteins having similar physical properties.

Affinity separation based on physicochemical properties includes the use of phenyl-Sepharose, heparin–Sepharose, and metal-chelate chromatography. Antibodies are highly biospecific but can still be categorized as recognizing the protein by its structure rather than by its enzymatic activity. Expression cloning has provided a convenient tool for enhancing the efficiency (but not the selectivity) of physical recognition by insertion of an oligopeptide at the terminus of the desired protein (e.g., polyarginine for cation exchange, polyphenylalanine for hydrophobic interaction).

Ligands immobilized for true affinity chromatography of enzymes may be substrates, inhibitors (active-site directed or allosteric), cofactors, or other proteins associated with the biochemical pathways of these enzymes. Subtractive methods (e.g., removing an albumin contaminant with an antibody while allowing the enzyme to pass into the "flow through") are also categorized as affinity chromatography. Specific examples of both kinds of separation are given below.

II. PROTEIN-STRUCTURE BASED

A. Blue-Sepharose CL-6B

Cibacron Blue F3GA, a monochlorotriazinyl dye covalently attached to Sepharose CL-6B (Pharmacia) (Figure 1), has been employed frequently for the purification of dehydrogenases and finds continued use for such purposes (1). For a procedure to attach the triazine dye to a matrix, see Baird et al. (2). Such immobilized dyes can serve as substrate, competitive inhibitor, coenzyme, or effector analog for a variety of enzymes. This is probably due to the fact that the dye can assume the required geometry or polarity. Thus the specificity of these dyes is broad, and nonspecific interactions can contribute significantly to the total binding energy. Not only are nucleotide-requiring enzymes [e.g., dehydrogenases (3,4), kinases (5,6), ligases (7)] bound to the matrix, but other enzymes [e.g., organophosphate phosphatase (8)], albumin, α_2-macroglobulin, and interferon are also retained by such an affinity matrix. On the other hand, immobilized triazine dyes display a higher binding capacity than that of the corresponding specific nucleotide affinity adsorbents (3). To give specificity to this type of affinity chromatography, it is essential to utilize selective elution with a substrate, inhibitor, and so on. It should also be noted that the affinity for such a dye is considerably lessened in the presence of buffers having an ionic strength in excess of 0.2 *M*.

B. Procion Red HE3B

While blue-Sepharose CL-6B is useful for isolating NAD$^+$-linked enzymes, Procion red HE3B (Pharmacia), another triazine dye, covalently attached to the

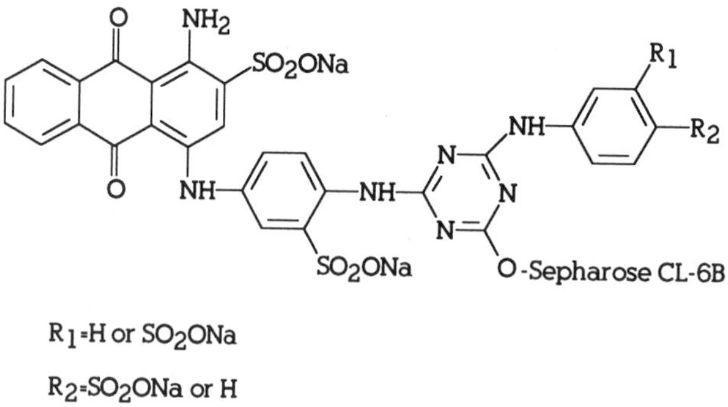

R_1=H or SO_2ONa

R_2=SO_2ONa or H

Figure 1 Partial structure of blue-Sepharose CL-6B.

Sepharose, is often preferred in the purification of $NADP^+$-requiring dehydrogenases. Matrix capacity is particularly high in the region of 70 nmol of immobilized ligand per milliliter of settled resin (3). Alcohol dehydrogenase from *Zymononas mobilis* has been purified by using in tandem Procion green-HE4BD and Procion HB blue-Sepharose CL-4B and eluting the enzyme with buffer containing 1 m*M* NAD^+ (4). This procedure resulted in a 36-fold purification and 80% yield of the enzyme.

C. Heparin–Sepharose CL-6B (Pharmacia)

Heparin-Sepharose CL-6B (Figure 2) has been of general value for the purification of restriction endonucleases and also of lipases. Owing to its polyanionic character, heparin interacts with many cationic molecules, possibly competing with nucleic acid substrates. Whereas the capacity of heparin to bind to lipases is not understood, the ability of heparin to release specific lipases from tissues and to stabilize their activities is well known. For a method of preparing heparin–agarose, see Davison et al. (9). Some specific applications for heparin–Sepharose are summarized in Table 1.

D. Lectin–Sepharose 6MB

Lectins exhibit the ability to react reversibly with specific sugar residues. Wheat germ lectin coupled to Sepharose 6B by the cyanogen bromide method is commercially available as lectin–Sepharose 6MB (Pharmacia). It binds to terminal *N*-acetylglucosaminyl residues and glycoproteins that possess such residues. Immobilized lectins are particularly useful for the isolation of extracellular components, detergent-solubilized cell membranes, subcellular particles, and glycoproteins in general. Similarly, concanavalin A (ConA) shows specificity for molecules that contain α-D-mannopyranosyl or α-D-glucopyranosyl residues. For

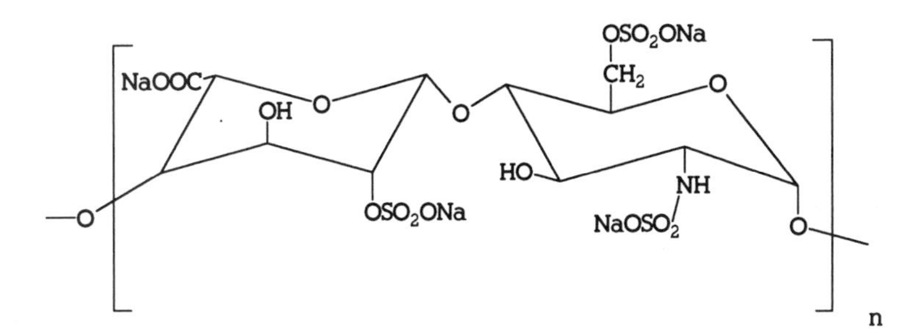

Figure 2 Partial structure of heparin.

Table 1 Enzymes Displaying Affinity for Immobilized Heparin

Enzyme	Buffer/elution condition	Fold purification[a]	Ref.
Hamster liver lipase	0.05 *M* phosphate 0.2 *M* NaCl (pH 7.2) 20% glycerol NaCl increased to 2 *M* NaCl	Not indicated	10
Human plasma lipoprotein lipase	5 m*M* sodium barbital–1.5 *M* NaCl	1500	11
Yeast mitochrondrial nuclease (has RNase, DNA endonuclease, and 5′-exonuclease activity)	0.25 → 0.6 *M* KCl in 0.01 *M* K_2PO_4–K_2HPO_4 (pH 7.4), 5 m*M* $MgCl_2$, 1 m*M* DTT, 15% glycerol, and 0.1% Nonidet P-40	20	12
Yeast DNA exonuclease V	0.03 → 0.6 *M* $(NH_4)_2SO_4$ in 0.025 *M* K_2HPO_4–KH_2PO_4 (pH 7.0), 2 m*M* EDTA and 5 m*M* DTT, 2 m*M* benzamidine, 2 *M* pepstatin A, and 10% glycerol	7.5	13
Bacteriophage N4-induced DNA polymerase (also has 3′ → 5′ exonuclease activity)	0.04 → 0.5 *M* NaCl in 0.1 *M* Tris–HCl (pH 8.0), 1 m*M* EDTA, 0.02 *M* 2-mercapto-ethanol, and 10% glycerol	2	14
Porcine liver mitochondrial DNA polymerase	0 → 0.8 *M* NaCl in 0.015 *M* Tris–HCl (pH 8.0), 5 m*M* 2-mercaptoethanol, 1 m*M* EDTA, and 10% glycerol	40	15
Sulfolobus acidocaldarius DNA polymerase	0.2 → 1 *M* NH_4Cl in 0.05 *M* Tris (pH 7.5), 1 m*M* DTT (0.5 m*M* EGTA)	3	16
Human lymphoblast DNA polymerase-α DNA primase	0.06 → 0.03 *M* K_2HPO_4–KH_2PO_4 (pH 7.8)	3	17
Human liver ribonuclease	0.2 → 1.0 *M* NaCl in 0.02 *M* MES (pH 6.0), 0.2 *M* NaCl	40	18

Table 1 (Continued)

Enzyme	Buffer/elution condition	Fold purification[a]	Ref.
A431 cells phosphatidylinositol kinase	5 m*M* Hepes (pH 7.2), 0.25% Triton X-100, 0.05 *M* NaCl, 2 m*M* phosphatidyl inositol	11	19
Yeast Cl tetrahydrofolate synthase (has 10-CHO-THF synthetase, 5,10-CH^+-THF cyclohydrolase, and 5,10-CH_2-THF dehydrogenase activity)	0.14 *M* KCl	66	20
Bacterial 10-formyltetrahydrofolate synthetase	0.25 *M* KCl	19	20
Tissue plasminogen activator (tPA)	2% SDS, 5% glycerol in phosphate-buffered saline	Not indicated	21

[a] A low "fold" in the last stages of enzyme purification may frequently still be very significant.

the reaction, the binding sugar requires hydroxyl groups at positions C-3, C-4, and C-5. ConA–Sepharose is prepared by coupling ConA to CNBr-activated Sepharose 4B (Pharmacia).

Isoenzymes of adenosine deaminase interact differently with wheat germ lectin and ConA affinity columns, suggesting that the various forms of the enzyme differ in the composition and accessibility of their sugar residues (22). Because many extracellular carboxypeptidases (such as the N-type) contain covalently bound carbohydrate, they can readily be purified by ConA–Sepharose and eluted by 0.1 *M* α-methyl-D-mannoside (23). Angiotensin-converting enzyme (ACE) has also been purified by Con A–Sepharose (24).

E. Thiopropyl-Sepharose 6B

Thiopropyl-Sepharose 6B has a short hydroxypropyl spacer arm and serves as a major tool of covalent chromatography. Ligands containing reactive thiol groups (cysteine residues) can be reversibly immobilized under relatively mild conditions. For instance, an acyl carrier protein (ACP)–Sepharose was employed to confirm that 2-acylglycerol phosphate ethanolamine acyl transferase and acyl ACP synthase activity of *E. coli* are expressed by one enzyme and that this enzyme binds ACP (25). ACP functions as a carrier in fatty acid biosynthesis by forming a thioester linkage between the acyl moiety and its 4′-phosphopantetheine prosthetic group.

For the production of ACP–Sepharose, ACP was reduced with dithiothreitol and treated with 2,2′-dithiodipyridine to yield the 2-thiopyridine derivative of ACP. Activated thiopropyl-Sepharose 6B (Pharmacia) (Figure 3) was also reduced with dithiothreitol and washed to remove the 2-thiopyridine groups. Then the gel and ACP were mixed in the presence of 0.2 *M* NaCl (26). Crude enzyme was applied to the ACP–Sepharose and the acyltransferase/synthetase was eluted as a single peak with 0.5 *M* LiCl (which significantly decreases the activity of the enzyme for ACP) (25).

Recombinant DNA techniques make possible gene fusion of enzymes with specific proteins or specific polyamino acids sometimes called *affinity tails.* Often the tail selected is a short homopolymer. For example, a synthetic DNA fragment encoding four additional cysteine residues was inserted at the 5′ end of the *Escherichia coli* galactokinase gene, *gal K.* After prepurification, the recombinant enzyme was incubated with thiopropyl-Sepharose 6B and subsequently eluted with 10 m*M* dithiothreitol. Such treatment enhanced the specific activity of the enzyme approximately fivefold (27).

F. L-Arginine– and L-Lysine–Agarose

Arginine– and lysine–agaroses (Figure 4) are amphoteric derivatives that can contribute both electrostatic and stereospecific forces. L–Arginine–Sepharose prepared according to Porath and Fornstedt (28) has been useful in the purification of B-type carboxypeptidases that recognize C-terminal arginine or lysine residues (29,30). The matrix was prepared by converting agarose to the oxirane derivative by treatment with epichlorohydrin. The oxirane–agarose was then reacted with the amino groups of arginine or lysine. Substitution of the gel amounted to 14 to 20 μmol/mL of gel.

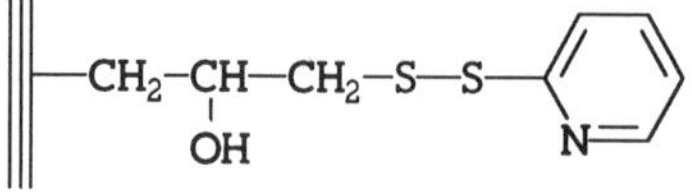

$-CH_2-CH(OH)-CH_2-S-S-$(2-pyridyl, N)

Figure 3 Partial structure of activated thiopropyl-Sepharose 6B.

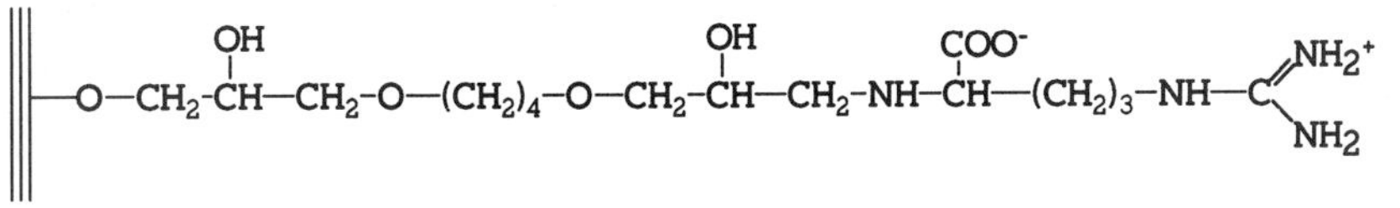

$$\text{Agarose}-O-CH_2-\underset{|}{\overset{OH}{C}}H-CH_2-O-(CH_2)_4-O-CH_2-\underset{|}{\overset{OH}{C}}H-CH_2-NH-\underset{|}{\overset{COO^-}{C}}H-(CH_2)_3-NH-C\begin{matrix}{\nearrow NH_2^+} \\ {\searrow NH_2}\end{matrix}$$

Figure 4 Structure of L-arginine–agarose.

The carboxypeptidase B was eluted from the L-arginine matrix with 1 m*M* guanidylthiopropionic acid (GEMSA), an inhibitor of the enzyme (10 μ*M* GEMSA completely inhibits carboxypeptidase M of human placental microvilli) (30). A carboxypeptidase B of human seminal plasma was purified nearly 1000-fold with a yield of 37% by this affinity matrix (29). A slightly modified ligand, *p*-aminobenzoyl L-arginine, was immobilized on epoxy-activated Sepharose and used to purify carboxypeptidase H, a B-type carboxypeptidase (31). When this matrix was employed as a terminal step in the purification and 1 m*M* GEMSA was used for elution, the enzyme was purified more than 100-fold with a yield of nearly 60%. Carboxypeptidase H of bovine pituitary was purified over 1300-fold, with a yield of 69% after elution from this same matrix with a gradient of arginine (0 to 800 m*M*) (32).

L-Arginine– or L-lysine–agarose was also useful in the isolation of serine proteinases that exhibit specificity for peptide bonds associated with basic amino acid residues. Prothrombin (33), plasminogen activator (34), and prekallikrein (35) have been isolated using this matrix. Plasminogen activators (PAs) are trypsin-like serine proteases involved in pathological processes and are of value in thrombolytic therapy. Two important plasminogen activators are urokinase and tissue-type activator (tPA). Lysine–Sepharose was used to isolate and purify tPA from murine mast cells (36). Elution was promoted by ε-aminocaproic acid, an inhibitor of plasminogen (37).

G. Hydrophobic Interaction Chromatography

Employment of the phenyl-substituted resin represents a good example of hydrophobic chromatography. The phenyl ligand is intermediate in hydrophobicity between *n*-pentyl and *n*-butyl ligands. The aromatic amino acid residues of an enzyme bind to the matrix through π–π interactions. In most instances, samples are loaded onto the column at high ionic strength to enhance hydrophobic interaction between a biomolecule and the matrix, and eluted by a decrease in salt concentrations. Addition of chaotropic molecules or of miscible organic solvents (e.g., ethylene glycol) to disrupt the hydrogen bonding of water or to change the polarity of the mobile phase can also be used to facilitate the attraction of apolar groups to the mobile phase. In the case of detergent-solubilized membrane proteins, the detergent is diluted for sample applications (e.g., to 0.2% cholate), but elution is promoted by increasing the detergent concentration of the elution buffer (e.g., to 0.4% cholate).

An affinity tail of 11 phenylalanine residues attached at the amino terminus by recombinant DNA techniques was used to increase the hydrophobic character of *E. coli* β-galactosidase. The partially purified enzyme was adsorbed onto a column of phenyl-Sepharose CL-4B (Pharmacia) and eluted with a gradient of ethylene glycol. The purification factor of this step was about 14 (27).

H. Metal-Chelate Affinity Chromatography

Metal ions are essential constituents of some protein structures. Ca^{2+} forms stabilizing complexes with carboxyl and carbonyl oxygens; Zn^{2+} coordinates to both imidazole nitrogens and carboxyl oxygens, and also bonds covalently to thiols; and Fe^{3+} interacts with phosphoamino acids. Metal-chelate columns are effective provided that such residues are favorably exposed in the native protein or have been genetically engineered in as a terminal domain. Metal-chelate affinity chromatography (IMAC) was first described by Porath et al. (38). Chelate-forming ligands for metal ions such as the disodium salt of iminodiacetic acid are attached to epoxy-activated amino-Sepharose 6B (Pharmacia) (Figure 5). Gel-bound chelates of the metal are prepared by passing a solution of the metal salt onto the column.

Nucleoside diphosphatase was purified starting with a specific activity of 47.5 to 185 units/mg utilizing a Ca^{2+}-chelate column and subsequently, from 185 to 310 units/mg employing a Zn^{2+} chelate column (39). The enzyme was eluted with buffer containing histidine. A Zn^{2+} chelate column (but not a Ca^{2+} column) has also been of value for the purification of collagenase, achieving a doubling in specific activity (40). In this case the enzyme was eluted by a low-pH, high-ionic-strength buffer.

Recombinant proteins possessing exposed histidine or cysteine residues, in particular, bind to metal chelate–Sepharose. Since such binding is pH dependent, the protein of interest can be eluted simply by reducing the pH of the buffer. For example, to isolate mouse dihydrofolate reductase, coding for six adjacent histidines at the amino terminus was fused to the gene for this enzyme (41). After

$$\text{AGAROSE—O—CH}_2\text{-CHOH—CH}_2\text{-O—(CH}_2)_4\text{-O—CH}_2\text{-}\underset{}{\overset{\text{OH}}{\overset{|}{\text{CH}}}}\text{—CH}_2\text{—N}\begin{matrix}\text{CH}_2\text{-COO}^-\\ \text{CH}_2\text{-COO}^-\end{matrix}$$

Figure 5 Iminodiacetic acid derivative of oxirane-activated agarose.

expression of the fusion protein in *E. coli*, the cell extract was purified by chromatography on a Ni^{2+}-nitrilotriacetate–containing matrix. The preparation of this metal-chelating adsorbent has been described by Hochuli et al. (42). Finally, the N-terminal histidine affinity peptide was removed by treatment of the enzyme with carboxypeptidase A. Since recombinant fusion proteins may fail to fold properly to produce a native and active molecule, the removal of the added peptide is important. If, however, enzyme activity and substrate specificity appear unaltered, the modified enzyme can be used directly without treatment. This novel approach has a general application. For instance, recombinant HIV-1 reverse transcriptase has been purified rapidly by "tailing" modifications permitting isolation by metal affinity chromatography (43).

III. PROTEIN-FUNCTION BASED

A. Use of Immobilized Small Substrate or Cosubstrate

An affinity matrix for the purification of norsolorinic acid dehydrogenase has been described (44). This NADPH-requiring enzyme is involved in aflatoxin biosynthesis. It catalyzes the conversion of norsolorinic acid to averantin (Figure 6). ω-Aminohexylagarose (Sigma) was treated with *p*-nitrobenzoyl azide in dimethylformamide to yield *p*-nitrobenzamidoalkylagarose. The *p*-nitrobenzamidoalkylagarose was suspended in 0.2 *M* sodium dithionite (in 0.5 *M* $NaHCO_3$) and sodium nitrite was added to produce the diazonium agarose derivative. This derivative was coupled to the hydroxyanthraquinone moiety of norsolorinic acid at position 4, 5, or 7 [i.e., as directed by the hydroxy (phenyl) groups] (Figure 7). This affinity matrix, however, exhibited little or no binding of the dehydrogenase. Since the enzyme is inactivated by high concentrations of the substrate (possibly because of the presence of phenol groups possessed by the substrate), the investigators methylated the affinity matrix with diazomethane to block the more acidic phenol groups at positions 6 and 8. The methylated matrix

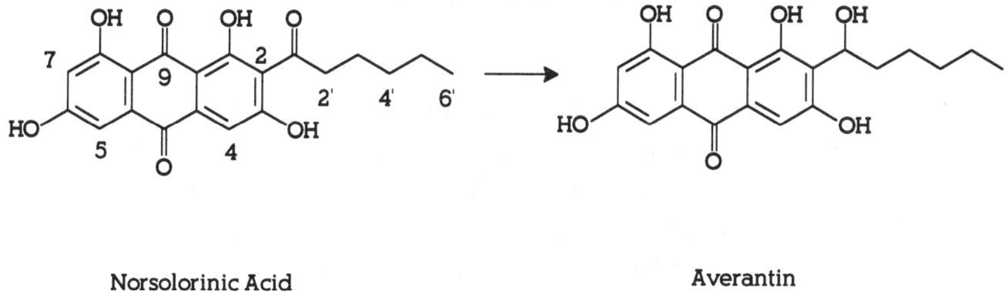

Norsolorinic Acid → Averantin

Figure 6 Conversion of norsolorinic acid to averantin by NA dehydrogenase.

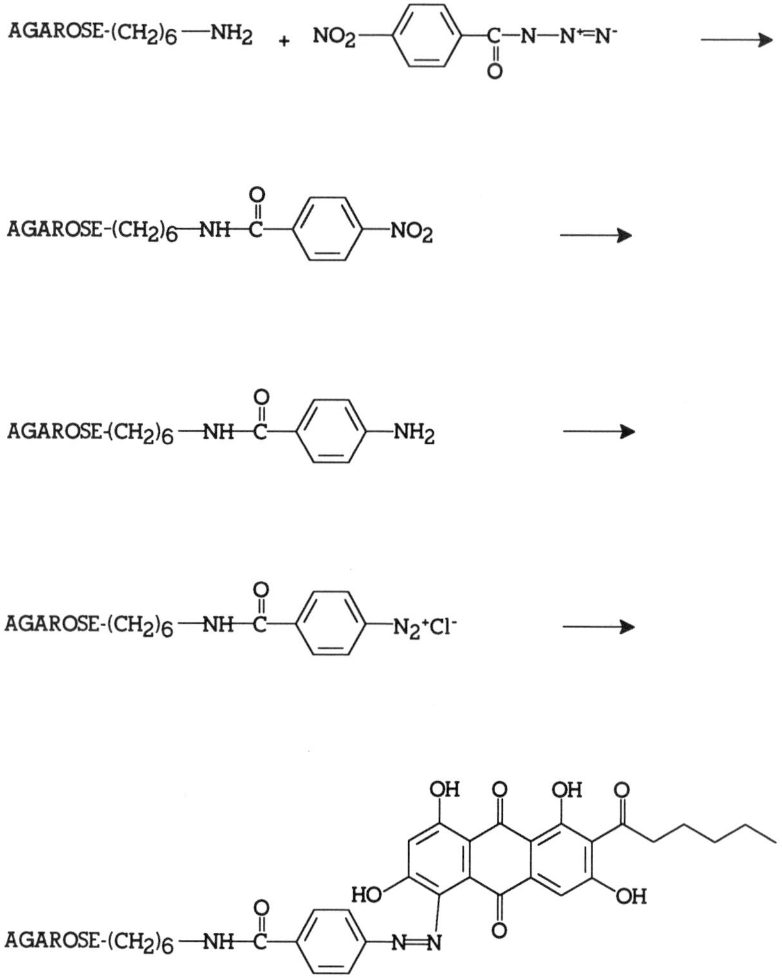

Figure 7 Preparation of the matrix for isolation of norsolorinic acid dehydrogenase.

allowed a one-step purification of the enzyme from a crude homogenate, resulting in 138-fold purification (44). Employing bioselective interaction is normally not recommended at an early stage of purification since nonspecific interaction is always a possibility in complex protein mixtures. Still, researchers should be encouraged to try affinity procedures with crude enzyme extracts.

Glutathione-S-transferase is a drug-metabolizing enzyme that transfers the entire glutathione (GSH) molecule via its sulfur to a wide spectrum of electrophiles [e.g., leukotriene A4 (LTA4) (Figure 8)]. Affinity matrices, commonly used for the purification of this enzyme, are glutathione–Sepharose 6B and *S*-hexylglutathione–Sepharose 7B. They are prepared by treating epoxy-activated Sepharose 6B (Pharmacia) with glutathione or *S*-hexylglutathione, respectively (45). The GSH-affinity column is prepared at pH 7, where the addition of GSH to the epoxy-activated Sepharose is presumed to occur by sulfhydryl addition to the epoxide (Figure 9). At this pH value the single amino group should be fully protonated. At higher pH, GSH also couples to epoxy-activated Sepharose, but the resulting material has very poor affinity for the transferases. Presumably, at higher pH, coupling proceeds by addition of the amino group of GSH to the epoxide. Such matrices, however, failed to bind an isoform known as α-glutathione transferase, subtype Ya1 Ya1. These transferases represent a complex-multigene family, comprising at least four distinct classes of proteins (designated cytosolic α, μ, π, and microsomal enzyme). Whereas some of these show little enzymatic activity when bromosulfophthalein is used as substrate, others, such as Ya1 Ya1, exhibit high activity. For the latter isoform, bromosulfophthalein–glutathione–Sepharose provided an effective matrix and the adsorbed protein was eluted with buffer containing glutathione (46).

Coenzyme A transferases catalyze transfer of the coenzyme A (CoA) moiety from the CoA thiol ester to carboxylic acids. Succinyl-CoA:3-hydroxyl-3-methylglutarate CoA transferase is an enzyme that catalyzes the reaction (Figure 10). The enzyme can be purified utilizing agarose–hexane–CoA (type 5 Pharmacia) with a resulting rise in specific activity from 31.2 nmol/min/mg to 248.5

LTA4 + GSH ⟶

Figure 8 Formation of LTA4–GSH complex.

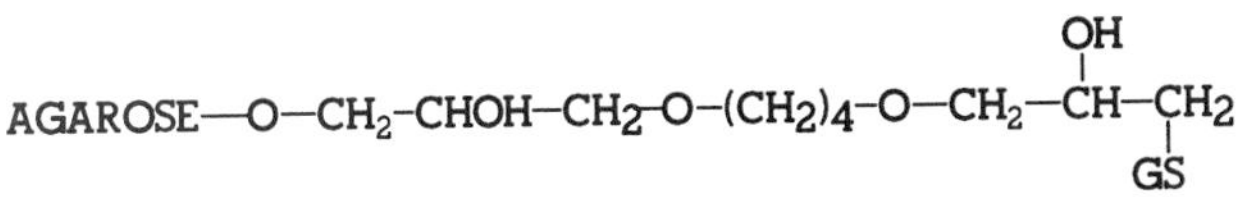

AGAROSE—O—CH_2-CHOH—CH_2-O-$(CH_2)_4$-O—CH_2—CH(OH)—CH_2(GS)

Figure 9 GSH coupled to epoxy-activated Sepharose 6B.

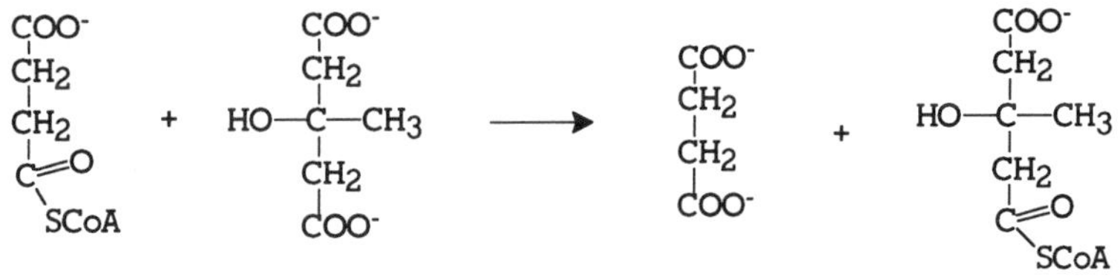

Figure 10 Succinyl-CoA + 3-hydroxy-3-methylglutarate → succinate + 3-hydroxy-3-methylglutaryl-CoA.

nmol/min/mg in this one step (47). Unfortunately, the amount of enzyme obtained by this type of affinity chromatography was limited and the agarose–hexane–CoA underwent rapid degradation.

The CoA–agarose is prepared by mixing 6-aminohexanoic acid with CNBr-activated Sepharose 4B to yield CH-Sepharose 4B (Pharmacia). The free carboxyl group at the end of the six-carbon spacer arm is activated by carbodiimide—coupling to the *N*-hydroxysuccinimide ester. When this ester is reacted with CoA, a thiol ester linkage between the spacer and the CoA is formed (Figure 11).

FAD synthetase (ATP:FMN adenyltransferase) has been purified using an FMN affinity matrix synthesized from CH-Sepharose 4B. FMN can be linked to CH-Sepharose 4B in the presence of the water-soluble *N*-ethyl-*N*′-(3-dimethyl-aminopropyl)carbodiimide, presumably by mixed acid anhydride linkage to the 5′-phosphate or by ester linkage to one of the hydroxyl groups of the ribitol moiety

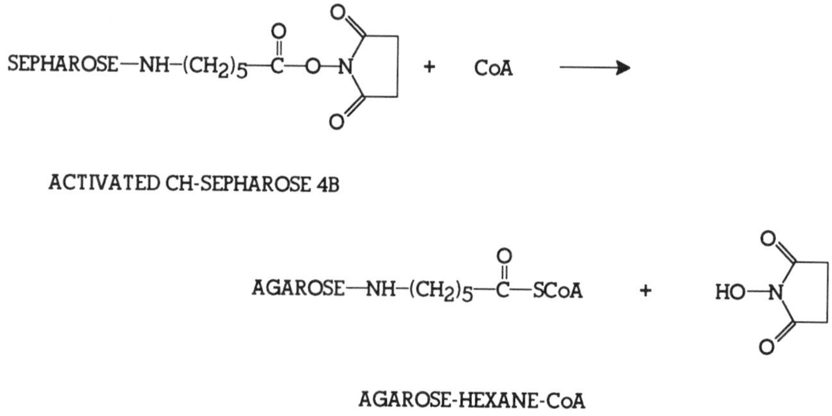

ACTIVATED CH-SEPHAROSE 4B

AGAROSE-HEXANE-CoA

Figure 11 Reaction of CoA with activated Sepharose.

of the FMN. For elution, FMN is added to the buffer. This one step of affinity chromatography increased the specific activity approximately 130-fold (2.8 units/mg to 374 units/mg) (48).

Sialyl transferases are a group of glycosyl transferases that catalyze the transfer of a sialic acid (Neu Ac) residue to the nonreducing terminal sugar of glycoproteins and glycolipids within the general reaction (Figure 12). CDP-Sepharose was employed to purify CMP-Neu Ac:GM3 sialyl transferase and also CMP-Neu Ac:lactosyl ceramide sialyl transferase present in developing rat brain. To separate the two enzymes from each other, a GM3 acid–Sepharose was prepared

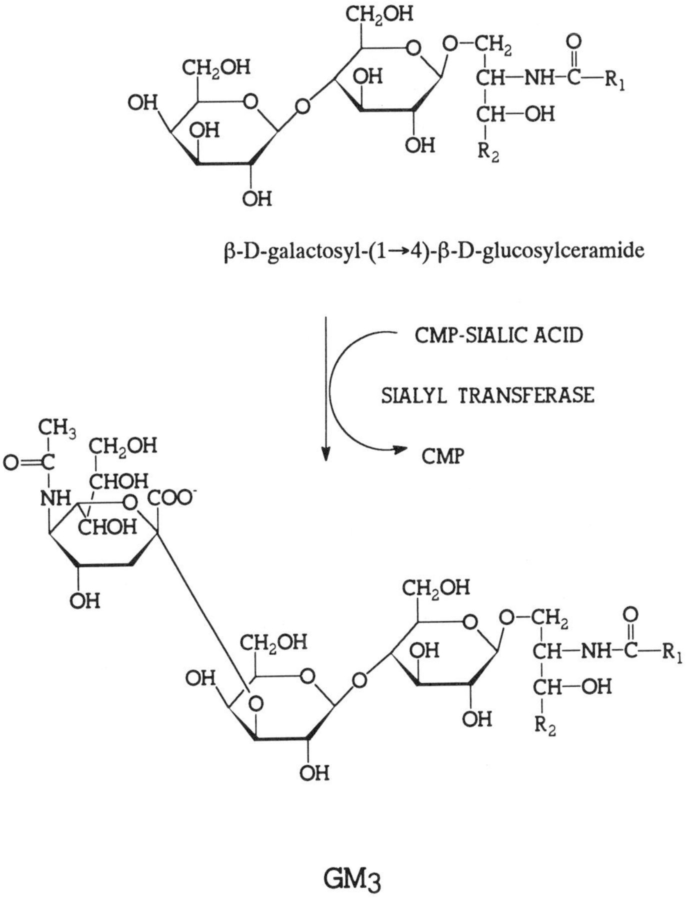

β-D-galactosyl-(1→4)-β-D-glucosylceramide

CMP-SIALIC ACID

SIALYL TRANSFERASE

CMP

GM3

Figure 12 CMP-Neu Ac + HO acceptor → CMP + Neu Ac-O-acceptor.

by nonspecific oxidation of GM3 with permanganate and coupling of the GM3 acid to aminohexyl-Sepharose 4B. For GM3 sialyl transferase, employment of CDP-Sepharose affinity chromatography led to an increase in specific activity from 0.016 unit/mg/protein to 0.79 unit/mg and that of the GM3-Sepharose affinity chromatography to 18.9 units/mg (49).

Boronic acid matrices for use in *cis*-diol affinity chromatography display high capacities for binding and separating low-molecular-weight compounds with coplanar *cis*-diol groups, including ribonucleotides, ribonucleosides, sugars, catecholamines, and coenzymes. Once a substrate has interacted, the column can serve for specific protein function–based affinity chromatography. For instance, poly(ADP-ribose)glycohydrolase catalyzes the cleavage of ribosyl–ribose bonds, liberating ADP-ribose. An affinity matrix for the purification of this enzyme can be prepared by allowing poly(ADP-ribose) to interact with dihydroxyboryl-Sepharose (DHB-Sepharose) (50). DHB-Sepharose is prepared by treating CNBr-activated Sepharose with 6-aminohexanoic acid, followed by coupling of the carboxy group of the spacer to *m*-aminophenylboronic acid hemisulfate (51) (Figure 13).

B. Use of Immobilized Protein Substrate or Cosubstrate

AMP deaminase has been shown to bind to the myosin heavy chain in vitro. When coupled to activated Sepharose (52), myosin can be used as affinity matrix for the deaminase. Myosin–Sepharose has also been utilized for preparation of the protein phosphatase, which catalyzes the dephosphorylation of isolated myosin light chains (53).

A fibrin–Sepharose affinity matrix has been of value in defining the mechanism of reaction between single-chain tissue plasminogen activator (sc-tPA) and the human tPA inhibitor-1 (PAI-1) (54). Isolation of 1:1 complexes of sc-tPA and PAI-1 on fibrin-agarose indicated cleavage of a "bait" peptide bond during complex formation. The bait peptide bond (i.e., Arg^{346}Met^{347}Ala–Pro–Glu–Glu–), which is located near the C terminus of inhibitor 1, resembled those found in other forms of serine protease inhibitors known as serpins.

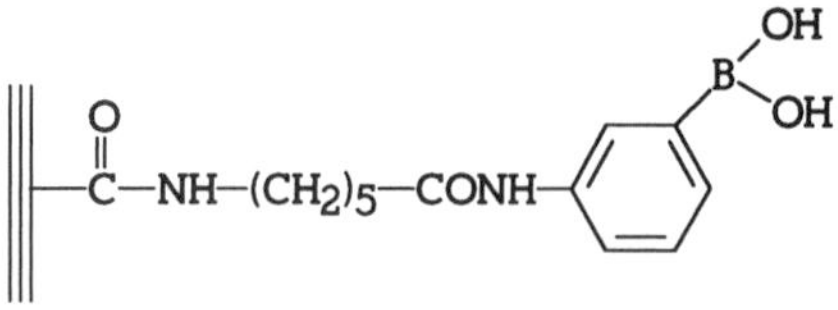

$$\|\!-\!\overset{O}{\overset{\|}{C}}\!-\!NH\!-\!(CH_2)_5\!-\!CONH\!-\!C_6H_4\!-\!B(OH)_2$$

Figure 13 Active group of DHB-Sepharose.

Ferredoxin:NADP$^+$ oxidoreductase, a flavoprotein, from plant tissue was bound by a ferredoxin–Sepharose 4B affinity column and eluted by increasing the concentration of phosphate buffer. Specific activity increased from 7.1 units/mg protein to 30.5 units/mg protein (55).

Pancreatic deoxyribonuclease I forms a very tight 1:1 complex with monomeric actin (K_d-10^{-10} *M*). The commercially available preparations of deoxyribonuclease I contain significant contamination from chymotrypsinogen, chymotrypsin, trypsin, and multiple ribonucleases. An affinity column prepared by adding 100 mg of G-actin to 25 mL of *N*-hydroxysuccinimide-substituted agarose (Affi-Gel 15, Bio-Rad) was used to bind the commercial DNase I. [Affi-Gel 15 possesses a cationic charge in its 15-atom spacer arm, which significantly enhances coupling efficiency for acidic proteins at physiological pH (Figure 14).]

Elution with buffer containing 10 *M* formamide yielded an enzyme with a specific activity of 2780 Kunitz units/mg protein (activity before this chromatographic step was 2651 units/mg), which was essentially free of protease and ribonuclease contamination (56).

Cytochrome c oxidase from heart mitochondria has been purified by linking cytochrome c directly to CNBr-activated Sepharose 4B (Pharmacia) (43). The cytochrome c molecule can be attached to the Sepharose through multiple lysine residues, which are located mainly on one side of the molecule (Figure 15). Hydrolysis of some of the CNBr-activated groups prior to the coupling can minimize multipoint attachment of cytochrome c via such lysine residues. Alternatively, the cytochrome c can be linked to activated thiol-Sepharose 4B (Pharmacia). In this procedure the protein is bound via a single cysteine residue at the opposite side of the molecule (Figure 15). Since the cytochrome c oxidase interacts with lysine residues located on only one side of the cytochrome c, the latter method is preferred. Another resin can be prepared by allowing Affi-Gel 102 (Bio-Rad), an amino alkyl agarose gel with a six-atom, hydrophilic spacer arm, to react with the heterobifunctional reagent succinimidyl-4-(*p*-maleimidophenyl) butyrate (SMPB). Reduced cytochrome c is then covalently coupled to the extended spacer arm of the gel (Figure 15). [This gel is stable under reducing conditions and in addition exhibits a high capacity for cytochrome c oxidase (57).]

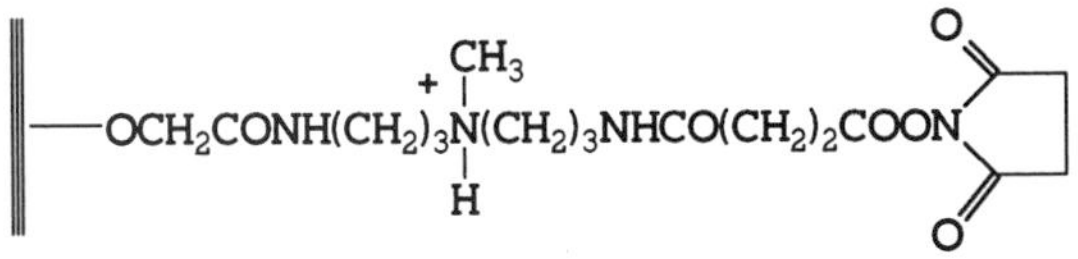

$$\text{—OCH}_2\text{CONH(CH}_2)_3\overset{+}{\underset{\text{H}}{\overset{\text{CH}_3}{\text{N}}}}\text{(CH}_2)_3\text{NHCO(CH}_2)_2\text{COON} \text{(succinimide ring, two C=O)}$$

Figure 14 Affi-Gel 15.

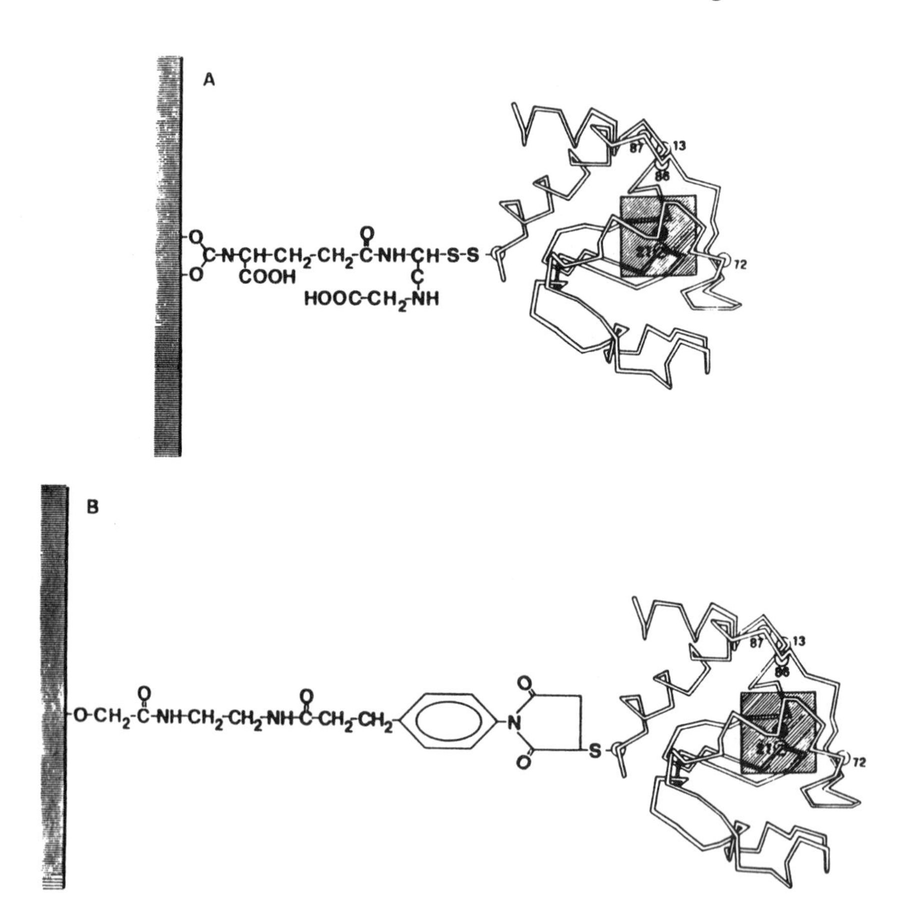

Figure 15 (A) Schematic drawing of yeast cytochrome c linked to activated thiol-Sepharose 4B. The hatched square in cytochrome c represents the heme plane and the numbers indicate the lysine residues, which are important in the interaction of this protein with cytochrome *c* oxidase. Notice that cytochrome c is attached through a cysteine that is located on the opposite side of the molecule. (B) Schematic drawing of yeast cytochrome c linked to Affi-Gel 102 through the heterobifunctional reagent SMPB. The spacer between the gel and cytochrome c is longer compared to the spacer of the gel above and not sensitive to reducing agents. (From Ref. 57.)

Restricting the number of attachment sites increases the available binding domains of immobilized proteins. Hadas et al. (58) have used dimethylmaleic anhydride to protect reversibly some of the free amino groups of a protein, thus reducing the number of attachment points between protein and matrix. Duerksen-Hughes et al. (59) used the glyoxal derivative 4-(oxoacetyl) phenoxyacetic acid (OAPA) (Figure 16) coupled with *N*-ethyl-*N'*-(3-dimethyl aminopropyl)-carbodiimide to amine-substituted polyacrylamide beads (Bio-Gel P-150) to immobilize ubiquitin via reactive arginine residues. [Phenyl glyoxal and its derivatives react under mild conditions (pH 7.9) primarily with the guanidino group of the arginine residues of proteins.] The immobilized OAPA bound primarily three closely spaced arginine residues in the ubiquitin (numbers 42, 72, and 74, all within a few angstroms of each other). Ubiquitin carboxyl terminal hydrolase (an enzyme that hydrolyzes various amide and ester leaving groups from the carboxyl terminus) was specifically and reversibly bound by this affinity support. In addition, three other proteins exhibiting ubiquitin carboxyl-terminal hydrolytic activities were isolated. These three proteins were not bound by a conventional matrix containing ubiquitin immobilized via lysine residues.

In eukaryotes, intracellular proteins targeted for degradation are first covalently linked to ubiquitin: An enzyme, E_1, catalyzes adenyl group transfer from ATP to the C-terminal glycine residue of ubiquitin, which is then moved to the sulfhydryl group of a cysteine residue within E_1. Next, ubiquitin is transferred from E_1 to another enzyme, E_2, by transthioesterification. A third enzyme, E_3, binds the target protein at a specific protein binding site and then catalyzes the transfer of ubiquitin from E_2 to the ε-amino group of a lysine residue in the target protein (Figure 17).

Ubiquitin coupled to activated CH-Sepharose (Pharmacia) allowed isolation of ubiquitin-activating enzyme (E_1) but also of the other components of the ubiquitin-protein ligase system, E_2 and E_3 (60). Sepharose-bound ubiquitin must be approximately 20 mg/mL of swollen gel for the isolation of E_3, whereas E_1 and E_2 were completely bound to columns containing much less ubiquitin (approximately 5 mg/mL). E_1 is covalently linked to the affinity column in the presence of adenosine triphosphate (ATP) and can be specifically eluted with AMP and

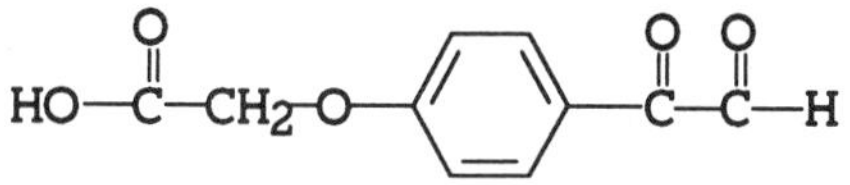

Figure 16 Glyoxal derivative of OAPA.

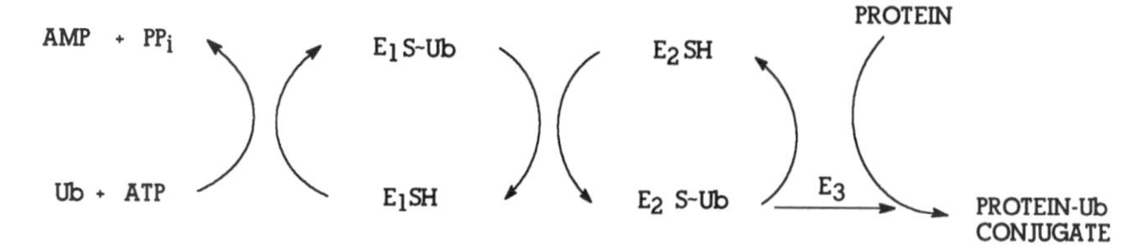

Figure 17 Suggested sequence of events in the ubiquitin–protein ligase system.

pyrophosphate. E_2 is bound when E_1 and ATP are present and is eluted with a thiol compound at high concentration. E_3 is noncovalently adsorbed and is eluted at high salt concentration or increased pH.

In an earlier review (61), the possibility of using affinity chromatography for isolation of the subunits of an enzyme was mentioned. If aldolase, covalently bound to Sepharose, is washed with 8 *M* urea solution, dissociation of the enzyme into subunits is evoked. All subunits not covalently linked to the matrix may be removed. It appears that subunits of rabbit muscle aldolase are capable of enzyme activity and thus may be isolated for studies on the function of subunits (62). This approach, however, has not been widely employed even though it could accomplish the isolation of a pure oligomeric enzyme from a crude extract (e.g., if octameric porphobilinogen synthase is immobilized on Sepharose 4B and treated with 4 *M* urea, four subunits per octamer are removed, which can be reassociated into a soluble octameric enzyme). The specific activity is about one-half that of the immobilized octameric structure (63). This is similar to the finding with immobilized aldolase. For such experiments to succeed the enzyme molecule has to be covalently linked to the support through a small fraction of its total subunits.

Employment of avidin affinity columns for the isolation of biotin containing enzymes was unsuccessful, at first, because the affinity of native avidin for biotin is too high ($K_d = 10^{-15}$ *M*). These enzymes could be recovered from such a matrix only under conditions of denaturation. Tetrameric avidin, however, can be dissociated into monomers by urea or guanidine HCl, and the monomeric protein exhibits reduced affinity for biotin. Monomeric avidin affinity matrixes (prepared by treating CNBr activated Sepharose 4B with avidin, followed by overnight exposure to 6 *M* guanidine HCl) will purify the biotin containing bacterial propionyl-CoA carboxylase. The enzyme was eluted with 0.2 m*M* biotin and 1 m*M* $MgCl_2$ in 0.05 *M* K_2HPO_4–KH_2PO_4, pH 7.0, and 0.015 ng/mL DTT, resulting in a sevenfold purification (64). A monomeric avidin affinity column was also employed for the purification of bacterial oxalacetate decarboxylase with 15- to 20-fold purification (65).

C. Cofactors as Ligands

Immobilized cofactors for the dehydrogenases may serve as affinity ligands. Because $NADP^+$ may undergo structural modification during coupling, it is usually not chosen. However, 2′,5′,ADP–Sepharose 4B (Pharmacia) has sufficient stability for a broad range of applications and can readily be prepared (66) from N^6-(6-aminohexyl)adenosine 2′,5′-bisphosphate (Figure 18). Crude bacterial luciferase was first chromatographed on a Sepharose–hexanoate gel to remove proteins that bind to the spacer and then applied to a FMN–Sepharose column and eluted by increasing the concentration of either NaCl or FMN. The activity increased from 8×10^{12} quanta s^{-1} mg^{-1} to 1.4×10^{14} quanta s^{-1} mg^{-1} (67).

D. Use of Enzyme Inhibitors as Ligands

For the preparation of a β-galactosidase-specific affinity column, the competitive inhibitor, *p*-aminophenyl-β-D thiogalactopyranoside, was linked to the hydrocarbon arm of succinyl-agarose using *N*-cyclohexyl-*N*′-(2 morpholinoethyl)carbodiimide metho-*p*-toluenesulfonate as coupling agent. [For a detailed description of the general method, see Steers and Cuatrecases (68).] Chromatography of β-galactosidase on this matrix increased specific activity from 1.27 units/mg to 41 units/mg (69).

Pig liver man_9-mannosidase (α-1,2-mannosidase), an enzyme involved in the processing of N-linked oligosaccharides, has been isolated on ConA–Sepharose

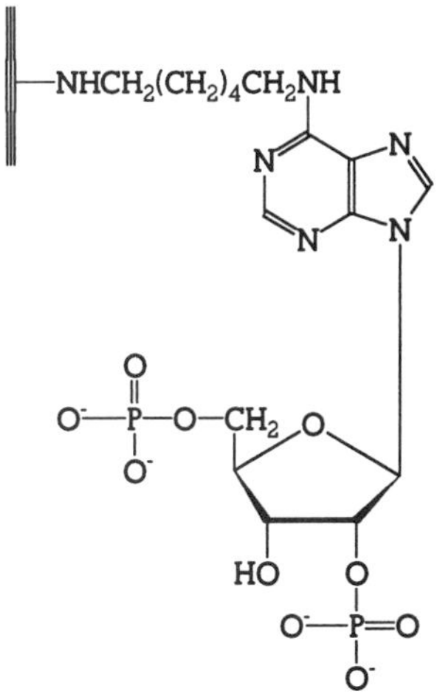

Figure 18 Structure of N^6-(6-aminohexyl) adenosine 2′,5′-bisphosphate.

columns. This step was preceded, however, by the far more efficient purification of the enzyme on a resin of synthetic *N*-5-carboxypentyl-1-deoxymannojirimycin (CP-dMM; Figure 19) linked to AH-Sepharose 4B, permitting a 280-fold increase in specific activity (70). 1-Deoxymannojirimycin is a basic sugar analog that is inhibitory to the enzyme activity. Thus the combination of ConA–Sepharose chromatography with an additional specific affinity matrix step is particularly effective in the purification of enzymes that are glycoproteins.

Novobiocin is a potent competitive inhibitor for the ATP-binding moiety of type II topoisomerases. (These are enzymes that introduce negative superhelical turns into relaxed covalently closed circular DNA.) Novobiocin–Sepharose prepared by coupling of this antibiotic to epoxy-activated Sepharose 6B (Figure 20) provided an efficient matrix for the purification of the enzyme from *E. coli* (71). More recently, the mitochondrial type II topoisomerase was applied as crude extract onto such a column and eluted with a buffer containing 20 m*M* ATP and 20 m*M* $MgCl_2$. This one step resulted in an increase in specific activity from 32 units/mg to 9600 units/mg (72).

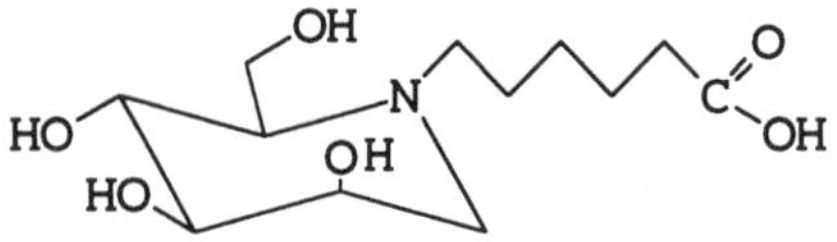

Figure 19 Structure of CP-dMM.

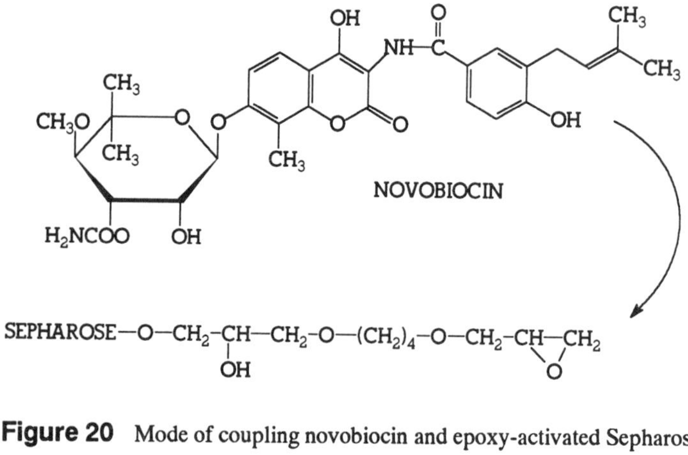

Figure 20 Mode of coupling novobiocin and epoxy-activated Sepharose.

Choline acetylase, which catalyzes the synthesis of acetylcholine from acetyl CoA and choline, has been isolated from the electric organ of *Torpedo mamorata*. The enzyme was first purified on a Sepharose–CoA affinity matrix prepared by coupling the CoA to Sepharose 4B (Pharmacia) via a six-carbon spacer. Next, an immobilized inhibitor matrix containing 2-[3-(2-ammonioethoxy)benzoyl] ethyltrimethylammonium bromide was employed. This ligand contains a primary amino group that reacts with the free carboxyl group of the CH-Sepharose (Pharmacia) in the presence of *N*-ethyl-*N'*-(3-dimethylaminopropyl)-carbodiimide at pH 4.8 to yield the structure shown in Figure 21. The bound material of the CoA–Sepharose 4B column was eluted with buffer containing acetyl-CoA, while that of the second column was eluted with buffer containing the inhibitor. Upon elution from the first column the specific activity had risen from 0.02 unit/mg to 0.11 unit/mg, and upon elution from the second column the specific activity was 73 units/mg (73).

Transglutaminases are a family of enzymes catalyzing the incorporation of primary amines into glutamine residues of proteins. When the primary amine donor is protein-bound lysine, an ε-(-glutamyl)lysine isopeptide bond is formed (Figure 22). These enzymes require Ca^{2+} and are differentially inhibited by GTP: Tissue-type transglutaminase is inhibited by GTP while epidermal transglutaminase is unaffected. The enzyme preparation was applied to a heparin–agarose column in Hepes–EGTA buffer and eluted with the same buffer containing GTP (purification about 10- to 17-fold). It may be assumed that the binding of the GTP to the enzyme changes its conformation. After supplementing the buffer with $CaCl_2$ the enzyme was adsorbed onto an α-casein–agarose column and eluted by addition of EGTA. Thus an additional fivefold purification was provided by this step (74).

Adenylosuccinate synthetase catalyzes the reaction

$$\text{IMP} + \text{asparate} + \text{GTP} \rightarrow \text{adenylosuccinate} + \text{GDP} + P_i$$

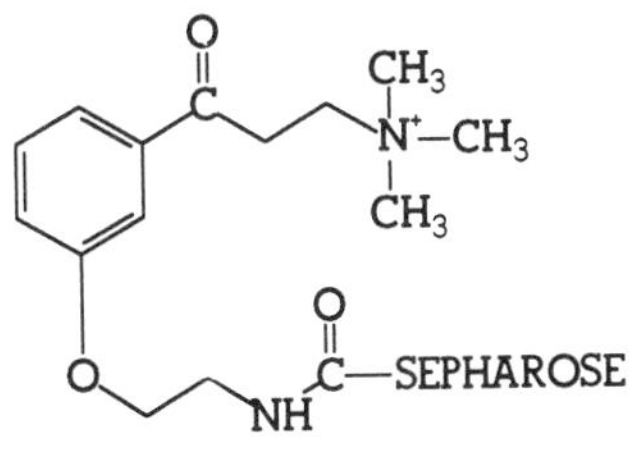

Figure 21 Sepharose-CONH–EtO (BzEt)–Me_3NBr.

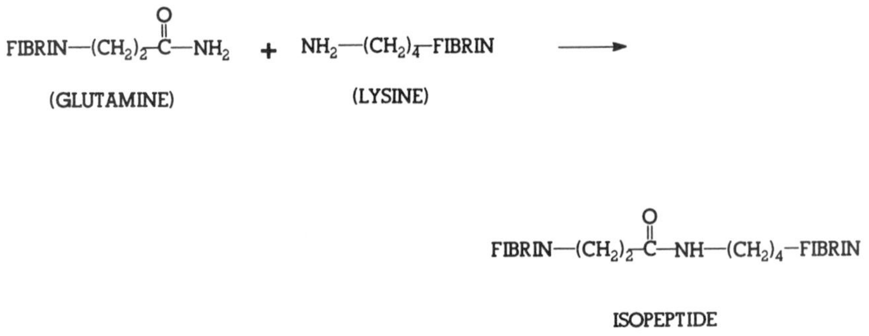

$$\text{FIBRIN—(CH}_2)_2\text{—C(=O)—NH}_2 + \text{NH}_2\text{—(CH}_2)_4\text{—FIBRIN} \longrightarrow \text{FIBRIN—(CH}_2)_2\text{—C(=O)—NH—(CH}_2)_4\text{—FIBRIN}$$

(GLUTAMINE) (LYSINE) ISOPEPTIDE

Figure 22 Isopeptide bond formation.

The enzyme was extracted from an *Ascites* sarcoma utilizing hadacidin–Sepharose 4B. Hadacidin (*N*-formyl hydroxyaminoacetic acid) is an antibiotic that acts as a powerful competitive inhibitor of this enzyme. The antibiotic is coupled to the aminohexyl-Sepharose 4B (Pharmacia) through its formyl group. This column absorbs many kinds of proteins, but the adenylosuccinate synthetase could be eluted specifically with aspartate. This step afforded a 15-fold purification (75).

The allosteric inhibitor, dATP, immobilized on Sepharose, was utilized in the isolation of the anaerobic ribonucleoside triphosphate reductase of *E. coli*, an enzyme that reduces CTP to dCTP and requires NADPH, dithiothreitol, Mg^{2+}, and ATP. dATP-Sepharose was prepared by coupling aminophenyl-γ-dATP to CNBr-activated Sepharose (Pharmacia) (76). Aminophenyl-γ-dATP was synthesized according to the procedure given by Knorre et al. (77).

Affinity chromatography for the isolation of the multiprotein complex DNA polymerase α-DNA primase from cherry salmon testes utilizes 5-(*E*)-(4 amino-styryl) araUTP as ligand (78) (see Figure 23). This compound is a strong and specific inhibitor of DNA polymerase. It is coupled to the Affi-Gel 10 matrix (Bio-Rad) via a reactive *N*-hydroxysuccinimide ester spacer arm. The succinimide ester forms a stable covalent bond with the primary amine of the araUTP. If the spacer is omitted, the enzyme is not bound. Furthermore, the enzyme is retained by the affinity resin only in the presence of a template or a template primer.

p-Aminobenzamidine (PAB) is an inhibitor of trypsin and trypsinlike serine proteases. A PAB column was used as a terminal step to separate single-chain urokinase from the two-chain urokinase (disulfide dimer), which binds quantitatively to the *p*-aminobenzamidine matrix (79). The sc-urokinase was recovered in the "through" fraction. In the preparation of the column the PAB is covalently bound to Sepharose 6B (Pharmacia) by a long hydrophilic spacer arm, in a stable ether linkage.

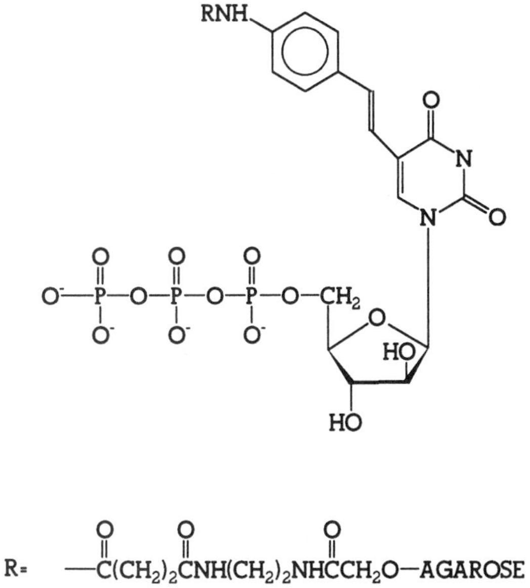

R= $-C(=O)(CH_2)_2C(=O)NH(CH_2)_2NHC(=O)CH_2O-AGAROSE$

Figure 23 Structure of AraUTP-Affi-Gel 10.

Amidine derivatives coupled to silica beads coated with DEAE–dextran have been used to purify serine proteases (thrombin, trypsin, and tPA) by high-performance affinity chromatography (80). Renal dipeptidase, which also has β-lactamase activity, was affinity purified using the antibiotic cilastatin (Figure 24) attached to Sepharose 4B (81). β-Lactamase activity has been shown to be selectively inhibited by cilastatin. The specific activity of the peptidase was enhanced from 1.04 μmol/min/mg to 65.7 μmol/min/mg.

The pepstatins, of which pepstatin A is the most common, are peptides that strongly inhibit certain proteases. They are produced by several *Streptomyces* species. Pepstatin A contains an unusual natural amino acid (4-amino-3-hydroxy-6-methyl heptanoic acid) which is essential for its inhibitory activity. Pepstatin is specific for the acid proteases; it very strongly inhibits pepsin, renin, and cathepsin D. Effective concentrations are 10^{-6} to 10^{-9} M. It does not inhibit trypsin, chymotrypsin, papain, or plasmin. Pepstatin–agarose prepared according to Huang et al. (82) has been used to separate lysosomal procathepsin D from active forms of cathepsin D (83). Procathepsin D was bound by the pepstatin matrix (maintained at 4°C) when the pH was 3.5, but not at pH 5.3. This contrasted with the active forms of the enzyme that were bound at both pH values. Active

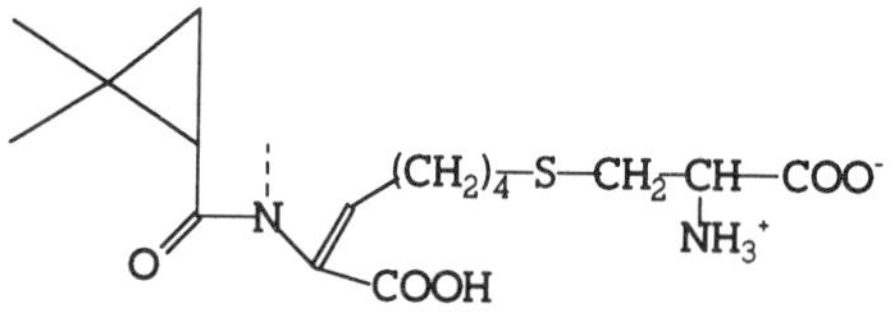

Figure 24 Cilastatin.

cathepsin was removed by pepstatin agarose at pH 5.3 and the unbound procathepsin was then reapplied to a second pepstatin column at pH 3.5 and finally, eluted with pH 8.0 buffer. The pure procathepsin D isolated by this technique appears to undergo an autocatalytic conversion to active enzyme in a manner analogous to other carboxyl proteinase, such as pepsinogen. In combination with ion-exchange chromatography, pepstatin–agarose has also been useful in obtaining homogeneous recombinant HIV-1 and HIV-2 retroviral proteases from bacterial lysates (84).

Lysosomal cysteine proteases (cathepsins B, H, and C) are important in cellular protein turnover. Peptide aldehyde inhibitors resembling leupeptin (Ac–leu–leu–argininal) are useful affinity ligands in the purification of cysteine proteases. An affinity matrix for the cysteine proteinase, cathepsin B, was produced by reacting gly–phe–glycinaldehyde semicarbazone with carboxyl-substituted agarose (CH-Sepharose 4B) (85). Elution was effected by incubation of the gel with 1.5 m*M* 2,2′-dipyridyl disulfide (DDS) at pH 4. DDS reacts selectively with cysteine residues of the active site to produce an inactive and protected enzyme that can be rapidly reactivated by reducing agents under assay conditions. Cathepsin B was purified to near homogeneity by a two-step procedure (acetone fractionation and the affinity matrix). Cathepsin L, papain, and actinidin were also bound by the column, whereas other proteinases (cathepsin H and chymopapain) were not (85). A similar matrix was used to purify cathepsin B from purulent sputum (86).

t-Butyloxycarbonyl–gly–phe–$NHCH_2CN$ is a potent inhibitor of papain, chymopapain, and papaya proteinase III. After deprotection of the N terminus, this inhibitor was reacted with CH-Sepharose 4B (Pharmacia) in the presence of carbodiimide to form Sepharose–6-aminohexanoyl–gly–phe–$NHCH_2CN$ (87). Hydroxyethyl disulfide was used to elute the cysteine proteases.

E. Use of Immobilized Transition-State Analogs as Ligands

Certain tripeptides in which the C-terminal arginine is substituted with argininal (arginine aldehyde) resemble transition-state analogs and are capable of forming a hemiacetal structure with the active-site serine of serine proteases (88). Agarose containing the peptide D-phe–D-phe–argininal was effective in binding tissue plasminogen activator (tPA) tightly in hemiacetal formation. The tripeptide was formed sequentially on a carboxyl-substituted matrix by carbodiimide coupling. Alternatively, preformed dipeptides were bound to succinimidyl-agarose followed by carbodiimide coupling of the argininal semicarbazone and subsequent removal of the semicarbazide group. A small percentage of the tPA bound by the matrix could be eluted with NaCl (0.2 to 1.0 *M*), but complete desorption from the immobilized peptide required low pH. For recombinant, large-scale purification, tPA from the conditioned media was captured on a Zn^{2+}–chelate column, applied to the argininal column and after serial washing was eluted with 0.1 *M* acetic acid. Purity of the tPA exceeded 95% with yields near 95% as well.

Kallikrein has been purified using aldehyde peptide analogs that are unable to form the hemiacetal linkage. Leuenkephalinargininal semicarbazone was covalently attached to succinimidyl-agarose and employed to purify plasma kallikrein. Unlike the immobilized aldehyde, the semicarbazone matrix interacted noncovalently and reversibly with kallikrein giving a yield of 71% and a purification of about 350-fold (89). The semicarbazone of peptide aldehyde inhibitors has also been used to purify cysteine proteases (see Section V.A).

IV. TANDEM METHODS

Only a few examples of tandem methods of affinity chromatography employed for the purification of enzymes are given here. It should be noted that immobilized dyes used in tandem where a specific positive-binding matrix is the last column in the chain are particularly effective in the selective purification of proteins. Preliminary screening on a small scale prior to fractionation is required to assess selectivity of dye adsorption.

The thermostable DNA polymerase from the archaebacterium *Sulfolobus acidocaldarius* fractionated in tandem first on phosphocellulose, next on heparin–Sepharose, and finally, on blue-Sepharose exhibited an increase in specific activity from 4000 units/mg to 11,800 units/mg during the heparin–Sepharose purification step and to 200,000 units/mg during a blue-Sepharose step (16). The DNA polymerase α-DNA primase complex (the two activities are tightly associated) from human lymphoblasts was similarly treated first by phosphocellulose, next by heparin–Sepharose, and finally, by an antibody column (17).

Partially purified arginyl tRNA synthetase applied to blue-Sephadex G-150 and eluated with 150 m*M* buffered KCl exhibited at least a 16- to 18-fold increase in activity (90). Purification of aminoacyl-tRNA provides an example of major inherent success when general affinity chromatography is coupled to biospecific steps. For isolation of yeast mitochondrial methionyl-tRNA synthetase, the partially purified extract was first placed onto heparin–Ultrogel, allowing for a rise in specific activity from 2.9 units/mg to 23 units/mg, next onto tRNAmet–Sepharose with an increase in activity to 500 units/mg, and finally, onto agarose–hexyl-AMP with a resulting activity of 1800 units/mg.

Glycosyltransferases from mammalian tissues have been purified on a support matrix to which nucleoside diphosphate was linked (via an arm) through the 5′-pyrophosphate group of ribose (e.g., example: UDP–hexanolamine–agarose) (91). Not all glycosyltransferases, however, were retained by this matrix. Hence 5-Hg-UDP *N*-acetylglycosamine linked through the 5-position of the uracil via a mercaptide linkage to a thiopropyl arm of the Sepharose 6B support matrix (Pharmacia) has been employed for the purification of UDP-*N*-acetylglucosamine α-D-mannoside β 1→ 2 *N*-acetylglucosaminyl transferase II. This purification step was repeated three times with the interphasing of one step utilizing Affi-Gel blue (Bio-Rad). Specific activity of the enzyme increased over 440-fold, from 0.0062 unit/mg to 2.75 units/mg (92). The method of preparation of the affinity matrix is given by Bendiak and Schachter (92). The structure of 5-Hg-UDP *N*-acetylglycosamine linked to the thiopropyl-Sepharose is shown in Figure 25.

V. PURIFICATION OF PEPTIDES WITH IMMOBILIZED ENZYMES

A. Isolation of Inhibitory Peptides

Affinity chromatography of proteolytic enzymes often requires information concerning their naturally occurring inhibitors. In turn, such inhibitors have provided

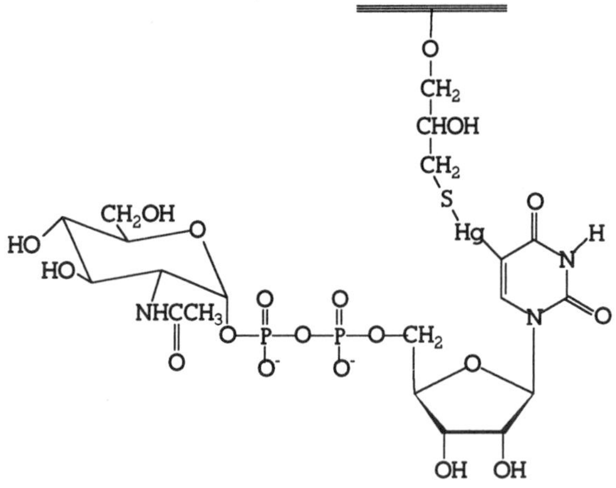

Figure 25 Chemical structure of 5-Hg-UDP-GlcNAc linked to thiopropyl-Sepharose.

stable and effective ligands that have been useful in the purification of proteolytic enzymes. Immobilized enzymes or their apoenzymes can be employed for inhibitor isolation. Commercial bovine carboxypeptidase A (without modification) was immobilized on CNBr-activated Sepharose and used to isolate carboxypeptidase inhibitors from the parasitic nematode *Ascaris suum* (93). Three isoforms of an 8-kDa carboxypeptidase inhibitor were released from the column by 1% formic acid.

Pepsin inhibitors from *A. suum* have also been isolated by a two-step procedure using native pepsin coupled to aminohexyl-Sepharose via carbodiimide (94). The peptide extracts of the parasite were applied to the column at pH 1.95 and eluted at pH 10. Chromatofocusing yielded one major homogeneous inhibitor suitable for sequencing and three minor forms.

An immobilized apoenzyme of carboxypeptidase B has also been used to isolate specific peptides (95). Native porcine carboxypeptidase was coupled to CNBr-activated agarose and then treated with *o*-phenanthroline and EDTA to produce the zinc-depleted, inactive apoenzyme, which has the same affinity for peptide substrates as the catalytically active enzyme (96). Peptides with C-terminal basic residues (e.g., metenkephalin–argarg or dynorphin 1–13) could be separated from nonreactive peptides by this apoenzyme procedure (80). Bound peptides were released by decreasing the pH of the elution buffer to 4.0. Although coupling efficiency was high (95%), only 2.7% of the immobilized carboxypeptidase (before conversion to the apoenzyme) retained activity.

B. Proteolytically Derived Peptides

The amino acid sequence of proteins deduced from the cDNA nucleotide sequence must be verified by direct protein sequencing of N- and C-terminal regions to determine if posttranslational processing occurs during biosynthesis. Anhydrotrypsin affinity chromatography has provided a simple and selective method of isolating the C-terminal fragment of proteins for sequencing (97). Anhydrotrysin is a catalytically inactive derivative of trypsin where serine 195 is converted to a dehydroalanine residue. Chemical conversion involves base elimination by alkaline treatment of phenylmethane–sulfonyl-trypsin. The affinity of the inactive anhydrotrypsin for peptides containing arginine or lysine at their C terminus is higher than the native enzyme (98). The anhydrotrypsin was immobilized on CNBr-activated Sepharose (7.6 mg/mL agarose), and the matrix was treated with tosyl-L-lysine chloromethyl ketone to inhibit any residual tryptic activity. All of the peptides from tryptic digests of proteins without lysine or arginine at the C terminus bind to this affinity matrix with the exception of the C-terminal peptide (subtractive method). If the protein has a lysine or arginine residue at the C terminus, only the C-terminal peptide from a chymotryptic digest will bind to the column. Aside from its utility in isolating C-terminal peptides, two classes of soybean trypsin inhibitors were separated and purified on a preparative scale by the anhydrotrypsin matrix (83). Inhibitors were applied at pH 7.5 and eluted with 0.05 *M* glycine–HCl buffer (pH 2.2) containing 0.2 *M* NaCl.

REFERENCES

1. D. Fischer, C. Ebenau-Jehle, and H. Grisebach, *Arch. Biochem. Biophys., 276*:390 (1990).
2. J. K. Baird, R. F. Sherwood, R. J. G. Carr, and A. Atkinson, *FEBS Lett., 70*:61 (1976).
3. D. H. Watson, M. J. Harvey, and P. D. G. Dean, *Biochem. J., 173*:591 (1978).
4. D. Neale, R. K. Scopes, J. M. Kelly, and R. E. H. Wettenhall, *Eur. J. Biochem., 154*:119 (1986).
5. A. Baxter, L. M. Currie, and J. P. Durham, *Biochem. J., 173*:1005 (1978).
6. T. P. Kasten, D. Naqui, D. Kruep, and G. H. Dunaway, *Biochem. Biophys. Res. Commun., 111*:462 (1983).
7. M. Sugiura, *Anal. Biochem., 108*:227 (1980).
8. S. B. Pai, *Biochem. Biophys. Res. Commun., 110*:412 (1983).
9. B. L. Davison, T. Keighton, and J. C. Rabinowitz, *J. Biol. Chem., 254*:9920 (1979).
10. H. Jansen, R. Lammers, M. G. Baggen, N. W. Wouters, and J. C. Birkenhäger, *Biochim. Biophys. Acta, 1001*:44 (1989).
11. I. Becht, O. Schrecker, G. Klose, and H. Greten, *Biochim. Biophys. Acta, 620*:853 (1980).
12. E. Dake, T. J. Hofmann, S. McIntire, A. Hudson, and H. P. Zassenhaus, *J. Biol. Chem., 263*:7691 (1988).
13. P. M. J. Burgers, G. A. Bauer, and L. Tam, *J. Biol. Chem., 263*:8099 (1988).
14. G. K. Lindberg, J. K. Rist, T. A. Kunkel, A. Sugino, and L.-B. Rothman-Denes, *J. Biol. Chem., 263*:11319 (1988).
15. D. W. Mosbaugh, *Nucleic Acids Res., 16*:5645 (1988).
16. C. Elie, A. M. DeRecondo, and P. Forterre, *Eur. J. Biochem., 178*:619 (1989).
17. G. Bialek, H. P. Nasheuer, H. Goetz, B. Behnke, and F. Grosse, *Biochim. Biophys. Acta, 951*:290 (1988).
18. S. Sorrentino, G. K. Tucker, and D. G. Glitz, *J. Biol. Chem., 263*:16125 (1988).
19. D. H. Walker, N. Dougherty, and L. Pike, *Biochemistry, 27*:6504 (1988).
20. C. Staben, T. R. Whitehead, and J. C. Rabinowitz, *Anal. Biochem., 163*:257 (1987).
21. P. L. Stein, A.-G. van Zonneveld, H. Pannekock, and S. Strickland, *J. Biol. Chem., 264*:15441 (1989).
22. D. M. Swallow, L. Evans, and D. A. Hopkinson, *Nature, 269*:261 (1977).
23. B. G. Grimwood, T. H. Plummer, Jr., and A. L. Tarentino, *J. Biol. Chem., 263*:14397 (1988).
24. M. Sharma and U. S. Singh, *J. Biochem., 104*:57 (1988).
25. S. Jackowski and C. O. Rock, *J. Biol. Chem., 258*:15186 (1983).
26. C. L. Cooper, L. Hsu, S. Jackowski, and C. O. Rock, *J. Biol. Chem., 264*:7384 (1989).
27. M. Persson, M. G. Son Bergstrand, L. Bulow, and K. Mosbach, *Anal. Biochem., 172*:330 (1988).

28. J. Porath and N. Fornstedt, *J. Chromatogr., 51*:479 (1970).
29. R. A. Skidgel, P. A. Deddish, and R. M. Davis, *Arch. Biochem. Biophys., 267*:660 (1988).
30. R. A. Skidgel, R. M. Davis, and F. Tan, *J. Biol. Chem., 264*:2236 (1989).
31. B. G. Grimwood, T. H. Plummer, Jr., and A. L. Tarentino, *J. Biol. Chem., 264*:15662 (1989).
32. D. Parkinson, *J. Biol. Chem., 265*:17101 (1990).
33. N. Sakuragawa, K. Takahashi, and T. Ashizawa, *Acta Med. Biol., 25*:119 (1977).
34. M. Wu, G. K. Arimura, and A. A. Yunis, *Biochemistry, 16*:1908 (1977).
35. T. Suzuki and H. Takahashi, *Methods Enzymol., 34*:432 (1974).
36. J. S. Bartholomew and D. E. Woolley, *Biochem. Biophys. Res. Commun., 153*:540 (1988).
37. A. K. Chibber, D. G. Deutch, and E. T. Mertz, *Methods Enzymol., 34*:424 (1974).
38. J. Porath, J. Carlson, I. Olsson, and G. Belfrage, *Nature, 258*:598 (1975).
39. T. Ohkubo, T. Kondo, and N. Taniguchi, *Biochim. Biophys. Acta, 616*:89 (1980).
40. T. E. Cawston and J. A. Tyler, *Biochem. J., 183*:647 (1979).
41. E. Hochuli, W. Bannwarth, H. Döbeli, R. Genz, and D. Stüber, *Biotechnology, 6*:1321 (1988).
42. E. Hochuli, H. Döbeli, and A. Schacher, *J. Chromatogr., 411*:177 (1987).
43. S. F. Le Grice and F. Grüninger-Leitch, *Eur. J. Biochem., 187*:307 (1990).
44. A. A. Chuturgoon, M. F. Dutton, and R. K. Berry, *Biochem. Biophys. Res. Commun., 166*:38 (1990).
45. P. C. Simons and D. L. Van der Jagt, *Anal. Biochem., 82*:334 (1977).
46. L. I. McLellan and J. D. Hayes, *Biochem. J., 263*:393 (1989).
47. M. A. Francesconi, A. Donella-Deana, V. Furlanetto, L. Cavallini, P. Palatini, and R. Deana, *Biochim. Biophys. Acta, 999*:163 (1989).
48. D. M. Bowers-Komro, Y. Yamada, and D. B. McCormick, *Biochemistry, 28*:8439 (1989).
49. X. B. Gu, T. J. Gu, and R. K. Yu, *Biochem. Biophys. Res. Commun., 166*:387 (1990).
50. H. Thomassin, M. K. Jacobson, J. Guay, A. Verreault, N. Aboul-ela, L. Menard, and G. G. Poirier, *Nucleic Acids Res., 18*:4691 (1990).
51. M. K. Jacobson, D. M. Payne, R. Alvarez-Gonzales, H. Jurez-Salinas, J. L. Sims, and E. L. Jacobson, *Methods Enzymol., 106*:483 (1984).
52. R. Marquetant, R. L. Sabina, and E. W. Holmes, *Biochemistry, 28*:8744 (1989).
53. M. D. Pato and E. Kero, *Arch. Biochem. Biophys., 276*:116 (1990).
54. T. L. Lindahl, P.-I. Ohlsson, and B. Wiman, *Biochem. J., 254*:109 (1990).
55. M. Hirasawa, K. T. Chang, and D. B. Knaff, *Arch. Biochem. Biophys., 276*:251 (1990).
56. B. Nefsky and A. Bretscher, *Eur. J. Biochem., 179*:215 (1989).
57. C. Broger, K. Bill, and A. Azzi, *Methods Enzymol., 126*:64 (1986).
58. E. Hadas, R. Koppel, F. Schwartz, O. Raviv, and G. Fleminger, *J. Chromatogr., 510*:303 (1990).
59. P. J. Duerksen-Hughes, M. M. Williamson, and K. D. Wilkinson, *Biochemistry, 28*:8530 (1989).
60. A. Hershko, H. Heller, S. Elias, and A. Ciechanover, *J. Biol. Chem., 258*:11992 (1983).
61. F. Friedberg, *Chromatogr. Rev., 14*:121 (1971).

62. W. W.-C. Chan, *Biochem. Biophys. Res. Commun., 41*:1198 (1970).
63. D. Grune, J. Chen, and D. Sherwin, *Proc. Natl. Acad. Sci. USA, 74*:1383 (1977).
64. K. P. Henrikson, S. G. Allen, and W. L. Maloy, *Anal. Biochem., 94*:366 (1979).
65. P. Dimroth, *FEBS Lett., 141*:59 (1982).
66. P. Brodelius, P.-O. Larsson, and K. Mosbach, *Eur. J. Biochem., 47*:81 (1974).
67. C. A. Waters, J. R. Murphy, and J. W. Hastings, *Biochem. Biophys. Res. Commun., 57*:1152 (1974).
68. E. Steers and P. Cuatrecaseas, *Methods Enzymol., 34B*:350 (1974).
69. F. M. Pisani, R. Rella, C. A. Raia, C. Rozzo, R. Nucci, A. Gambacorta, M. De Rosa, and M. Rossi, *Eur. J. Biochem., 187*:321 (1990).
70. J. Schweden and E. Bause, *Biochem. J., 264*:347 (1989).
71. W. L. Stundenbauer and E. Orr, *Nucleic Acids Res., 9*:3589 (1981).
72. T. Melendy and D. S. Ray, *J. Biol. Chem., 264*:1870 (1989).
73. A. J. Raeber, G. Riggio, and P. G. Waser, *Eur. J. Biochem., 186*:487 (1989).
74. C. Y. Dadabay and L. J. Pike, *Biochem. J., 264*:487 (1989).
75. S. Matsuda, K. Shimura, H. Shiraki, and S. Nakagawa, *Biochim. Biophys. Acta, 616*:340 (1980).
76. R. Eliasson, M. Fontecave, H. Jörnvall, M. Krook, and E. Pontis, *Proc. Natl. Acad. Sci. USA, 87*:3314 (1990).
77. D. G. Knorre, V. A. Kurbatov, and U. V. Samukov, *FEBS Lett., 70*:105 (1976).
78. S. Izuta and M. Saneyoshi, *Anal. Biochem., 174*:318 (1988).
79. L. Holmberg, B. Bladh, and B. Astedt, *Biochim. Biophys. Acta, 445*:215 (1976).
80. S. Khamlichi and D. Muller, *J. Chromatogr., 510*:123 (1990).
81. B. J. Campbell, L. J. Forrester, W. L. Zahler, and M. Burks, *J. Biol. Chem., 259*:14586 (1984).
82. J. S. Huang, S. S. Huang, and J. Tang, *J. Biol. Chem., 254*:601 (1989).
83. G. E. Conner, *Biochem. J., 263*:601 (1989).
84. J. Rittenhouse, M. C. Turon, R. J. Helfrich, K. S. Albrecht, D. Weigl, R. L. Simmer, F. Mordini, J. Erickson, and W. E. Kohlbrenner, *Biochem. Biophys. Res. Commun., 171*:60 (1990).
85. D. H. Rich, M. A. Brown, and A. J. Barrett, *Biochem. J., 236*:731 (1986).
86. D. J. Buttle, B. C. Bonner, D. Burnett, and A. J. Barrett, *Biochem. J., 254*:693 (1988).
87. D. J. Buttle, A. A. Kembhavi, S. L. Sharp, R. E. Shjute, D. H. Rich, and A. J. Barrett, *Biochem. J., 261*:469 (1989).
88. A. Patel, M. O'Hara, J. E. Callaway, D. Greene, J. Martin, and A. H. Nishikawa, *J. Chromatogr., 510*:83 (1990).
89. A. Basak, Y. T. Gong, J. A. Cromlish, J. A. Paquin, F. Jean, N. G. Seidah, C. Lazure and M. Chretien, *Int. J. Protein Res., 36*:7 (1990).
90. S. X. Lin, J. P. Shi, X. D. Cheng, and Y. L. Wang, *Biochemistry, 27*:6343 (1988).
91. Y. Nishikawa, W. Pegg, H. Paulsen, and H. Schachter, *J. Biol. Chem., 263*:8270 (1988).
92. B. Bendiak and H. Schachter, *J. Biol. Chem., 262*:5775 (1987).
93. G. A. Homandberg, R. D. Litwiller, and R. J. Peanasky, *Arch. Biochem. Biophys., 270*:153 (1989).
94. M. R. Martzen, B. A. McMullen, N. E. Smith, K. Fujikawa, and R. J. Peanasky, *Biochemistry, 29*:7366 (1990).

95. T. Yasuhara and A. Ohashi, *Biochem. Biophys. Res. Commun., 166*:330 (1990).
96. J. E. Coleman and B. L. Vallee, *J. Biol. Chem., 237*:3430 (1962).
97. T. Kumazaki, K. Teraswa, and S.-I. Ishii, *J. Biochem., 102*:1539 (1987).
98. H. Yokosawa and S. Ishii, *J. Biochem., 81*:647 (1977).

4

Affinity Chromatography of Regulatory and Signal-Transducing Proteins

Allen R. Rhoads and Felix Friedberg

Howard University, Washington, D.C.

I. INTRODUCTION

The highly selective binding and concentrating ability of affinity chromatography are of considerable value in the purification of proteins involved in signal transduction pathways. Signal transduction occurs when messages are communicated across the plasma membrane. In some cases integral membrane receptor proteins are triggered, causing the production of a cellular second messenger such as cyclic AMP or inositol phosphates. Other transduction mechanisms include the interaction of extracellular factors with ion-channel receptors or with transmembrane receptors regulating cytoskeletal organization and direct interaction of hormones with cytosolic receptors (1). Enzymes and proteins of the transduction pathway may be relatively low in cellular concentration and readily susceptible to proteolysis, leading to a loss in regulatory function during isolation. The specificity and rapidity of affinity chromatography can be invaluable in these situations. Affinity chromatography also offers a rapid method of studying specific and sequential reactions between biomolecules in these pathways.

Signal-transducing proteins often possess strikingly different affinities for target proteins or associated subunits in the presence of messenger or activating factors compared to affinities measured in their absence. This dependency can often be exploited in the development of an affinity chromatography procedure. Proteins such as calmodulin or second messengers that interact with different proteins in the signal transduction system can serve as a general affinity ligand for isolating several proteins associated with the pathway.

In contrast to small biomolecules, proteins can be immobilized through a wide variety of functional groups and may offer a higher degree of specificity. The stability of immobilized peptides and proteins can also be increased by incorporation of D-amino acids, cross-linking, or chemical modification as long as the binding specificity is not compromised. The effectiveness of a particular affinity procedure often depends on the method of immobilization. Currently, refined techniques of affinity chromatography based on site-specific immobilization procedures to ensure the retention of biospecificity and high binding capacity are being developed. For example, immobilization of immunoglobulins by covalent coupling to the carbohydrate portion of the molecule results in a more effective orientation of the antibody and preserves antigen-binding ability (2,3).

Immunoaffinity purification using immobilized monoclonal antibodies directed against specific peptide regions of the regulatory proteins has been invaluable in elucidating and characterizing the protein components of signal-transducing pathways. Selective elution of the protein with a synthetic peptide antigen obviates the need for elution by chaotropic agents that denature proteins. The development of special monoclonal antibodies or elution procedures permitting release of protein antigens under relatively mild, undenaturing conditions has also enhanced the effectiveness of immunoaffinity techniques.

The field of signal transduction represents a wide range of systems and proteins. Practical applications of affinity chromatography to selected areas of signal transduction research will be examined. Sources that are more comprehensive in scope have appeared and should be useful in obtaining a broader understanding of affinity chromatography (4–7).

II. CALCIUM REGULATION

A. Purification of Calmodulin Target Proteins

1. Structural Basis of Interaction

The calmodulin (CaM) molecule can be divided into four homologous domains (Figure 1). The level of homology is greatest when domain 1 (residues 8 to 40) is aligned with domain 3 (residues 81 to 113) and domain 2 (residues 44 to 75) is aligned with domain 4 (residues 117 to 148) (8). The four calcium binding sites fall into two classes of high and low affinity. Regulation of CaM or troponin C is viewed as being controlled by the binding of an additional two calcium ions to domains 1 and 2 on the amino terminal half of molecule. These sites are calcium-specific but low-affinity sites (9,10).

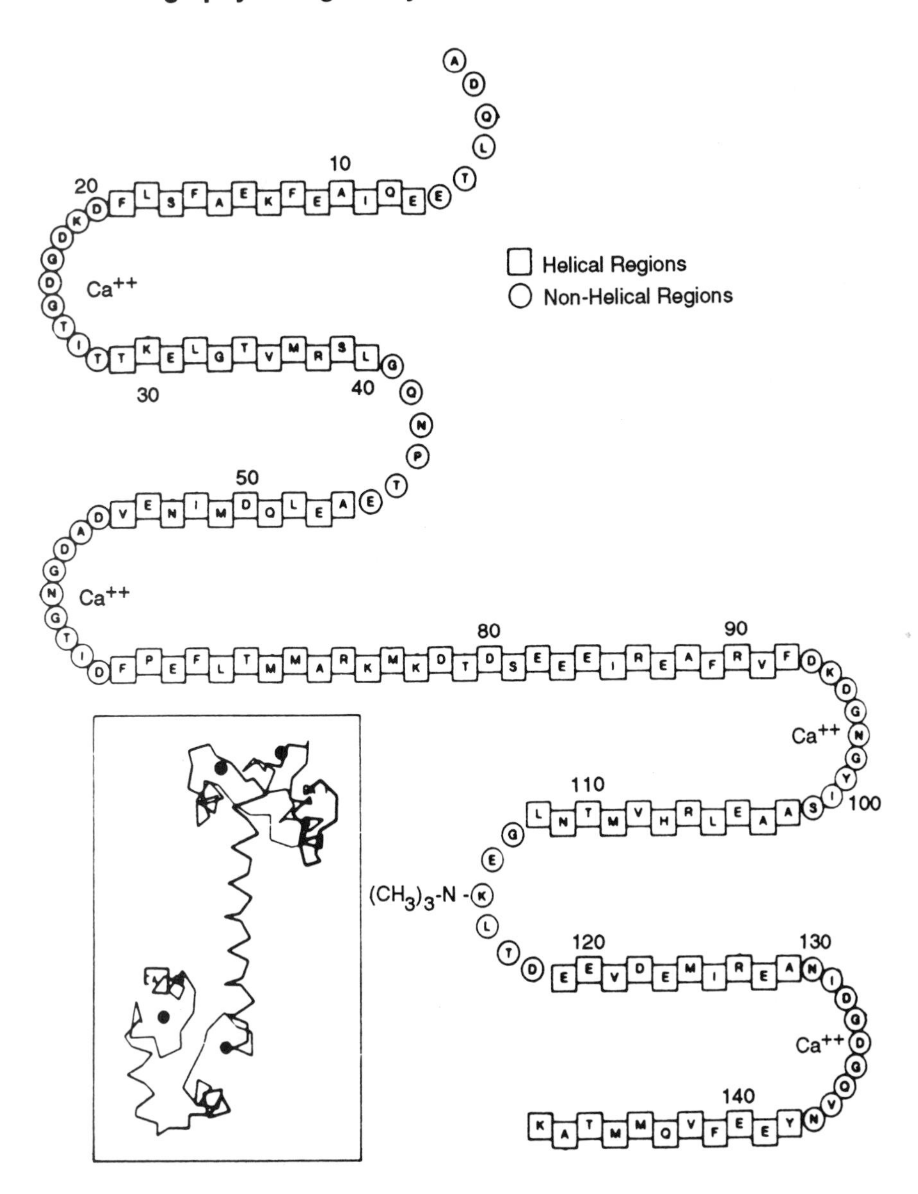

Figure 1 Primary amino acid sequence of vertebrate calmodulin. The sequence is derived from chicken CaM DNA [Putkey et al., *J. Biol. Chem. 258*:11864 (1983)]. Helical and nonhelical regions were assigned based on x-ray crystallographic data of Babu et al., *Nature 315*:37 (1985). [After Putkey et al., *J. Biol. Chem. 261*:9896 (1986).]

The CaM target proteins that bind in a calcium-dependent manner to CaM are characterized by a basic amphipathic helix (11). This helical region is composed of positively charged residues (Lys, Arg, and His) interspersed among hydrophobic residues in a sequence spanning approximately 20 amino acid residues. Polypeptide fragments of target proteins containing the CaM recognition domain (12,13), and certain analogous peptides from toxic components such as melittin (14) interact strongly with the CaM. The CaM recognition domain may occur near important regulatory regions of target proteins. A 10-kDa CaM-binding domain resulting from the CNBr cleavage of caldesmon was isolated by CaM–Sepharose chromatography (15). Cosedimentation with F-actin indicted that the peptide was bound by both actin and CaM and thus may be important in regulating the interaction of caldesmon with actin and CaM.

2. *Applications of CaM Affinity Chromatography*

Watterson and Vanaman (16) and Miyake et al. (17) were the first workers to employ immobilized CaM to purify CaM target enzymes. Miyake (17) reported that the CaM-dependent phosphodiesterase of rat cerebrum was purified over 50-fold by the CaM affinity column. These groups utilized cyanogen bromide and *N*-hydroxysuccinimide-activated Sepharose 4B to covalently couple primary amino groups of CaM. Covalent attachment by either procedure could be expected at any of seven lysines (Lys 13 in the first helix, Lys 21 and 30 in the first calcium-binding domain, Lys 75 and 77 in the central helix, Lys 94 in the third calcium-binding domain, and Lys 148 at the C terminal). Lysine 115 is trimethylated in most species and would thus be unreactive.

CaM-matrix affinity chromatography has been useful in the purification (18–27) of numerous target proteins (Table 1). Due to the high degree of conservation in sequence (28), CaM is a useful affinity matrix for target proteins from different species. The utility of the CaM affinity matrix lies in the fact that in the presence of calcium, most proteins bind with relatively high affinity (K_d is in the nanomolar region), whereas in the absence of calcium, there is little affinity for CaM. Included in this sizable group of CaM-regulated proteins is the Ca^{2+}-independent target protein, neuromodulin. The protein binds specifically to CaM in the absence of calcium and is released by addition of calcium (22). Neuromodulin is a major protein of the neuronal growth cone, a prominent feature of axonal growth. Neuromodulin is also a major target of protein kinase C action and has been shown to stimulate GTPγS binding to G_o in the growth cone membrane (29). CaM-matrix affinity chromatography can also be advantageous in separating CaM-sensitive and CaM-insensitive enzymes such as adenylate cyclase (26). Although CaM affinity chromatography is highly selective, lysozyme of egg white binds to the matrix in a calcium-dependent manner (30). The significance of this interaction is not known since CaM has no influence on the lytic activity of lysozyme.

Table 1 Applications of CaM Affinity Chromatography

Target protein	Ref.
Cyclic 3′,5′-nucleotide phosphodiesterase	18
Myosin-light-chain kinase	19
CaM-dependent protein kinase	20
CaM-dependent protein phosphatase (2B)	21
Neuromodulin (GAP-43)	22
CaM-dependent histone 3-kinase	23
Inositol 1,4,5-trisphosphate 3-kinase	24
Major CaM-binding protein of cardiac muscle	25
CaM-dependent adenylate cyclase	26
Guanylate cyclase–activating factor synthase (nitric oxide synthase)	27

Prior to the application of the sample, the column is usually equilibrated with Tris buffer containing 100 m*M* NaCl, 0.5 m*M* calcium, and calpain inhibitors. Protease inhibitors (e.g., leupeptin, soybean trypsin inhibitor, etc.) are included to prevent the loss of CaM sensitivity from proteolysis. The CaM affinity matrix is eluted with equilibration buffer containing 5 m*M* EGTA in place of calcium.

Studies by Klee and others (31–33) on HPLC-purified CaM fragments have demonstrated that two fragments (N terminal, 1–77 or C terminal, 78–148) can act as antagonists or agonists of target enzymes. Employing proteolytic fragments of CaM coupled to Sepharose 4B, Ni and Klee (34) developed affinity matrixes exhibiting some selectivity for target enzymes. CaM fragments were coupled through cyanogen bromide–activated Sepharose utilizing lysine residues. Using this procedure, nearly all of the CaM-dependent enzymes could bind to fragment 78–148, but only phosphodiesterase and cyclic AMP-dependent protein kinase were bound by fragment 1–77. A rapid procedure for the selective purification of CaM-stimulated phosphodiesterase on CaM fragment 1–77 immobilized on Sepharose was developed from these studies (35). This technique requires careful purification to obtain homogeneous peptide free from intact CaM.

The extracellular adenylate cyclase of *Bordetella pertussis* is activated by CaM in both the presence and absence of calcium (36,37). The difference in affinity of CaM for *B. pertussis* adenylate cyclase is only slightly higher in the presence of Ca^{2+} than its absence (38). Because of this calcium-independent high binding affinity, the elution of the enzyme from the normal CaM matrix requires chaotropic agents. An engineered CaM (VU-8) in which three glutamic acid residues (positions 82 to 84) have been substituted with three lysine residues possesses 1000-fold lower affinity for *B. pertussis* adenylate cyclase in the absence of calcium compared to the normal CaM. Immobilized VU-8 CaM thus provides an affinity matrix for *B. pertussis* adenylate cyclase that permits elution by the removal of calcium. A nearly homogeneous enzyme in high yield was obtained from this VU-8 matrix by elution with 2 m*M* EGTA (38). However, the holotoxin of *B. pertussis* has been purified to apparent homogeneity by unmodified CaM–Sepharose chromatography using detergent-supplemented buffer containing 8 *M* urea for elution (39). The procedure effectively separated proteolyzed low-molecular-weight catalytic subunits without toxin activity from the 216-kDa holotoxin.

3. *Preparation of CaM for Immobilization*

a. Conventional Purification

Bovine brain CaM is conveniently purified by the procedure of Dedman and Kaetzel (40) from the acetone powder of the tissue. The purity of the CaM can be assessed by comparison of the dry weight of the protein with the amount of protein determined using an extinction coefficient of 0.18 absorbance units at 276 nm for a solution of 1 mg/mL (8). CaM should be examined for its ability to activate bovine brain CaM-dependent phosphodiesterase (41) or any of the other target enzymes. The purity of CaM over the range 10 to 80 μg should be assessed by SDS–polyacrylamide gel electrophoresis to ensure column specificity (42).

b. Phenothiazine Matrix

In the presence of calcium, CaM exhibits two hydrophobic, phenothiazine-binding sites on each of the two halves of the molecule (43–45). Thus fragments 1–77 and 78–148 and intact CaM can be purified by phenothiazine- and phenyl-agarose chromatography. Fluphenazine, a hydroxyphenothiazine, can readily be coupled to epoxy-activated Sepharose 4B and employed to isolate CaM (46). Elution of CaM is accomplished by using EGTA-containing buffers of moderate ionic strength. Several other phenothiazines and CaM antagonists differing in stability and capacity have also been employed as affinity ligands. The phenothiazines are subject to photochemical degradation, but with proper precautions the technique provides a rapid and effective means of purifying CaM from acidic proteins of similar molecular weight. Although possessing lower capacity for CaM than the phenothiazine matrix, phenyl-Sepharose is a very stable matrix for the isolation of CaM from reactive and complex protein extracts of plants (47).

c. Melittin Matrix

Melittin has been considered a model peptide resembling the recognition sequence or "calmodulin-binding domain" of many target proteins that interact in a calcium-dependent manner with CaM. Melittin is a basic amphipathic peptide consisting of 26 amino acids present in bee venom of *Apis mellifera*. It forms a 1:1 calcium-dependent high-affinity complex with CaM (K_d: 3 n*M* and 10 μ*M* with and without calcium, respectively) (14). Melittin immobilized on CNBr-activated

Sepharose still displays high-affinity, calcium-dependent binding to CaM (48). The melittin matrix can be used to purify CaM and to measure the affinity of different peptides for radiolabeled CaM in a competitive binding assay.

Melittin–Sepharose has been employed to purify polyclonal antibody against melittin (49). The antibody to melittin was then used to prepare antimellitin–Sepharose to demonstrate that myosin light-chain kinase, CaM-dependent cyclic 3′,5′-nucleotide phosphodiesterase and other CaM-binding proteins can interact with antibody against the model peptide. Results from studies with antimellitin–Sepharose and with monoclonal antibody (50) suggest that structural similarities exist among the CaM recognition domains of several target proteins.

4. *Preparation of the CaM-Affinity Matrix*

CaM can be coupled through primary amino groups to Sepharose 4B by the procedure of Klee (51). Sepharose 4B is activated by suspending 25 g of the moist agarose cake in 25 mL of 2 *M* sodium carbonate (pH 11.5) and placing the mixture on a stirrer in a well-ventilated hood (52). A CNBr solution (3.45 g in 3 mL of acetonitrile) is then added dropwise while the suspension is stirred. (*Caution*: CNBr is volatile and toxic.) After 5 min of additional stirring (pH remains constant), the mixture is placed on a 250-mL fritted glass funnel and washed with 500 mL of cold deionized water and 1 L of cold 0.2 *M* sodium bicarbonate, pH 9.5. The activated Sepharose is then transferred to 25 mL of 0.1 *M* sodium borate buffer pH 8.2, and 27 mg of CaM in 4.4 mL of borate buffer is added. After 10 to 18 h, the coupled gel is washed and reacted with 0.5 *M* ethanolamine pH 8.2 as described (51). The gel should contain approximately 0.8 mg CaM/mL. This is best determined by measuring the absorbance of the unreacted CaM in the filtrate at 276 nm. Alternatively, the amount of uncoupled CaM in the filtrate can be determined by the ability to activate CaM-deficient phosphodiesterase of bovine brain or by protein measurement using either fluorescamine (53) or Coomassie Blue dye–binding (54) assays.

The effect of immobilizing CaM via primary amino groups on its affinity for target enzymes is not clear. Chemical modification studies indicate that after carbamoylation of all seven lysines with isocyanate, CaM showed enhanced activation (152%) of brain adenylate cyclase with no change in the apparent K_m value or concentration required for half-maximal stimulation (55). Complete lysine modification of CaM had no effect on maximal activation of cyclic 3′,5′-nucleotide phosphodiesterase but increased the apparent K_m value to two- to threefold above that of native CaM. However, monoacetylation of any of the lysines of CaM was found to reduce its affinity for calcineurin by 5- to 10-fold (56).

Although standard conditions for coupling CaM to CNBr-activated Sepharose do not involve addition of calcium (51), calcium in laboratory reagents approaches 1 μM unless special precautions are taken. Thus under standard conditions, coupling probably includes significant calcium, and thus Lys 75 and Lys 94 would be the most reactive of the seven lysines in CaM (57). In the presence of calcium, CaM can be monomodified on Lys 75 by carbamoylation with a nitrosourea (58). Phosphodiesterase required a sevenfold higher amount of the mono-modified CaM to reach normal maximal activation, whereas the activation profile of Ca^{2+}, Mg^{2+}-ATPase was unaffected. This suggests that stereochemical factors resulting from the modification of the flexible central helix may be important, depending on the target enzyme. From these studies it is difficult to predict the effect of coupling lysines of CaM during the immobilization procedure. Cyanogen bromide-activated coupling of intact CaM to Sepharose via amino groups results in lower than calculated capacities. Relatively, large columns are routinely used for purification of milligram amounts of target proteins (59).

5. *Preliminary Purification of CaM Target Proteins*

CaM-binding proteins can be conveniently prepared according to Hathaway et al. (60) as described by Ni and Klee (34). Bovine brain cortex, a rich source of many of the CaM-dependent enzymes, should be obtained at the time of slaughter and placed immediately on ice for transportation.

The brain cortex is homogenized for 1 min at 4°C in a Waring Blendor with 2.5 volumes of 25 m*M* Tris–HCl (pH 7.5) containing 2.5 m*M* dithiothreitol, 10 m*M* EDTA, 5 m*M* EGTA, and protease inhibitors: phenylmethylsulfonylfluoride (PMSF), 75 μg/mL; leupeptin, 1 μg/mL; soybean trypsin inhibitor: 50 μg/mL; α-*N*-benzoyl-L-arginine methyl ester, 100 μg/mL; and 1 m*M* benzamidine. Experience indicates that proteolysis of highly labile CaM target proteins in brain occurs rapidly. Most CaM-binding proteins are sensitive targets of the Ca^{2+}-dependent neutral proteases (calpains) (61). Due to the close proximity or overlap between the autoinhibitory and CaM-binding domains of some target proteins, proteolysis often results in activation of target enzyme activity with concomitant loss of CaM binding. Inclusion of cysteine protease inhibitors such as 1 μg/mL leupeptin and antipain can minimize proteolysis, which prevents many CaM target proteins from binding to the CaM affinity matrix.

Before CaM affinity chromatography, endogenous CaM is removed from the target enzymes using DEAE–cellulose chromatography and gradient elution with NaCl. This step must be performed with high resolution and care; otherwise, the trace amounts of CaM remaining in the enzyme fraction will prevent target enzymes from binding to the immobilized CaM. Often an additional preliminary step such as gel filtration or blue dextran chromatography (18,26) can ensure complete elimination of endogenous CaM. Endogenous CaM can be detected by measuring the activity of the CaM-dependent enzyme being purified in the presence and absence of EGTA or directly by immunoassay. CaM can also be assayed by heating a sample at 100°C to eliminate endogenous CaM-dependent enzymes and assaying the supernatant based on the degree of activation of a purified CaM-deficient reagent enzyme.

B. Purification of Other Calcium-Dependent Proteins

Several calcium-dependent phospholipid- or membrane-binding proteins have been identified in tissues. These include calpactins, chromobindins, lipocortins, calcimedins, synexin, calelectrins, endonexin, and others, broadly termed annexins (62). The cellular elements responsible for binding these proteins, such as phospholipid, membranes, and actin, have all been immobilized on agarose and employed in both the isolation of the proteins and to study specific properties. Glutaraldehyde-stabilized F-actin prepared according to Herman and Pollard (63) has been employed to isolate annexins from bovine aorta (64) (Table 2). Stabilization of F-actin by glutaraldehyde, phalloidin (65), or suberimidate cross-linking (66) prevents dissociation of the actin monomer from the immobilized filament during chromatography. Although the F-actin affinity matrix has been used primarily to study myosin interaction, a number of additional actin-binding proteins have been isolated by this technique (67). Liposomes entrapped within polyacrylamide gel particles or covalently linked to agarose have also been used to isolate members of the annexin family and to study their Ca^{2+}-binding properties (Table 2). Proteolytic fragmentation of a 36-kDa annexin that normally contains four calcium phospholipid–binding regions into peptides containing only one or two calcium phospholipid–binding regions led to a decrease (10- to 20-fold) in calcium-dependent affinity for the liposome matrix (68).

Cytosolic proteins known as chromobindins can be bound to immobilized chromaffin granule membranes in the presence of calcium (Table 2) and isolated by elution with EGTA (69–71). The majority of the isolated chromobindins were also bound by the immobilized liposomes prepared from total lipid extracts of chromaffin granule membranes via CNBr-activated Sepharose (69). A total of 23 soluble chromobindins have been isolated by binding to the chromaffin granule membrane matrix. These proteins appear to play important roles in such membrane events as exocytosis, membrane recycling, and cytoskeletal interactions.

Chromaffin granule proteins binding to microtubules in a calcium-independent manner have been isolated using taxol-stabilized microtubules. Taxol (Figure 2) stabilizes microtubules by inhibiting their depolymerization. Severin et al. (72) employed taxol-stabilized microtubules immobilized on agarose derivatized with *N*-hydroxysuccinimide ester to demonstrate that MAP 2 was present in chromaffin granules and may be responsible for their interaction with microtubules.

Table 2 Purification and Identification of Calcium-Binding Proteins

Protein	Ligand/Matrix	Binding/elution conditions	Results	Refs.
Annexins of bovine aorta	Glutaraldehyde-stabilized F-actin	Tris buffer (pH 7.5), 100 m*M* NaCl, 0.5 m*M* dithiotreitol (DTT); eluted with 5 m*M* EGTA	Five proteins of 67, 36, 34, 32, and 10 kDa identified	64
Actin-binding proteins of chicken smooth muscle	F-actin (phalloidin stabilized and suberimidate cross-linked actin)	Hepes buffer (pH 7.5), 0.05% Nonidet P40, 10% glycerol, protease inhibitors; batch elution with 0.1, 0.5, and 1.0 *M* KCl with 1 m*M* ATP/3 m*M* Mg^{2+}	Several proteins identified, including villin, fimbrin, spectrin, TW260/240, and filamin	67
Annexins of porcine intestine	Liposomes of dipalmitoyl-glycerophosphoserine entrapped in polyacrylamide gel	Applied sample at 2 m*M* Ca^{2+} and eluted at 8 m*M* EGTA	Proteolytic fragments of one or two annexin repeats have decreased (10- to 20-fold) affinity	68
Chromobindins of adrenal medullary cytosol	Immobilized chromaffin granule membrane or immobilized liposomes	Equilibrated with 240 m*M* sucrose, 30 m*M* KCl, 1 m*M* Mg^{2+} in Hepes buffer pH 7.3 at 37°C, and 4 m*M* Ca^{2+}/2 m*M* EGTA; batch elution with free Ca^{2+} at 40 and 0.1 μ*M*	23 chromobindins identified, including CaM, protein kinase C, caldesmon, synexin, calelectrins (32 and 68 kDa), and protein substrates for protein kinase C and protein tyrosine kinases	69–71

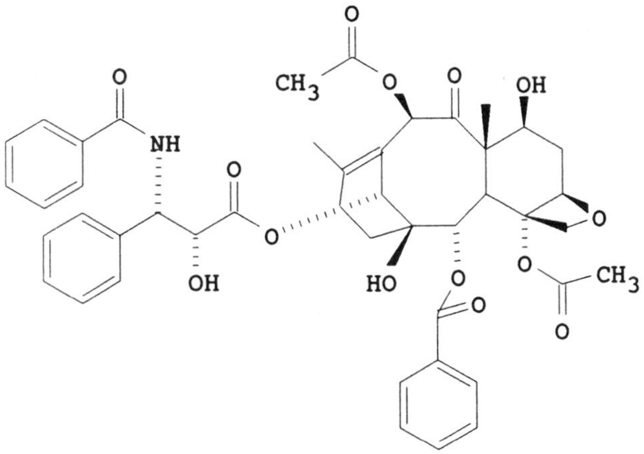

Figure 2 Taxol.

III. CYCLIC NUCLEOTIDE REGULATION

A. Adenylyl Cyclase

Signal transduction by receptors of hormones and neurotransmitters coupled to cyclic AMP production often involves mediation through a GTP-binding protein that subsequently stimulates or inhibits effector enzyme, adenylyl cyclase (73). The adenylyl cyclase regulated by GTP-binding proteins may be sensitive or insensitive to stimulation by CaM (74). Classical GTP-binding proteins are heterotrimeric complexes consisting of α, β, and γ subunits. The binding of GTP or guanosine 5′-O-3-thiotriphosphate (GTPγS) to the α subunit of the αβγ complex in exchange for GDP leads to dissociation of the activated α subunit from the βγ dimer. Activation of the α subunit can also occur through the interaction of AlF_4– with bound GDP, causing a mimicry of GTP binding (75). Due to its low concentration in membranes and its lability in the detergent-solubilized state, adenylyl cyclase was one component of this signal transduction triad that proved difficult to purify by conventional techniques. Initially, adenylyl cyclase of cardiac muscle was purified over 40-fold by N^6-ATP–Sepharose but remained associated with the GTP-binding protein (G_s) even after extensive washing of the bound enzyme with 5 m*M* GTP (76). However, GTP–Sepharose chromatography proved effective in separating the G_s from adenylyl cyclase (77) and demonstrated for the first time that membrane-associated GTP-binding and adenylyl cyclase activity were separable entities (78).

Forskolin, a major diterpene isolated from the roots of the Indian plant *Coleus forskohlii* (Figure 3), was known to be a potent activator of adenylyl cyclase from different sources, and this activation explains the basis for many of its pharmacological actions (79,80). Forskolin, specifically the 7-*O*-hemisuccinyl-7-deacetyl derivative, which is biologically active, proved exceedingly useful as an affinity matrix in the purification and subsequent characterization and cloning of adenylyl cyclase (81–84). The deacetylforskolin provides a reactive derivative with an activation constant (K_{act}) and maximum activation more closely related to the parent compound (K_{act}, 0.3 μM) than to any obtained by direct succinylation of the acetylated oligohydroxyderivative. For the preparation of the matrix, forskolin was deacetylated by alkaline hydrolysis and reacted with succinic anhydride to form 7-*O*-hemisuccinyl-7-deacetylforskolin (82). The carboxy derivative was then coupled to the aminoethyl-Sepharose activated via the formation of the *N*-hydroxysuccinimide ester (85). Agarose substitution measured by including radiolabeled forskolin was estimated to be 0.4 to 0.8 μmol of forskolin per milliliter of packed resin.

Approximately 90% of the Lubrol-solubilized adenylyl cyclase from cardiac muscle was bound by this column, and 35 to 55% of the retained enzyme could be eluted with 100 μM forskolin in buffers containing Tween-60. Detergent alone did not lead to release of the enzyme but was essential for elution with forskolin. A 300- to 500-fold enrichment of adenylate cyclase was obtained by this single-step procedure. Adenylate cyclase was also readily separated from GTP-binding proteins, and thus was unresponsive to nonhydrolyzable derivatives of GTP or to NaF. The isolated enzyme was depleted of bound forskolin simply by Sephadex

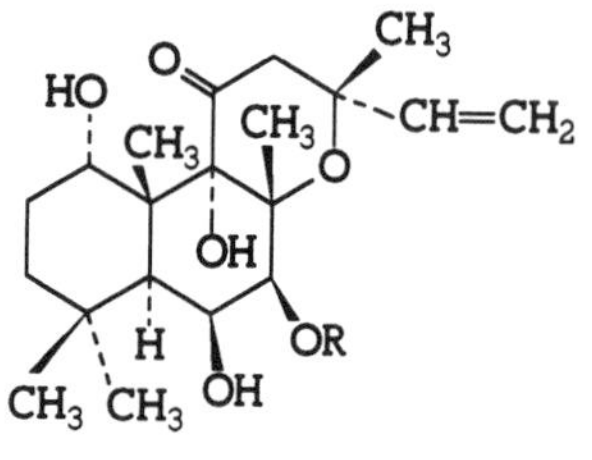

FORSKOLIN : R = C(=O)—CH_3

FORSKOLIN, 7-O-HEMISUCCINYL-7-DEACETYL : R = C(=O)—CH_2–CH_2–COOH

Figure 3 Forskolin and derivative.

G-25 chromatography. The forskolin affinity column could be regenerated by treatment with buffered solutions containing urea. The ratio of basal to Mn^{2+}- or forskolin-stimulated activity was essentially unchanged, indicating that the purified enzyme maintains responsiveness to direct activators while losing responsiveness to guanine nucleotides and NaF acting via the GTP-binding protein. Activation by GTP analogs and NaF could be fully restored by addition of purified GTP-binding proteins to the purified enzyme.

Under mild conditions (low $MgCl_2$ and NaCl concentrations), a GppNHp-activated adenylate cyclase was purified over 3000-fold by a series of steps involving forskolin–Sepharose (86). The stoichiometry between the α_s subunit, $\beta\gamma$ subunit, and catalytic subunit was near unity, suggesting that $\beta\gamma$ dimer remains associated with the α subunit during the activation cycle of adenylyl cyclase under physiological conditions. The forskolin affinity procedure has also been used to purify adenylate cyclase to homogeneity from other sources (87,88), including a novel 180-kDa form of the enzyme from olfactory cilia (89).

Adenylyl cyclase from yeast has been purified by a recombinant tailing technique in which the expressed fusion protein of adenylate cyclase was linked at the N terminus to a peptide epitope (YPYDVPDYA) of the hemagglutinin of influenza virus (90). This short peptide represents the complete epitope for the monoclonal antibody. The expressed adenylate cyclase was adsorbed and then released from the immobilized antibody by competitive elution with peptide. Competitive elution avoided deleterious extremes in pH normally employed to elute antigen from the immunoaffinity matrix.

A CaM-independent adenylyl cyclase of bovine brain was isolated in the unbound fraction from CaM–Sepharose (91). This enzyme was used to produce a monoclonal antibody that was immobilized on protein A–Sepharose to provide an immunoaffinity matrix. The antibody column specifically bound only the CaM-independent enzyme of brain. The N-terminal amino acid sequence of the purified protein permitted synthesis of oligonucleotide probes that were employed to screen a cDNA library of brain.

B. GTP-Binding Proteins

GTP-binding proteins were first identified by their modulation of adenylyl cyclase activity. These proteins are now known to participate in the activation of other target systems, including ion channels and phospholipases (92,93). Isolation procedures for membrane-associated GTP-binding proteins have relied mainly on classical techniques and hydrophobic chromatography on heptyl- or phenyl-Sepharose (94). However, in some cases GTP–Sepharose affinity chromatography has been employed (95). For the preparation of GTP–Sepharose, phenylethylenediamine is coupled to the γ-phosphate of GTP and the resulting product was immobilized on carboxypropylamino-Sepharose via carbodiimide coupling (76).

Recently, affinity chromatography with immobilized βγ subunits has been applied to isolate and purify a new family of α subunits (96). Because the different β and γ subunits are closely related in both structure and function (97), βγ subunits were employed to isolate different α subunits that bind to them in a reversible fashion. Binding was stabilized by GDP and dissociation was promoted by GTP or AlF_4^- The matrix bound more than 60% of the α_s, α_{i1}, α_{i2}, and α_0 from the Lubrol extracts of brain and also bound a new α subunit protein of 42 kDa (α42). Binding was specific since heat-inactivated βγ-agarose was not effective (96). Extracts were routinely added slowly to the column to allow the α subunit to dissociate from endogenous βγ subunits and associate with the immobilized βγ subunits (98). GTPγS (50 μ*M*) was found to elute all α subunits except the α42. Thus the column was sequentially eluted with GTPγS and then AlF_4^- to obtain a fraction enriched in the α42 subunit. This novel 42-kDa α subunit was not a substrate for ADP ribosylation by *B. pertussis* toxin. The new α subunit was shown to stimulate polyphosphoinositide-specific phospholipase C from bovine brain (99).

C. Guanylate Cyclase

Cyclic GMP is formed by the action of guanylate cyclase. Two major forms (one soluble and the other membrane-bound) of the enzyme exist. The soluble enzyme, which is composed of two subunits of 82,000 and 70,000 kDa, is activated by free radicals such as nitric oxide (100) and also certain porphyrins (101). The soluble enzyme has been isolated by immunoaffinity chromatography using antibody developed against a synthetic peptide of the C terminus of the 70-kDa subunit (102). Competitive elution with the synthetic peptide successfully preserved enzyme activity in contrast to elution with chaotropic agents. A soluble guanylate cyclase was purified from rat lung using monoclonal antibody immunoaffinity procedure (103). The 70-kDa subunit isolated by denaturing polyacrylamide gel electrophoresis was subjected to N-terminal amino acid sequencing and an oligonucleotide probe prepared for cDNA screening. Sequence information indicated that the 70-kDa subunit has a regulatory rather than a catalytic function.

The membrane-bound guanylate cyclase consists of only one polypeptide chain and can be stimulated by atriopeptins (104) and *Escherichia coli* entereotoxin (105). GTP–agarose and wheat germ agglutinin–agarose have been employed to purify the membrane-bound enzyme approximately 600-fold (106). The purified protein was then used to immunize mice for monoclonal antibody production. When immobilized on Sepharose, the monoclonal antibodies bound a 180-kDa protein exhibiting both atrial natriuretic factor (ANF) binding and the guanylate cyclase activity.

D. Cyclic 3′,5′-Nucleotide Phosphodiesterase

Purification of cyclic 3′,5′-nucleotide phosphodiesterases has proved difficult due to the presence of multiple forms of the enzyme in cytosol and membranes of cells and the relatively low amount of the enzyme. Multiplicity is due primarily to the expression of discrete enzymes that differ in substrate specificity, response to CaM, and allosteric modulation by cyclic GMP (107). Proteolysis leading to alteration in enzyme properties has also contributed to the difficulties of clearly defining the physical and inhibitory properties of the different phosphodiesterases (108).

Cilostamide (Figure 4A), a quinoline compound, specifically inhibits an insulin-sensitive, low-K_m cyclic AMP phosphodiesterase of fat cells and hepatocytes. This specific enzymic form has also been termed a cyclic GMP-inhibited isozyme. The carboxylic acid analog of cilostamide (Figure 4B) was converted to the *N*-(2-isothiocyanato)ethyl derivative and coupled to aminoethyl agarose (Figure 4C) (109). The enzyme solubilized by 150 m*M* NaBr and nonionic detergent was applied to the affinity gel and eluted by incubation with 100 m*M* NaBr and 50 m*M* cyclic AMP for a period of 5 h at 4°C. This single step resulted in selective purification (300-fold) of a low-K_m isozyme with a 60% yield. A major 63.8-kDa protein representing the native form of cyclic GMP–inhibited phosphodiesterase of adipose tissue was obtained by this affinity matrix.

A number of distinct isozymes of phosphodiesterase from the visual system have been identified by affinity chromatography. Phosphodiesterase of the visual transduction system of the cone photoreceptor was purified approximately 15,000-fold to homogeneity by a procedure using cyclic GMP–Sepharose affinity chromatography (110). Epoxy-activated Sepharose 6B prepared according to Sundberg and Porath (111) was reacted with cyclic GMP by the method of Martins et al. (112) to form the affinity matrix. The coupling density of the cyclic GMP matrix was 0.5 to 1.0 μmol/mL of agarose. The matrix was regenerated with 4 *M* guanidine, 2 *M* NaCl, and 1 m*M* EDTA. An 8-(6-aminohexylamino) cyclic AMP–Sepharose column was placed in tandem with the cyclic GMP–Sepharose column to eliminate minor contaminants of the R subunit of cyclic AMP–dependent protein kinase and also cyclic GMP–dependent protein kinase, which also binds to the cyclic AMP matrix. After washing, the cyclic GMP column was removed and the cyclic GMP phosphodiesterase was eluted with 5 m*M* cyclic GMP.

A soluble rod phosphodiesterase was purified 2600-fold to homogeneity by immunoaffinity chromatography using a monoclonal antibody (ROS 1a) affinity

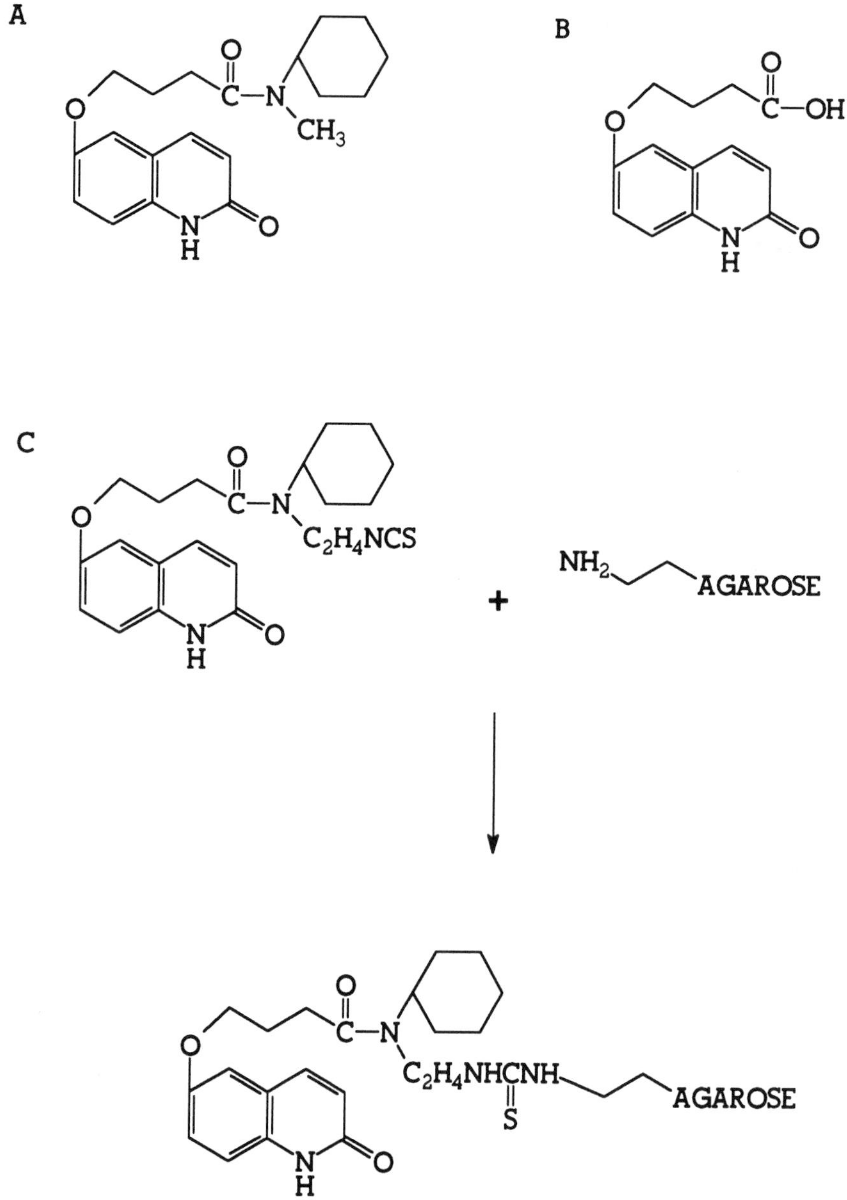

Figure 4 Structure of cilostamide: (A) cilostamide; (B) carboxylic acid analog of cilostamide; (C) coupling of the isothiocyanatoethyl analog of cilostamide to amine-substituted agarose. (After Ref. 109.)

column (113). The enzyme was eluted using a pH 10.7 buffer, and fractions were collected in tubes preloaded with buffer below pH 7.0 to neutralize the eluted proteins rapidly.

In another study, a distinct form of phosphodiesterase from *Neurospora crassa* was purified by monoclonal antibody immunoaffinity chromatography (114). Hybridomas producing specific antibody were selected based on their capacity to inhibit cyclic AMP phosphodiesterase activity. The phosphodiesterase was eluted from the immobilized monoclonal antibody by increasing the ionic strength of the elution buffer.

A unique soluble calcium-stimulated phosphodiesterase was isolated from the cytosolic fraction of bovine lung (115). The phosphodiesterase associates with CaM even in the presence of 0.1 m*M* EGTA. The enzyme was dissociated from CaM by first binding the holoenzyme to an immunoaffinity matrix prepared from antibody against phosphodiesterase, washing with buffers containing 0.1 m*M* EGTA and 0.5 *M* NaCl to remove CaM and then eluting the CaM-dependent enzyme with 2.5 *M* $MgCl_2$. CaM affinity chromatography was employed to purify the enzyme further. The enzyme was similar to *B. pertussis* toxin adenylate cyclase, neuromodulin, and phosphorylase kinase in its ability to bind CaM at very low levels of calcium. In vivo the enzyme may be expected to retain calcium sensitivity without being altered by acute changes in cellular CaM concentrations.

E. Cyclic Nucleotide–Dependent Protein Kinase

Regulation of proteins by cyclic AMP in cells occurs through covalent modification via phosphorylation of serine and threonine residues (116). Cyclic AMP–dependent protein kinase is a tetramer consisting of two regulatory (R) and two catalytic (C) subunits. Cyclic AMP binds to the inhibitory R subunit, leading to dissociation of an R-subunit dimer and two active C subunits. Two different forms of the C subunit and R subunit have been identified (116).

Affinity chromatography of these protein kinases has exploited interactions with ATP, allosteric modulators, pseudosubstrate domains, and recognition sequences in the phosphorylation domains of the target proteins. A synthetic peptide of 18 amino acids corresponding to the inhibitory domain of a heat-stable protein kinase inhibitor (117) was used to construct an affinity column for the purification of the catalytic subunit of cyclic AMP-dependent protein kinase (118). The heat-stable inhibitor is a highly specific inhibitor of cyclic AMP–dependent protein kinase. The inhibitory peptide, which is homologous to the pseudosubstrate domain of the R subunit of cyclic AMP–dependent kinase, causes half-maximal inhibition at 200 n*M*. Immobilization was accomplished by linking the peptide to *N*-hydroxysuccinimide–agarose. Both C_α and C_β catalytic subunits of the enzyme were isolated by this method from 3T-3 cells transfected with expression vectors containing a metallothionein promoter. The vectors encoded either the C_α or the C_β isoforms of the catalytic subunit. Kinase activity of transfected cells was eluted from the peptide affinity column with 200 m*M* of L-arginine, an inhibitor of kinases. Both purified isoforms had similar K_m values for ATP (4 μ*M*) and for Kemptide (5.6 μ*M*), a synthetic phosphate-acceptor peptide that has the sequence LRRASLG.

In the yeast *Saccharomyces cerevisiae*, three genes encode the catalytic subunits of cyclic AMP–dependent protein kinase (119). The holoenzyme containing the regulatory and the specific catalytic subunit expressed by the TPK1 gene was bound to an immunoaffinity column prepared from a monoclonal antibody against the regulatory subunit of the yeast enzyme. The monoclonal antibody was immobilized on protein A–Sepharose. A homogeneous 52-kDa catalytic subunit was released by elution with 0.8 m*M* cyclic AMP.

Cyclic AMP-8-(6-aminohexylamino)–agarose has been used to isolate the regulatory subunit of cyclic AMP–dependent protein kinase of *Trypanosom cruzi* (120). Unbound fractions contained the catalytic subunit, and fractions eluted with cyclic AMP contained a 56-kDa regulatory component that cross-reacted with polyclonal antibody against mammalian regulatory subunits of type I or II.

Cyclic GMP protein kinases have been purified using cyclic AMP–agarose (121,122). During the initial affinity chromatography of this enzyme from bovine aorta (121) on a cyclic GMP column, a major portion of a specific isozyme (form I) and a small amount of form II were eluted by 1 *M* NaCl. Following this with elution by 0.1 m*M* cyclic GMP yielded mainly form II. Upon cyclic AMP–agarose affinity chromatography of the same enzyme preparation, form I exchanged more rapidly with cyclic GMP during elution than form II. Proteolytic treatment of the separated enzymes yielded unique peptides suggesting differences in either the amino acid sequence or posttranslational modification. Cyclic AMP–agarose has also been employed to purify the cyclic GMP–dependent protein kinase from *Acaris suum* (122). The 8-(6-aminohexylamino)-cyclic AMP can be synthesized by the method of Jergil and Mosbach (123) by reacting 8-bromo-cyclic AMP with 1,6-diaminohexane to form 8-(6-aminohexylamino)-cyclic AMP that can be immobilized on CNBr-activated Sepharose. The yield on this cyclic AMP affinity step was 75% with a purification of about 10-fold. Another useful immobilized cyclic AMP analog for the purification of protein kinase is N^6-(2-aminoethyl)-cyclic AMP synthesized from 6-chloropurine riboside 3′,5′-monophosphate and 1,2-aminoethane as described by Dills et al. (124).

F. Other Protein Kinases

Affinity chromatography of protein kinases has utilized distinct recognition sequences within the phosphorylation domains of the target proteins (125).

Specific peptide sequences derived from glycogen synthase were synthesized and immobilized to provide a high density of binding sites for both glycogen synthase kinase-3 and casein kinase II (126). The synthetic peptide was coupled to *N*-hydroxysuccinimide–agarose. Within the recognition sequence, phosphorylation by casein kinase II was required for the binding of glycogen synthase kinase-3, providing a highly selective separation procedure. Columns were eluted by increasing the concentration of NaCl to 50 m*M* (phosphopeptide ligand) or 200 m*M* (dephosphopeptide ligand). The K_m value of the free peptides for the kinases was 10 μ*M* or less. A histone–tubulin protein kinase of *Leishmania donovani* was purified approximately fivefold with a yield of 90% by histone affinity chromatography and elution with 0.5 *M* NaCl (126). Calf thymus histones were covalently attached to Sepharose 4B as described by Wilchek et al. (127).

G. Protein Phosphatases

Protein phosphatases bear the reciprocal function to the kinases in the regulation of proteins by phosphorylation (128). Protein phosphatase inhibitor 1 (129) and an analogous protein, DARPP-32, when phosphorylated by cyclic AMP–dependent protein kinase are potent and specific inhibitors of protein phosphatase 1 (130,131). Ingebritsen and Ingebritsen (132) have used immobilized inhibitor 1 to study the mechanisms of protein phosphatase 1 inhibition (Table 3). Protein phosphatase 1 did not bind to dephosphorylated inhibitor 1 or BSA coupled to Sepharose. Protein phosphatase 1 of brain has been purified utilizing heparin-Sepharose at two steps in the purification scheme (133), and protein phosphatase 2C of muscle and liver by employing immobilized thiophosphorylated-myosin light chains (134).

The major portion of protein phosphatase 1 activity in mammalian tissues is tightly associated with glycogen. β-Cyclodextrin was immobilized on epoxy-activated Sepharose 6B and used to purify both the catalytic and noncatalytic subunits of the rat liver enzyme (135). The protein phosphatase holoenzyme was transferred to immobilized cyclodextrin by repeated cycling of isolated rat liver glycogen through the column. Selective elution resolved the 37-kDa catalytic from the noncatalytic (161- and 54-kDa) subunits. The noncatalytic subunits appear to function as the glycogen-binding domain of the hepatic protein phosphatase.

The tumor promoter okadaic acid inhibits mainly protein phosphatase 1 and 2A. A photoaffinity probe, methyl 7-*O*-(4-azidobenzoyl)okadaate has been used to selectively label the catalytic subunit of the heterotrimeric protein phosphatase 2A (136). Because of the functional significance of the carboxyl group of okadaic acid, another inhibitor of the phosphatase 1 and 2A, microcystin LR (Figure 5),

Table 3 Affinity Purification of Protein Phosphatases

Enzyme	Ligand/Matrix	Binding/elution	Results	Refs.
Protein phosphatase 1 (rabbit skeletal muscle)	Phosphorylated or dephosphorylated inhibitor 1 coupled to CNBr-activated Sepharose	Tris buffer (pH 7.0, 0.1 m*M* EGTA, 10% glycerol, 0.01% Brij 35, and protease inhibitors; elution with 3 *M* NaSCN or 1 *M* NaCl	Phosphatase activity (74%) was bound by the phosphorylated inhibitor matrix	132
Protein phosphatase 1 (bovine brain)	Heparin–Sepharose	Tris buffer pH 7.0 containing 0.2 m*M* EGTA, 0.1% mercaptoethanol, and protease inhibitors; elution with 120 to 400 m*M* NaCl	Heparin–Sepharose was used at two steps in the purification; enzyme purified to near homogeneity	133
Protein phosphatase 2C (rabbit muscle/liver)	Thiophosphorylated myosin-*p*-light-chain Sepharose	Triethanolamine buffer (pH 7.0), 1 m*M* EGTA, 0.1% mercapto-ethanol, 5% glycerol, 10 m*M* Mg^{2+}, 25 m*M* NaCl; elution by replacing Mg^{2+} with 2 m*M* EDTA	Single activity peak of two isozymes with 44-fold purification and 81% yield was obtained	134
Protein phosphatase 1 (glycogen-associated, rat liver)	β-Cyclodextrin–Sepharose	Glycylglycine buffer (pH 7.4), 1 m*M* EGTA, 0.1% mercaptoethanol and protease inhibitors; elution with 2 *M* NaCl and the noncatalytic subunit with 1 mg/mL β-cyclodextrin	Catalytic subunit of 37 kDa and noncatalytic subunits of 161 and 54 kDa	135
Protein phosphatase 2A (mouse brain)	Microcystin LR-Sepharose	Tris buffer (pH 7.4), 10% glycerol, 2 m*M* EDTA, 2 m*M* EDTA; elution with 1 *M* NaCl	$<$ 50% protein phosphtase 1 bound; column has low binding capacity	136

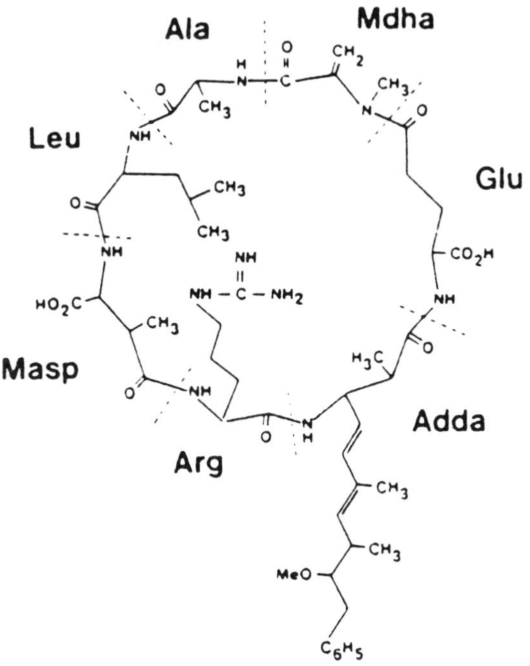

Figure 5 Structure of microcystin-LR. Masp, β-methylaspartic acid; Mdha, *N*-methyldehydroalanine; Adda, 3-amino-9-methoxy-2,6,8-trimethyl-10-phenyldeca-4,6-dienoic acid. (From Ref. 137.)

had to be employed as an affinity ligand for the purification of protein phosphatase 2A (137).

IV. INOSITOL PHOSPHATE AND DIACYLGLYCEROL REGULATION

A. Phosphatidylinositol and Inositol Metabolism

Activation of phospholipase C by calcium-mobilizing hormones promotes the hydrolysis of phosphatidylinositol 4,5-bisphosphate to form inositol 1,4,5-trisphosphate [Ins(1,4,5)P_3 or IP_3] and diacylglycerol (138,139). The released IP_3 acts as a second messenger to mobilize intracellular calcium and can undergo phosphorylation to Ins (1,3,4,5) P_4 or IP_4 (140). The diacylglycerol formed by phospholipase C action can activate protein kinase C (141) and also serve as a source of arachidonic acid for the production of numerous icosanoid mediators (142).

An analog of IP_3 immobilized on agarose has been employed in the isolation of inositol-metabolizing enzymes (143). CH-Sepharose 4B containing the succinimide ester of aminocaproic acid as an activated spacer arm was reacted with a primary amino group of tyramine (Figure 6A). The diazotization product of 2-*O*-(4-aminobenzoyl)-1,4,5-tri-*O*-phosphono-*myo*-inositol was then coupled to the ortho position of the phenyl ring of the immobilized tyramine to form the affinity matrix.

In an alternative procedure, the aminocyclohexanecarbonyl IP_3 analog was attached directly to the activated support via the primary amino group (i.e., without tyramine addition and diazotization) (Figure 6B). Such affinity supports were employed in the isolation of Ins(1,4,5)P_3 5-phosphatase, Ins(1,4,5)P_3 3-kinase, and Ins (1,4,5)P_3 binding protein (143). Chromatography buffers contained 2 m*M* EDTA to protect the immobilized inositol derivatives from Mg^{2+}-dependent IP_3 5-phosphatase activity. The enzymes were eluted from the matrix with 0.2 and 0.5 *M* KCl. The IP_3 5-phosphatase was bound to the IP_3 analog coupled via the azo linkage but not to the analog with the amide linker arm.

Four distinct groups ($\alpha,\beta,\gamma,\delta$) of phospholipase C enzymes (PLC) occur in cytosolic and membrane fractions of tissues (144). Some groups have multiple isoforms. A specific PLC isozyme of 145 kDa (PLCγ), a major target of EGF-induced tyrosine phosphorylation, was purified by immunoaffinity chromatography using an immobilized monoclonal antibody directed against phosphotyrosine residues (145) (Table 4). PLCγ was eluted with 10 m*M* phenylphosphate and isolated by precipitation using an anti-PLCγ monoclonal antibody and cells of *Staphylococcus aureus*.

Phosphatidylinositol 4-phosphate kinase has been purified from bovine brain membranes using ATP-agarose as a final step in the purification scheme (146). A 55-kDa phosphatidylinositol 4-phosphate kinase of A431 cells was purified to near homogeneity by a combination of conventional procedures involving heparin and reactive green-agarose (147). Quercetin, an oligohydroxy benzopyran inhibitor of the kinase, has also been used to purify phosphatidylinositol 4-phosphate kinase from rat liver plasma membrane (148).

Attenuation of the diacylglycerol signal can occur by phosphorylation of diacylglycerol by diacylglycerol kinase (DGK) or lipase degradation of diacylglycerol, leading to the production of arachidonic acid. ATP–agarose, phenyl-Superose, and heparin–Sepharose have all been employed in the purification and separation of the different forms of diacylglycerol kinase (149,150).

A soluble CaM-dependent IP_3 kinase that converts Ins(1,4,5)P_3 to Ins(1,3,4,5)P_4 has been isolated from several sources by procedures that employ CaM and adenosine 2,3-diphosphate agarose (24,151–153). The IP_3 kinase from smooth muscle was purified about 25-fold with a 20% yield by CaM–Sepharose chromatography (151). The enzyme from rat brain was purified to a specific

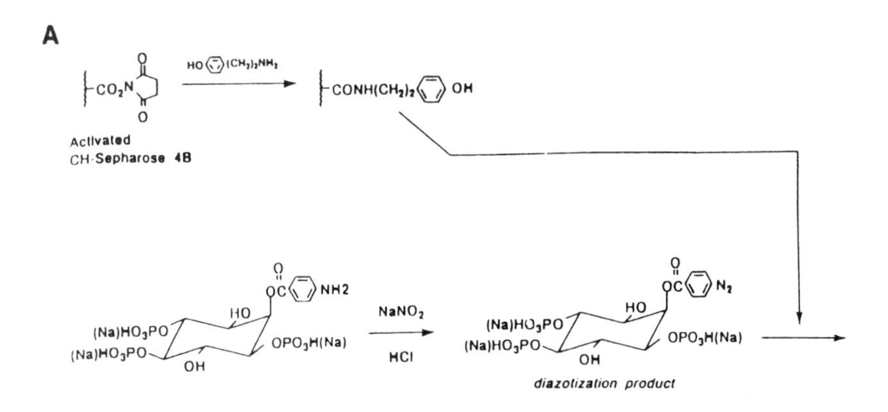

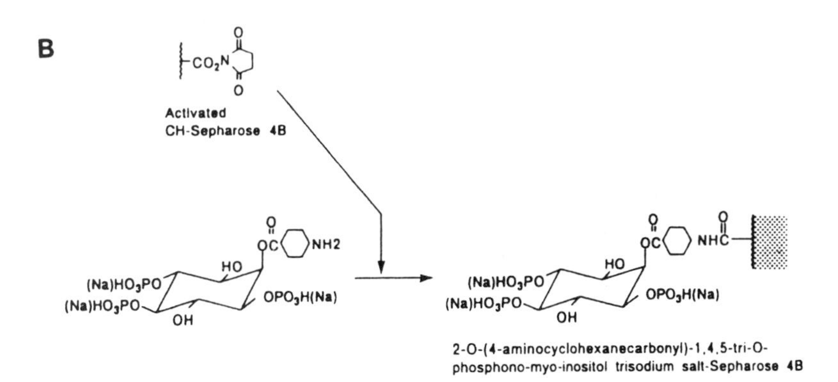

Figure 6 Preparation of immobilized inositol analogs: (A) preparation of 2-*O*-[4-(5-aminoethyl-2-hydroxyphenylazo)benzoyl]-1,4,5-tri-*O*-phosphono-myo-inositol–agarose; (B) preparation of 2-*O*-(4-aminocyclohexanecarbonyl)-1,4,5-tri-*O*-phosphono-myo-inositol–agarose. (From Ref. 143.)

Table 4 Affinity Purification of Enzymes in Inositol Metabolism

Enzyme	Ligand/Matrix	Binding/elution conditions	Results	Refs.
Phospholipase C (γ-type)	Antiphosphotyrosine monoclonal antibody–Sepharose	Buffered elution with 0.075% Triton X-100 and 10 m*M* phenylphosphate	Large-scale immunoaffinity procedure for the isolation of phosphorylated phospholipase C (γ)	145
Phosphatidylinositol 4-phosphate kinase of brain	ATP–agarose	Elute kinase with 300 to 400 m*M* NaCl in buffer containing 0.1% Triton X-100 and 0.11% PEG 20,000	ATP maintained at 50 μM in binding buffer to elute diacylglycerol kinase	146
Phosphatidylinositol 4-kinase of A431 cells	Heparin– and reactive green–Sepharose	Column buffers supplemented with 0.25% Triton X-100, heparin column eluted with salt gradient ± phosphoinositol; reactive green eluted with 1 *M* KCl	A nearly homogeneous 55-kDa protein was isolated	147
Phosphatidylinositol 4-kinase of rat liver	Quercetin–Sepharose	Elute with buffered 0.15 to 2.0 *M* NaCl containing 0.02% Triton X-100	26-fold purification with 92% yield	148
Diacylglycerol kinase (rat brain)	Affi-Gel Blue and ATP–agarose	Affi-Gel Blue enzyme eluted with 1.0 and 2.0 *M* NaCl; ATP–agarose: gradient of 0 to 0.5 *M* NaCl	Two isozymes were purified from cytosolic and membrane fractions	149; see also Ref. 150
Phosphatidylinositol 3-kinase and p85 (3T3 cells)	PDGF receptor–antibody complex immobilized on protein A–Sepharose	Immunocomplex washed extensively with buffers containing 1% Triton and 0.5 *M* LiCl	A 110-kDa kinase and and a 85-kDa subunit that associates with PDGF receptor isolated	155
Myoinositol monophosphate phosphatase (brain)	Phenyl-Superose	Decreasing ammonium sulfate concentration	29-kDa protein isolated	156

activity of 2.3 μmol/min/mg protein (2700-fold) from the cytosol by chromatography on phosphocellulose, Orange A dye (Amicon Corp.), CaM–agarose, and hydroxyapatite (153). The enzyme from bovine brain has also been purified using CaM–Sepharose affinity chromatography (154). Inclusion of calpain inhibitors during purification of the enzyme from rat brain was essential to obtain an intact 53-kDa enzyme (152).

The ligand-activated PDGF-β receptor that associates with phospholipase C-γ, GAP, and phosphatidylinositol 3-kinase has provided an affinity matrix to purify p85, a protein that links the kinase to the activated tyrosine-phosphorylated receptor (155). This unique approach may be useful with other PDGF receptor–associated proteins as well as other tyrosine kinase receptors. A lithium-sensitive, 29-kDa, myoinositol monophosphate phosphatase catalyzing the hydrolysis of Ins 1-P and Ins 4-P was purified to homogeneity from bovine brain using phenyl-Superose chromatography (156) and by a monoclonal immunoaffinity procedure (157) using low-ionic-strength buffers containing ethylene glycol to elute the enzyme in good yield (54%) and also to preserve the binding capacity of the immunoaffinity column. Enzyme elution with chaotropic agents (4 *M* $MgCl_2$, 3 *M* NaSCN) or low pH caused a total loss in activity.

B. Protein Kinase C

At least seven closely related phenotypes of protein kinase C (PKC) exist in tissues with different regional and cellular distributions (158). Threonine–Sepharose prepared according to Kitano et al. (159) in combination with phenyl-Sepharose chromatography and high-resolution hydroxylapatite chromatography have been used in purification procedures developed to resolve the isozymes of PKC (160,161). L-Threonine was coupled to an amine-substituted agarose (AH-Sepharose 4B) by addition of carbodiimide. The L-threonine matrix provided a 30-fold increase in PKC specific activity with 40% yield. The PKC was eluted with a linear gradient of NaCl (0 to 0.1 *M*). PKC-ε expressed in insect cells via the baculovirus vector has been purified to homogeneity by a procedure involving serine–Sepharose chromatography and NaCl gradient (0.15 to 1.0 *M*) elution (162). Serine–Sepharose was prepared by the same procedure as that described for threonine-Sepharose (159).

In contrast to other protein kinases, PKC is inhibited by a number of specific agents, including melittin, polymixin B, trifluoperazine, dibuccaine, compound W-7 [*N*-(6-aminohexyl)-5-chloro-1-naphthalenesulfonamide] (Figure 7A), and verapamil (141). Most of these same agents also inhibit calmodulin action by interaction with hydrophobic sites on the molecule. With PKC, these agents may in some cases cause inhibition by preventing the enzyme from interacting with essential phospholipids.

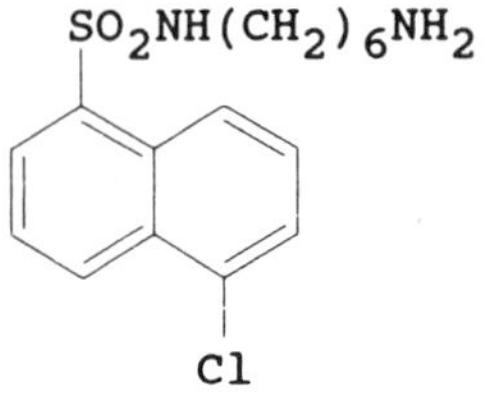

$SO_2NH(CH_2)_6NH_2$

Cl

A W-7

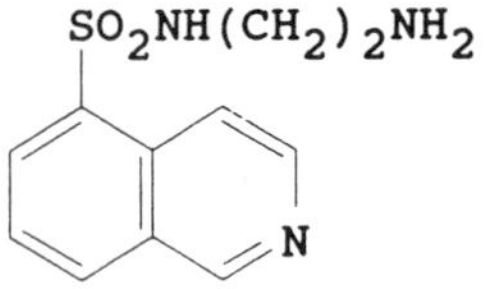

$SO_2NH(CH_2)_2NH_2$

N

B H-9

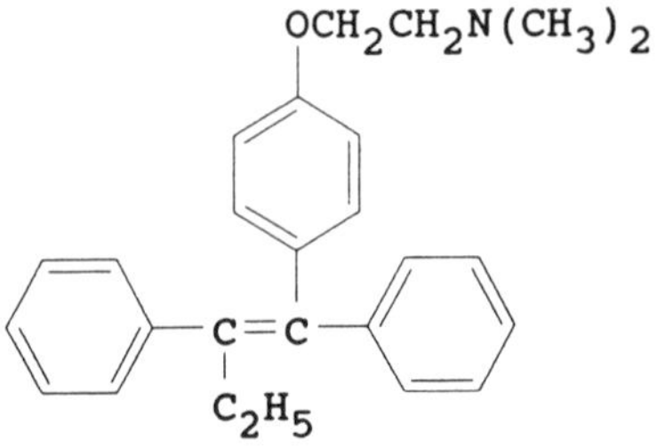

$OCH_2CH_2N(CH_3)_2$

C=C

C_2H_5

C TAMOXIFEN

Figure 7 Structures of W-7, H-9, and tamoxifen.

The sulfonamide, *N*-(2-aminoethyl)-5-isoquinolinesulfonamide (H-9, Figure 7B), and other analogs (H-7 and H-8) have been shown to act as competitive inhibitors of PKC with respect to ATP (163). PKC and other kinases bind to the immobilized H-9 and can be eluted with 30 m*M* ATP or L-arginine (164). H-9 was coupled directly to CNBr-activated agarose at a substitution of 10 to 12 μmol/mL of gel. Protein kinase C and cyclic GMP- and cyclic AMP-dependent protein kinase were eluted separately by a gradient of 0 to 1.5 *M* L-arginine, which is an inhibitor of all three kinases.

W-7-agarose has been used to investigate aspects of the interaction of W-7 with PKC (165). PKC was bound by the immobilized W-7 and could be eluted with 0.1% Triton X-100 and MgATP. It was suggested that the PKC interacted mainly with the ATP-binding region of PKC in the absence of phospholipid. In a similar study using mellitin–agarose chromatography, PKC was bound and released by Triton X-100–supplemented buffer containing MgATP (166). Based on ATP-sensitive binding and other results, mellitin, like W-7, appears to interact specifically with the catalytic domain of PKC. The same group of investigators also examined the binding of PKC to immobilized tamoxifen (Figure 7C), an anti-estrogen that is effective in the treatment of certain human breast cancers (167). PKC was found to bind specifically and reversibly to *N*-didesmethyltamoxifen when coupled to CNBr-activated agarose through a primary amine group. PKC activity was eluted from the tamoxifen matrix with 0.2% (v/v) Triton X-100. A derivative of similar hydrophobicity, 4-hydroxytamoxifen, immobilized on epoxy-activated Sepharose did not bind PKC, suggesting specific binding to the didesmethyltamoxifen matrix.

Since PKC has an absolute requirement for calcium and acidic phospholipids, phosphatidylserine coupled to Affi-Gel 102 (an amine-substituted agarose derivative, Bio-Rad Laboratories) has also been used in the purification of PKC (168). Emulsified phosphatidylserine was coupled to the amine-substituted agarose using a water-soluble carbodiimide, 1-ethyl-3-(3-dimethylaminopropyl)carbodiimide. The enzyme was eluted with Tris buffer containing 2 m*M* EDTA, 10 m*M* EGTA, 1 *M* NaCl, and 50 m*M* mercaptoethanolamine. In this terminal step of the isolation procedure, PKC was purified approximately fivefold but with a low recovery (5%). A Ca^{2+}-dependent phospholipid affinity column consisting of phosphatidylserine immobilized in polyacrylamide has also been used to purify protein kinase C (169).

REFERENCES

1. H. R. Bourne and A. L. DeFranco, in *Oncogenes and the Molecular Origins of Cancer* (R. A. Weinberg, ed.), Cold Spring Harbor Laboratory Press, Cold Spring Harbor, N.Y., 1989, p. 97.
2. P. L. Domen, J. R. Nevens, A. K. Mallia, G. T. Hernanson, and D. C. Klerk, *J. Chromatogr., 354*:292 (1990).
3. D. J. O'Shannessey, *J. Chromatogr., 510*:13 (1990).
4. P. Mohr and K. Pommerening, *Affinity Chromatography: Practical and Theoretical Aspects*, Marcel Dekker, New York, 1985.
5. P. D. G. Dean, W. S. Hohnson, and F. A. Middle, *Affinity Chromatography: A Practical Approach*, IRL Press, Oxford, 1985.
6. S. Ostrove, Affinity chromatography: general methods, *Methods Enzymol., 182*:357 (1990).

7. I. Parikh and P. Cuatrecasas, Affinity chromatography, *Chem. Eng. News*, Aug. 26, 1985, pp. 17–32.
8. D. M. Watterson, F. Sharief, and T. C. Vanaman, *J. Biol. Chem., 255*:962 (1980).
9. J. D. Potter and J. Gergely, *Biochim. Biophys. Acta, 405*:40 (1975).
10. M. Yazawa, K. Yagi, H. Toda, K. Kondo, K. Narita, R. Yamazaki, K. Sobue, S. Kakiuchi, S. Nagao, and Y. Nozawa, *Biochem. Biophys. Res. Commun., 99*:1051 (1981).
11. K. T. O'Neil and W. F. DeGrado, *Trends Biol. Sci., 15*:59 (1990).
12. D. K. Blumenthal, H. Charbonneau, A. M. Edelman, T. R. Hinds, G. B. Rosenberg, D. R. Storm, F. F. Vincenzi, J. A. Beavo, and E. G. Krebs, *Biochem. Biophys. Res. Commun., 156*:860 (1988).
13. B. Buschmeier, H. E. Meyer, and G. W. Mayr, *J. Biol. Chem., 262*:9454 (1987).
14. M. Comte, Y. Maulet, and J. A. Cox, *Biochem. J., 209*:269 (1983).
15. A. Bartegi, A. Fattoum, J. Derancourt, and R. Kassab, *J. Biol. Chem., 265*:15231 (1990).
16. D. M. Watterson and T. C. Vanaman, *Biochem. Biophys. Res. Commun., 73*:40 (1976).
17. M. Miyaki, J. W. Daly, and C. R. Creveling, *Arch. Biochem. Biophys., 181*:39 (1977).
18. R. K. Sharma, T. H. Wang, E. Wirch, and J. H. Wang, *J. Biol. Chem., 255*:5916 (1980).
19. R. S. Adelstein and C. B. Klee, *J. Biol. Chem., 256*:7501 (1981).
20. G. Thiel, A. J. Czernik, F. Gorelick, A. C. Nairn, and P. Greengard, *Proc. Natl. Acad. Sci. USA, 85*:6337 (1988).
21. E. A. Tallant and R. W. Wallace, *J. Biol. Chem., 260*:7744 (1985).
22. T. J. Andreasen, C. W. Leutje, W. Heideman, and D. R. Storm, *Biochemistry, 22*:4615 (1983).
23. B. T. Wakim, M. M. Picken, and R. J. DeLange, *Biochem. Biophys. Res. Commun., 171*:84 (1990).
24. T. J. Biden, M. Comte, J. A. Cox, and C. B. Wollheim, *J. Biol. Chem., 262*:9437 (1987).
25. R. K. Sharma, *J. Biol. Chem., 265*:1152 (1990).
26. K. R. Westcott, D. C. LaPorte, and D. R. Storm, *Proc. Natl. Acad. Sci. USA, 76*:204 (1979).
27. H. H. H. W. Schmidt, J. S. Pollock, M. Nakane, L. D. Gorsky, U. Forstermann, and F. Murad, *Proc. Natl. Acad. Sci. USA, 88*:365 (1991).
28. F. Friedberg, *Protein Sequences Data Anal., 3*:335 (1990).
29. S. M. Strittmatter, D. Valenzuela, T. E. Kennedy, E. J. Neer, and M. C. Fishman, *Nature, 344*:836 (1990).
30. J. F. Fierro, M. A. Pajares, and C. Hardisson, *FEBS Lett., 247*:22 (1989).
31. D. L. Newton, M. D. Oldewurtel, M. H. Krinks, J. Shiloach, and C. B. Klee, *J. Biol. Chem., 259*:4419 (1984).
32. J. Wolff, D. L. Newton, and C. B. Klee, *Biochemistry, 25*:7950 (1986).
33. D. Guerini, J. Krebs, and E. Carafoli, *J. Biol. Chem., 259*:15172 (1984).
34. W.-C. Ni and C. B. Klee, *J. Biol. Chem., 260*:6974 (1985).
35. G. Draetta and C. B. Klee, *Methods Enzymol., 159*:523 (1988).
36. L. Wolff, H. G. Cook, A. H. Goldhammer, and S. A. Berkowitz, *Proc. Natl. Acad. Sci. USA, 77*:3841 (1980).
37. D. V. Greenler, T. J. Andreasen, and D. R. Storm, *Biochemistry, 21*:2759 (1982).

38. J. Haiech, R. T. Predeleaner, D. M. Watterson, D. Ladant, J. Bellalou, A. Ullmann, and O. Barzu, *J. Biol. Chem., 263*:4259 (1987).
39. E. L. Hewlett, V. M. Gordon, J. D. McCaffery, W. M. Sutherland, and M. C. Gray, *J. Biol. Chem., 264*:19379 (1989).
40. J. R. Dedman and M. A. Kaetzel, *Methods Enzymol., 102*:1 (1983).
41. N. Nibhanupudy, F. Jones, and A. R. Rhoads, *Biochemistry, 27*:2212 (1988).
42. U. K. Laemmli, *Nature (London), 227*:680 (1970).
43. H. Brzcska, J. Szynkiewicz, and W. Drabikowski, *Biochem. Biophys. Res. Commun., 115*:87 (1983).
44. J. F. Head, H. R. Masure, and B. Kaminer, *FEBS Lett., 137*:71 (1982).
45. H. J. Vogel, L. Lindahl, and E. Thulin, *FEBS Lett., 157*:241 (1983).
46. H. Charbonneau, R. Hice, R. C. Hart, and M. J. Cormier, *Methods Enzymol., 102*:17 (1983).
47. J. M. Anderson, *Methods Enzymol., 102*:9 (1983).
48. J. A. Cox, M. Comte, A. J. E. Fitton, and W. F. DeGrado, *J. Biol. Chem., 260*:2527 (1985).
49. M. A. Kaetzel and J. R. Dedman, *J. Biol. Chem., 262*:3726 (1987).
50. K. C. Wang, H. Y. Wong, J. H. Wang, and H.-Y. P. Lam, *J. Biol. Chem., 258*:12110 (1983).
51. C. B. Klee, *Methods Enzymol., 102*:233 (1983).
52. S. C. March, I. Parikh, and P. Cuatrecasas, *Anal. Biochem., 60*:149 (1974).
53. S. Udenfriend, S. Stein, P. Bohlen, W. Darma, W. Leimgruber, and M. Weigele, *Science, 178*:871 (1972).
54. M. M. Bradford, *Anal. Biochem., 72*:248 (1976).
55. P. Thiry, A. Vandermeers, M.-C. Vandermeers-Piret, J. Rather, and J. Christophe, *Eur. J. Biochem., 103*:409 (1980).
56. A. S. Manalan and C. B. Klee, *Biochemistry, 26*:1382 (1987).
57. D. P. Giedroc, S. K. Sinha, K. Brew, and D. Puett, *J. Biol. Chem., 260*:13406 (1985).
58. D. M. Mann and T. C. Vanaman, *J. Biol. Chem., 263*:11284 (1988).
59. R. K. Sharma, W. A. Taylor, and J. H. Wang, *Methods Enzymol., 102*:210 (1983).
60. D. R. Hathaway, R. S. Adelstein, and C. B. Klee, *J. Biol. Chem., 256*:8183 (1981).
61. K. K. W. Wang, A. Villalobo, and B. W. Roufogalis, *Biochem. J., 262*:693 (1989).
62. C. B. Klee, *Biochemistry, 27*:6645 (1988).
63. I. M. Herman and T. D. Pollard, *J. Cell. Biol., 80*:509 (1979).
64. F. Martin, J. Derancourt, J.-P. Capony, A. Watrin, and J.-C. Cavadore, *Biochem. J., 251*:777 (1988).
65. A. Grandmont-Leblanc and J. Gruda, *Can. J. Biochem., 55*:949 (1977).
66. O. Ohara, S. Takahashi, T. Ooi, and T. Fujiyoshi, *J. Biochem., 91*:1999 (1982).
67. K. G. Miller and B. M. Alberts, *Proc. Natl. Acad. Sci. USA, 86*:4808 (1989).
68. N. Johnson and K. Weber, *Eur. J. Biochem., 188*:1 (1990).
69. W. H. Martin and C. E. Creutz, *J. Neurochem., 54*:612 (1990).
70. C. E. Creutz, W. J. Zaks, H. C. Hamman, S. Crane, W. H. Martin, K. L. Gould, K. M. Oddie, and S. J. Parsons, *J. Biol. Chem., 262*:1860 (1987).
71. C. E. Creutz, L. G. Dowling, J. J. Sando, C. Villar-Palasi, J. H. Whipple, and W. J. Zaks, *J. Biol. Chem., 258*:14664 (1983).

72. F. F. Severin, N. A. Shanina, S. A. Kuznetsov, and V. I. Gelfand, *FEBS Lett., 282*:65 (1992).
73. E. M. Ross and A. G. Gilman, *Annu. Rev. Biochem., 49*:533 (1980).
74. Wei-Jen Tang, J. Krupinski, and A. G. Gilman, *J. Biol. Chem., 266*:8595 (1991).
75. G. Bigay, P. Deterre, C. Pfister, and M. Chabre, *FEBS Lett., 191*:181 (1985).
76. C. Homcy, S. Wrenn, and E. Haber, *Proc. Natl. Acad. Sci., USA, 75*:59 (1978).
77. T. Pfeuffer, *J. Biol. Chem., 252*:7224 (1977).
78. L. Birnbaumer, *FASEB J., 4*:3178 (1990).
79. H. Metzger and E. Landner, *IRC Med. Sci., 9*:99 (1981).
80. K. B. Seaman, W. Padgett, and J. W. Daly, *Proc. Natl. Acad. Sci. USA, 78*:3363 (1981).
81. T. Pfeuffer, B. Gaugler, and H. Metzger, *FEBS Lett., 164*:154 (1983).
82. T. Pfeuffer and H. Metzger, *FEBS Lett., 146*:369 (1982).
83. E. Pfeuffer, R.-M. Drehev, H. Metzger, and T. Pfeuffer, *Proc. Natl. Acad. Sci. USA, 82*:3086 (1985).
84. J. Krupinski, F. Coussen, H. A. Bakalyar, Wei-Jen Tang, P. G. Feinstein, K. Orth, C. Slaughter, R. R. Reed, and A. G. Gilman, *Science, 244*:1558 (1989).
85. I. Porath, E. Nola, G. A. Puca, and P. Cuatrecasas, *Methods Enzymol., 34*:670 (1974).
86. I. Marbach, A. Bar-Sinai, M. Minich, and A. Levitzki, *J. Biol. Chem., 265*:9999 (1990).
87. F. Coussen, M. Guermah, J. d'Alayer, A. Monneron, J. Haiech, and J.-C. Cavadore, *FEBS Lett., 206*:213 (1986).
88. A. M. Minocherhomjee, S. Selfe, N. J. Flowers, and D. R. Storm, *Biochemistry, 26*:44 (1987).
89. E. Pfeuffer, S. Mollner, D. Lancet, and T. Pfeuffer, *J. Biol. Chem., 264*:18803 (1989).
90. J. Field, J.-I. Nikawa, D. Brock, B. MacDonald, L. Rodgers, I. A. Wilson, R. A. Lerner, and M. Wigler, *Mol. Cell. Biol., 8*:2159 (1988).
91. V. M. Lipkin, N. V. Khramtsov, S. G. Andreeva, M. V. Moshnyakov, G. V. Petukhova, T. V. Rakitina, E. A. Feshchenko, K. A. Ishchenko, S. F. Mirzoeva, M. N. Chernova, and S. M. Dranytsyna, *FEBS Lett., 254*:69 (1989).
92. H. R. Bourne, D. A. Sanders, and F. McCormick, *Nature, 348*:125 (1990).
93. M. Freissmith, P. J. Casey, and A. G. Gilman, *FASEB J., 3*:2125 (1989).
94. P. C. Sternweiss and J. D. Robishaw, *J. Biol. Chem., 259*:13806 (1984).
95. J. K. Northrup, P. C. Sternweiss, M. D. Smiget, L. S. Schleifer, E. M. Ross, and A. G. Gilman, *Proc. Natl. Acad. Sci. USA, 77*:6516 (1980).
96. I.-H. Pang and P. C. Sternweiss, *Proc. Natl. Acad. Sci. USA, 86*:7814 (1989).
97. L. Birnbaumer, *Annu. Rev. Pharmacol. Toxicol., 30*:675 (1990).
98. Pang, I.-H. and P. C. Sternweiss, *J. Biol. Chem., 265*:18707 (1990).
99. A. V. Smrcka, J. R. Hepler, K. O. Brown, and P. C. Sternweiss, *Science, 251*:804 (1991).
100. W. P. Arnold, C. K. Mitlal, S. Katsuki, and F. Murad, *Proc. Natl. Acad. Sci. USA, 74*:3203 (1977).
101. P. A. Craven, F. R. DeRubertis, and D. W. Pratt, *J. Biol. Chem., 254*:8213 (1979).
102. P. Humbert, F. Niromand, G. Fischer, D. Koesling, D.-D. Hinsch, H. Gausepohl, R. Frank, G. Schultz, and E. Bohme, *Eur. J. Biochem., 190*:273 (1990).

103. M. Nakane, S. Saheki, T. Kuno, K. Ishii, and F. Murad, *Biochem. Biophys. Res. Commun., 157*:1139 (1988).
104. S. A. Waldman, R. M. Rapoport, and F. Murad, *J. Biol. Chem., 259*:14332 (1984).
105. M. Field, L. H. Graf, W. J. Laird, and P. L. Smith, *Proc. Natl. Acad. Sci. USA, 75*:2800 (1978).
106. M. Takada, H. Takeuchi, and M. Shino, *Biochem. Biophys. Res. Commun., 164*:653 (1989).
107. J. A. Beavo, *Adv. 2nd Messenger Phosphoprotein Res., 22*:1 (1988).
108. R. L. Kincaid, I. E. Stith-Coleman, and M. Vaughan, *J. Biol. Chem., 260*:9009 (1985).
109. E. Degerman, P. Befrage, A. H. Newman, K. C. Rice, and V. C. Manganiello, *J. Biol. Chem., 262*:5797 (1987).
110. P. G. Gillespie and J. A. Beavo, *J. Biol. Chem., 263*:8133 (1988).
111. L. Sundberg and J. Porath, *J. Chromatogr., 90*:87 (1974).
112. T. J. Martins, M. C. Mumby, and J. A. Beavo, *J. Biol. Chem., 257*:1973 (1982).
113. P. G. Gillespie, R. K. Pnisti, E. D. Apel, and J. A. Beavo, *J. Biol. Chem., 264*:12187 (1989).
114. R. M. Ulloa, C. P. Rubinstein, L. Molinc y Vedia, H. N. Torres, and M. T. Tellez-Inon, *FEBS Lett., 241*:219 (1988).
115. R. K. Sharma and J. H. Wang, *J. Biol. Chem., 261*:14160 (1986).
116. S. S. Taylor, *J. Biol. Chem., 264*:8443 (1989).
117. J. G. Demaille, K. A. Peters, and E. H. Fischer, *Biochemistry, 16*:3080 (1970).
118. S. R. Olsen and M. D. Uhler, *J. Biol. Chem., 264*:18662 (1989).
119. M. J. Zollen, J. Kuret, S. Cameron, L. Levin, and K. E. Johnson, *J. Biol. Chem., 263*:9142 (1988).
120. R. M. Ulloa, E. Mesri, M. Esteva, H. N. Torres, and M. T. Tellez-Inon, *Biochem. J., 255*:319 (1988).
121. T. M. Lincoln, M. Thompson, and T. L. Cornwell, *J. Biol. Chem. 263*:17633 (1988).
122. J. P. Thalhofer and H. W. Hofer, *Arch. Biochem. Biophys., 273*:535 (1989).
123. B. Jergil and K. Mosbach, *Methods Enzymol., 34*:261 (1974).
124. W. L. Dills, C. D. Goodwin, T. M. Lincoln, J. A. Beavo, P. J. Bechtel, J. D. Corbin, and E. G. Krebs, *Adv. Cyclic Nucleotide Res., 10*:203 (1979).
125. J. R. Woodgett, *Anal. Biochem., 180*:237 (1989).
126. N. K. Mukhopadmyay, K. Shome, A. K. Saha, J. R. Hassell, and R. H. Glew, *Biochem. J., 264*:517 (1989).
127. M. Wilchek, T. Miron, and J. Kohn, *Methods Enzymol., 104*:3 (1984).
128. P. Cohen and P. T. W. Cohen, *J. Biol. Chem., 264*:21435 (1989).
129. T. S. Ingebritsen and P. Cohen, *Science, 221*:331 (1983).
130. H. C. Hemmings, Jr., A. C. Nairn, and P. Greengard, *J. Biol. Chem., 259*:14491 (1984).
131. H. C. Hemmings, Jr., P. Greengard, H. Y. L. Tung, and P. Cohen, *Nature, 310*:503 (1984).
132. V. M. Ingebritsen and T. S. Ingebritsen, *Biochim. Biophys. Acta, 1012*:1 1989.
133. H. Y. Tung and L. Reed, *J. Biol. Chem., 264*:2985 (1989).
134. P. Cohen, S. Klumpp, and D. L. Schelling, *FEBS Lett., 250*:596 (1989).
135. S. Wera, M. Bollen, and W. Stalmans, *J. Biol. Chem., 266*:339 (1991).

136. S. Nishiwaki, H. Fujiki, M. Suganuma, M. Ojika, K. Yamada, and T. Sugimura, *Biochem. Biophys. Res. Commun., 170*:1359 (1990).
137. S. Nishiwaki, H. Fujika, M. Suganuma, R. Nishiwaki-Matsushima, and T. Sugimura, *FEBS Lett., 279*:15 (1991).
138. M. J. Berridge, *Annu. Rev. Biochem., 58*:159 (1987).
139. P. W. Majerus, T. S. Ross, T. W. Cunningham, K. K. Caldwell, A. B. Jefferson, and V. S. Bansal, *Cell, 63*:459 (1990).
140. M. J. Berridge and R. F. Irvine, *Nature, 341*:197 (1989).
141. Y. Takai, U. Kikkawa, K. Kaibuchi, and Y. Nishizuka, *Adv. Cyclic Nucleotide Protein Phosphorylation Res., 18*:119 (1984).
142. J. H. Exton, *J. Biol. Chem., 265*:1 (1990).
143. M. Hirata, Y. Watanabe, T. Ishimatsu, F. Yanaga, T. Koga, and S. Ozaki, *Biochem. Biophys. Res. Commun., 168*:379 (1990).
144. M. Wahl and G. Carpenter, *BioEssays, 13*:107 (1991).
145. M. I. Wahl, S. Nishibe, J. W. Kim, H. Kim, S. G. Rhee, and G. Carpenter, *J. Biol. Chem., 265*:3944 (1990).
146. A. Moritz, P. N. E. DeGraan, P. F. Ekhart, W. H. Gispen, and K. W. A. Wirtz, *J. Neurochem., 54*:351 (1990).
147. D. H. Walker, N. Dougherty, and L. J. Pike, *Biochemistry, 27*:6504 (1988).
148. T. Urumov and O. H. Wieland, *Biochem. Biophys. Acta, 1052*:152 (1990).
149. M. Kata and T. Takenawa, *J. Biol. Chem., 265*:794 (1990).
150. Y. Yada, T. Ozeki, H. Kanoh, and Y. G. Nozawa, *J. Biol. Chem., 265*:10367 (1990).
151. K. Yamaguchi, M. Hirata, and H. Kuriyama, *Biochem. J., 244*:787 (1987).
152. S. S. Sim, J. W. Kim, and S. G. Rhee, *J. Biol. Chem., 265*:10367 (1990).
153. R. A. Johnson, C. A. Hansen, and J. R. Williamson, *J. Biol. Chem., 263*:17465 (1988).
154. K. Takazawa, H. Passareiro, J. E. Dumont, and C. Erneux, *Biochem. J., 261*:483 (1989).
155. J. A. Escobedo, S. Navankasattusas, W. M. Kavanaugh, D. Milfay, V. A. Fried, and L. T. Williams, *Cell, 65*:75 (1991).
156. N. S. Gee, C. I. Ragan, K. J. Watling, S. Asplay, R. G. Jackson, G. G. Reid, D. Gani, and J. K. Shute, *Biochem. J., 249*:883 (1988).
157. N. S. Gee, S. Howell, G. Ryan, and C. I. Ragan, *Biochem. J., 264*:793 (1989).
158. Y. Nishizuka, *Nature, 334*:661 (1988).
159. T. Kitano, M. Go, U. Kikkawa, and Y. Nishizuki, *Methods Enzymol., 124*:349 (1986).
160. R. M. Marais and P. J. Parker, *Eur. J. Biochem., 182*:129 (1989).
161. M. B. Wheeler and J. D. Veldhuis, *Mol. Cell. Endocrinol., 61*:117 (1989).
162. D. Schaap and P. J. Parker, *J. Biol. Chem., 265*:7301 (1990).
163. H. Hidaka, M. Inagaki, S. Kwamoto, and Y. Sasaki, *Biochemistry, 23*:5036 (1984).
164. M. Inagaki, M. Watanabe, and H. Hidaka, *J. Biol. Chem., 260*:2922 (1985).
165. C. A. O'Brian and N. E. Ward, *Biochem. Pharmacol., 38*:1737 (1989).
166. C. A. O'Brian and N. E. Ward, *Mol. Pharmacol., 36*:355 (1989).
167. C. A. O'Brian, G. M. Housey, and I. B. Weinstein, *Cancer Res., 48*:3626 (1988).
168. B. C. Wise, R. L. Raynor, and J. F. Kuo, *J. Biol. Chem., 257*:8481 (1982).
169. T. Uchida and C. R. Filburn, *J. Biol. Chem., 259*:12311 (1984).

5

Purification of Membrane Transport Proteins and Receptors by Immobilized-Ligand Affinity Chromatography

Malcolm G. P. Page

F. Hoffmann–La Roche Ltd., Basel, Switzerland

I. INTRODUCTION

The purification of membrane-bound proteins has a complexity that is not present with other proteins. The normal state of the protein is to be embedded in the phospholipid bilayer, where it interacts with the acyl chains in the hydrophobic core, the phosphate groups in the head-group region, the aqueous medium in the compartments on either side of the membrane, and possibly, other proteins. The aim of purification is essentially to disrupt all these interactions and to transfer the protein into a homogeneous and well-defined medium.

Affinity chromatography exploits the highly specific interactions involved in recognition of substrate and inhibitors. These facilitate the isolation and purification of hormone receptors and selective transporters, which are frequently present in very low copy numbers in cell membranes, while preserving the intrinsic activity of the protein and removing inactive, denatured material.

The covalent attachment of ligands to solid supports was discussed by Cuatrecasas in 1970 (1) and the remarks remain highly pertinent to the selection of immobilization strategies. Pharmacological studies of receptor and channel function have made available a great diversity of small molecule antagonists and venom toxins that bind adventiously to other sites on the proteins with such high affinity that they can be used for affinity chromatography. Capture of ligand–protein complexes either by immobilized antibodies or, for biotinylated ligands,

Table 1 Useful Detergents for Solubilization of Membrane Proteins for Affinity Chromatography

Detergent	Receptor	Refs.
3-[(3-Cholamidopropyl) dimethylammonio]-1-propanesulfonic acid (CHAPS)	A_1 adenosine receptor	37, 40
	D_2 dopamine receptor	29
	GABA receptor (+ asolectin)	17
	K^+-channel protein	8
	Neurotensin receptor	53
	NMDA receptor (+0.1% polyoxyethylene 10-tridecyl ether)	15
	Prostaglandin receptor	45
	Vasoactive intestinal peptide receptor	65
Cholate	Adenosine-binding proteins	44
	D_2 dopamine receptor	27
Deoxycholate	GABA receptor	18
	5-HT_3 receptor (+ Lubrol)	30
	H^+-lactose transporter	5
Digitonin	A_1 adenosine receptor	37, 40
	A_2 adenosine receptor	42
	α_1-Adrenergic receptor	20
	β-Adrenergic receptor	21
	Cholecystokinin receptor	57
	D_2 dopamine receptor	24
	Kappa-opioid receptor	75
	Melatonin receptor	31
	Muscarinic acetylcholine receptor	35
	Na^+–K^+–$2Cl^-$ cotransporter	9
Lubrol-PX	H^+-lactose transporter	5, 6
	Nictinic acetylcholine receptor	32
	Ovarian leuteinizing hormone receptor	71
Nonidet P-40	Collagen receptor (+ deoxycholate)	51, 52
	Fibronectin receptor	47
	Thyrotropin receptor	80
Octyl-β,D-glucoside	Adhesion receptors	46
	Collagen receptors	52
	Na^+–K^+–$2Cl^-$ cotransporter	10

Table 1 (Continued)

Detergent	Receptor	Refs.
Triton X-100	Collagen receptor	52
	Kainate receptor (+ digitonin)	16
	Lactotransferrin receptor	68
	Na^+-D-glucose transporter	11
	Neurotoxin binding proteins	69
	Pyruvate and α-ketoglutarate transporters	12
	Placental lactogen receptor	60
	Somatostatin receptor	79
Triton X-114	Dicarboxylate transporter	13
	Tricarboxylate carrier	14

by immobilized streptavidin are methods that have not been so widely used but are becoming more important. Biotin–avidin techniques have been the subject of a monograph (2).

Detergents play an essential role in the purification scheme, and the choice of detergent is known to be critical for maintenance of protein structure and function (3–6). This has also become clear in studies with receptors: Some detergents have proved better than others for maintaining the protein in solution (3,4,31), and in at least one case it has been necessary to add lipid to the solubilized receptor to maintain binding activity (37). The D_2 dopamine receptor appears to retain binding activity in crude extracts but not in a purified state, which could be due to a requirement for lipid (29–32). Certainly, the purified protein recovered its typical binding properties after being reconstituted into proteoliposomes. The detergents that have commonly been used for purification are listed in Table 1; whether these represent optimal detergents for study of all aspects of receptor structure and function is unclear. The efficacy of a particular detergent or combination of detergents is tested several times: for efficient solubilization, for maintaining binding activity, for maintaining stability during purification, for efficiency of reconstitution, for structural studies by spectroscopy, or most ambitiously, for crystallization and x-ray diffraction analysis. For many proteins (e.g., lactose transporter from *Escherichia coli*), detergents with high aggregation number and low critical micelle concentration (e.g., dodecyl maltoside or Lubrol-PX) have proved to the most efficacious for solubilization, stability, and binding. On the other hand, detergents with low aggregation number and high critical micelle concentration (e.g., octyl glucoside or decyl maltoside) have proved more efficacious for reconstitution (5,6) and crystallization attempts. Starting the purification in one detergent, which is best for activity and stability, and ending by exchanging the detergent for one that is more suitable for the end goal will become a more frequent strategy as more is known about protein–detergent interactions.

Robinson and colleagues (7) have observed that nonionic detergents can incorporate hydrophobic ligands into their micelles, thus rendering affinity chromatography unproductive. In an investigation involving anionic dye affinity chromatography, it was found that the detergent micelles could be dispersed by the addition of anionic detergents such as deoxycholate and sodium dodecyl sulfate. Azzi and colleagues (12,13) have reported that the addition of deoxycholate to carboxylate transporters solubilized in Triton detergents improved the performance of affinity chromatography columns with immobilized aromatic carboxylic acids as ligands. Similarly, addition of deoxycholate to the *E. coli* lactose transporter in solution in Lubrol-PX helps to stabilize the protein and improves recovery after affinity chromatography (Ref. 5 and unpublished observations).

II. MEMBRANE TRANSPORT PROTEINS

A. Channel Proteins

A calcium-activated K^+-channel protein was purified from pig kidney outer medulla using calmodulin immobilized by cross-linking to CNBr-activated Sepharose (8). The channel protein, solubilized in 3-[(3-cholamidopropyl) dimethylammonio]-1-propanesulfonic acid (CHAPS), was bound to the column in the presence of calcium and could be specifically eluted with EGTA.

B. Ion Transporters

The Na^+–K^+–$2Cl^-$ cotransporters from epithelial (9) and Ehrlich ascites cells (10) have been purified by affinity chromatography using the selective inhibitors furosemide and bumetanide, respectively. Furosemide was immobilized by carbodiimide-induced cross-linking to Affi-Gel 102 (Figure 1), while the bumetanide was cross-linked in a novel way by the light-induced reaction between 4′-azidobumetanide (Figure 2) and Pharmacia-activated thiol Sepharose 4B (reduced glutathione–Sepharose). The chemical nature of the linkage is not certain but involves reaction at the thiol group of the glutathione (10).

C. Ion-Coupled Sugar Transporters

Although the H^+-lactose transporter from *E. coli* is soluble in a number of detergents (5,6), it exhibits high-affinity substrate binding only in Lubrol-PX. Attempts

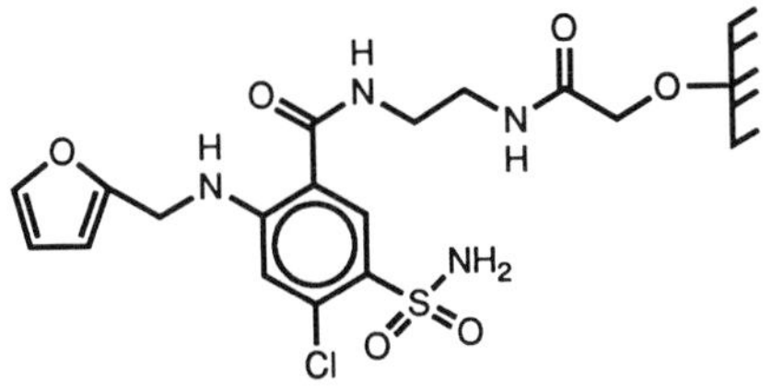

Figure 1 Immobilized furosemide.

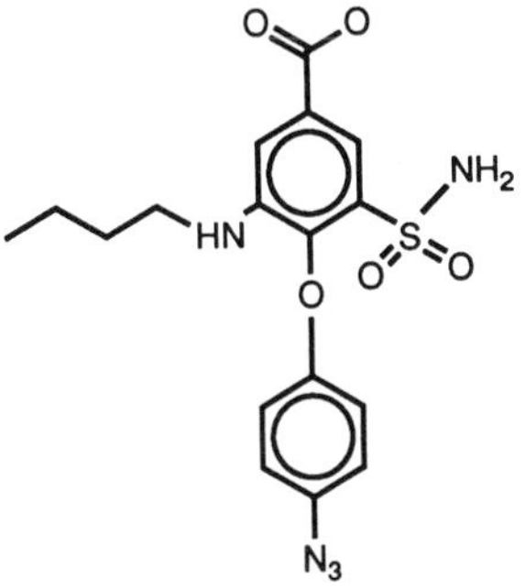

Figure 2 4′-Azidobumetanide.

at affinity purification using immobilized galactosides, such as lactose-agarose, have generally been unsuccessful, probably because the affinity of the transporter for such sugars is rather low (5). A spacer of several carbon atoms length was required for binding even to a high-affinity ligand. Thus, no significant binding occurred to aminoethylaminophenyl-α,D-galactopyranoside immobilized on CNBr-activated Sepharose (Figure 3a), while weak affinity for aminohexyl-aminophenyl-α,D-galactopyranoside (Figure 3b) immobilized in the same way (Page, unpublished observation). The transporter could be quantitatively bound to aminophenyl-α,D-galactopyranoside immobilized on Affi-Gel 15 (Figure 3c) and eluted either with 4-nitrophenyl-α,D-galactopyranoside or by exchanging the Lubrol-PX with the detergent octyl-β,D-glucoside, which does not support binding. The requirement for a relatively long spacer between ligand and matrix may indicate that the sugar-binding site on the protein is buried somewhat in the molecule, or that the ligand must be occluded before high-affinity binding can occur (although recognition occurs at the surface).

(a)

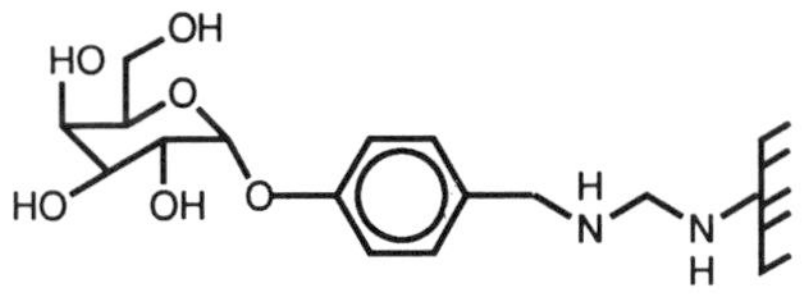

(b)

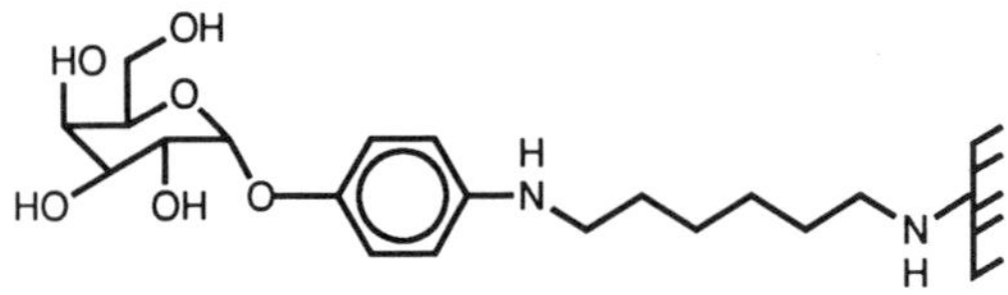

(c)

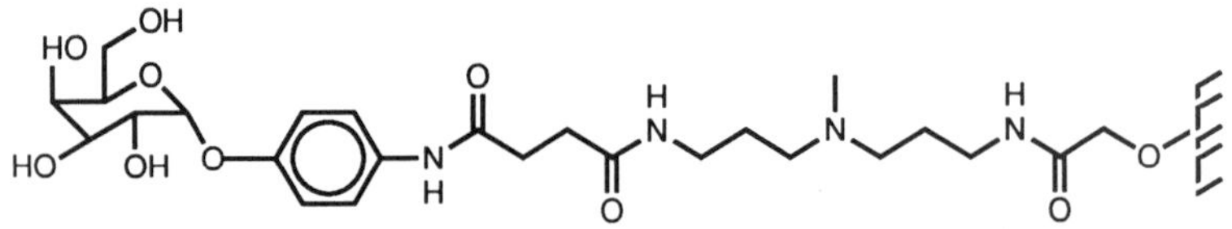

Figure 3 Immobilized ligands of the *E. coli* lactose transporter.

A protein that has the expected activities of the Na^+-glucose transporter from kidney (11) was isolated by renal medulla membranes using 3-aminophloridzin (Figure 4).

D. Mitochondrial Carboxylic Acid Transporters

The monocarboxylate transporter and the α-ketoglutarate transporter have been purified using immobilized 2-cyano-4-hydoxycinnamate (Figure 5), which is a potent inhibitor of pyruvate and α-ketoglutarate transport in mitochondria (12). The transporters were solubilized in Triton X-100, but deoxycholate was added to the mixture to improve the affinity chromatography step, possibly by increasing the availability of the rather hydrophobic ligand (see Ref. 7).

The dicarboxylate transporter was purified in a very similar manner, using immobilized phenylsuccinate (Figure 6) (13). A rather different strategy was adopted for the tricarboxylate transporter (14). This protein was purified by dye affinity chromatography using Matrix Gel Blue B (Amicon) and then sulfhydryl

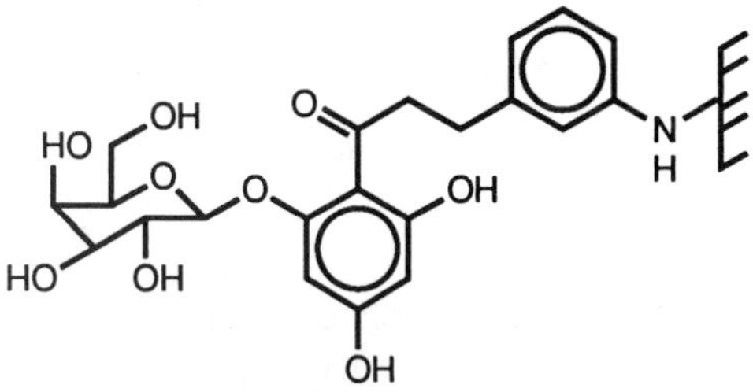

Figure 4 Immobilized 3-aminophloridzin.

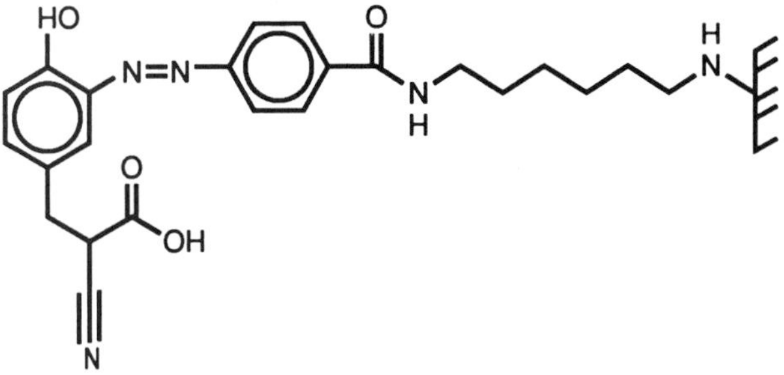

Figure 5 Immobilized 2-cyano-4-hydroxycinnamate.

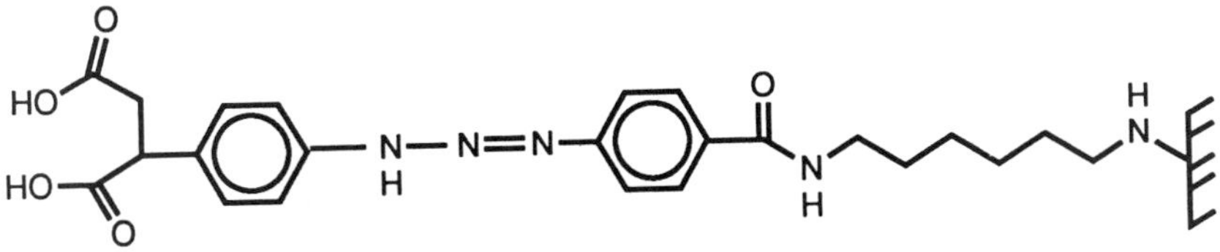

Figure 6 Immobilized 4-aminophenylsuccinate.

affinity chromatography using immobilized 4-hydroxymercuriphenylamine (Affi-Gel 501, Bio-Rad).

III. NEUROTRANSMITTER AND HORMONE RECEPTORS

A. Amino Acid Receptors

N-Methyl-D-aspartate receptor from rat brain, solubilized in a mixture of CHAPS and polyoxyethylene10-tridecyl ether, was purified more than 20,000-fold using an affinity matrix having the agonist ibotenic acid (Figure 7) cross-linked to Affi-Gel 15 (15). Kainate receptor was purified to apparent homogeneity using domoic acid, an analog of kainate, immobilized on Sepharose (Figure 8) (16). Immobilized kainate was also able to bind the receptor, although with lower affinity. Thus it appears that cross-linking may be achieved through any of the carboxyl groups, but the precise nature of the cross-linking and its affect on affinity remains to be studied. γ-Aminobutyric acid/benzodiazepine receptor has been purified using immobilized antagonists such as baclofen (Figure 9) (17) or the high-affinity benzodiazepine Ro-7-1986 (Figure 10) (18,19).

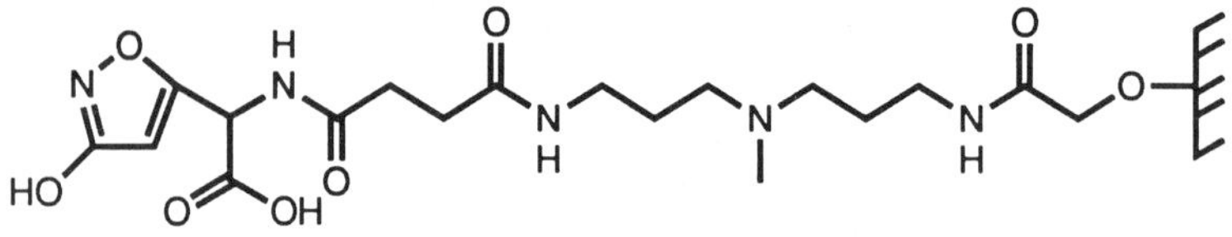

Figure 7 Immobilized ibotenic acid.

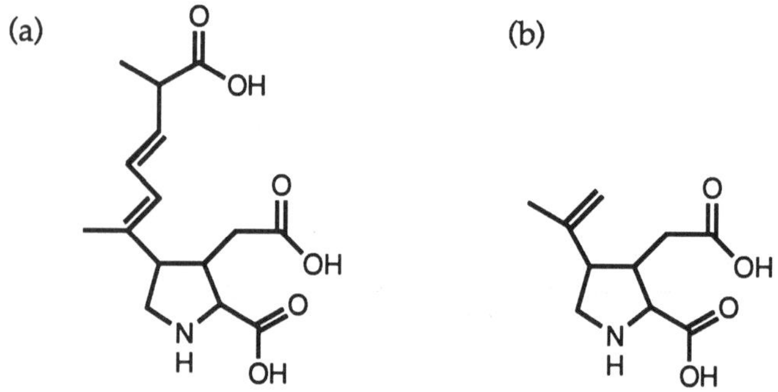

(a) (b)

Figure 8 Domoic acid (a) and kainic acid (b).

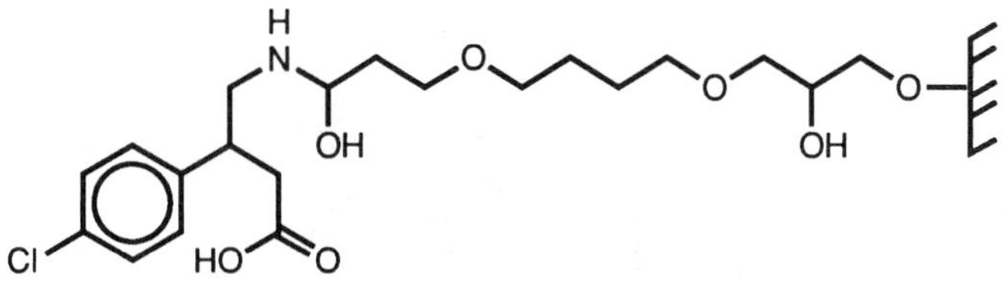

Figure 9 Immobilized baclofen.

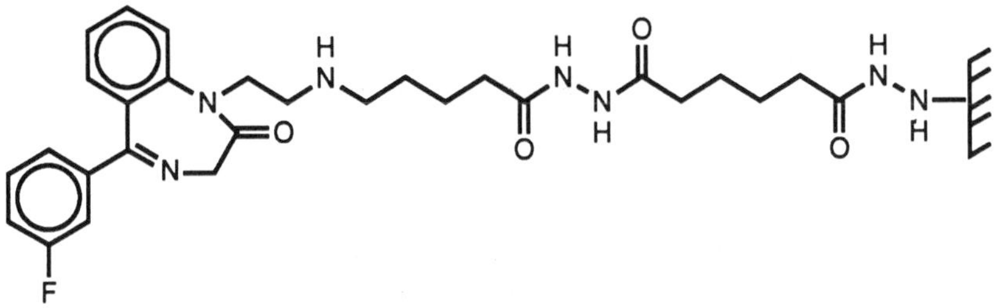

Figure 10 Immobilized Ro-7-1986.

B. Aminergic Receptors

Adrenergic receptors have been divided into α_1, α_2, β_1, and β_2 subclasses according to their interactions with ligands. An α_1 receptor has been purified using immobilized antagonist (Figure 11) (20), while β-type receptors have been purified using the antagonist alprenolol immobilized on Sepharose 4B (21,22).

Dopamine D_1 receptor has been purified by a novel method that involved reaction of membranes with *N*-ethylmaleimide in the presence of a selective antagonist, then purification on a mercury–agarose column. More than 8000-fold purification could be achieved with 33% yield of binding sites (23). Dopamine D_2 receptors have proved difficult to purify because of the low amounts of material expressed in source tissue. Estimates suggest that 50,000-fold purification would be necessary to reach homogeneity (24). The receptor has been effectively solubilized in digitonin, CHAPS, or sodium cholate for affinity chromatography using high-affinity antagonists (Table 2). Purification of 500- to 2000-fold was achieved in the affinity steps (25–27), and this has been combined with other methods to effect complete purification (28,29). The purified receptor proved very sensitive to its environment after the affinity step, and it was necessary to reconstitute the protein into liposomes before assaying binding activity.

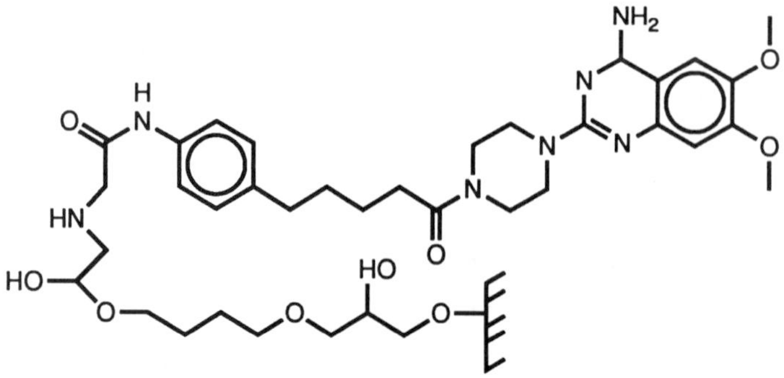

Figure 11 Immobilized quinazoline antagonist of the α_1-adrenergic receptor.

A 5-hydroxytryptamine $5HT_3$ receptor has been purified using the 3-carboxypropyl-1-azacyclic derivative of 2′-(1-methyl-1*H*-indol-3-yl)spiro[1-azabicyclo (2,2,2)octane-3,5′(4′*H*)oxazole] (Figure 12), a potent antagonist of 5-hydroxytryptamine, immobilized on an agarose support (30). A melatonin receptor from lizard brain was purified on 6-hydroxymelatonin Sepharose (Figure 13) (31). The receptor was effectively solubilized in a mixture of Lubrol and deoxycholate, but the affinity chromatography had to be performed with Triton X-100 because the protein was not stable in the solubilization buffer.

C. Acetylcholine Receptors

The nicotinic acetylcholine receptor, a cation-specific ion channel, is one of the most intensively and best studied membrane-bound receptors. High-affinity ligands in the form of agonists and antagonists are numerous and well characterized, and the highly specific interaction with cobra venom toxin is well established. It is therefore not surprising that several approaches to affinity chromatography have been taken. The earliest exploited the cobratoxin, immobilized on Sepharose (32,33), but more recently acetylcholine immobilized on Affi-Gel (34), and several antagonists (35) have been employed. The muscarinic acetylcholine receptor, a G-protein-linked receptor, has been purified using an immobilized antagonist (Figure 14) (36).

D. Adenosine Receptors

Adenosine receptors are classified in two subtypes according to their ability to inhibit (A_1) or stimulate (A_2) adenylate cyclase activity. The adenosine A_1

Table 2 Antagonist Affinity Matrices for Purification of D2 Dopamine Receptors

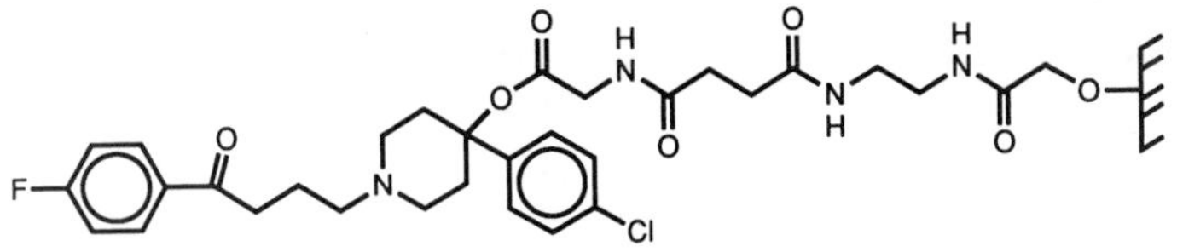

Purification	Recovery
500-fold purification Ref. 28, 29	40% recovery after reconstitution
Two pools: 350-fold purification 460-fold purification Ref. 25	 22% recovery after reconstitution 27% recovery after reconstitution
1000-fold purification Ref. 26	5% recovery 50% recovery after reconstitution
2000-fold purification Ref. 24	12% recovery

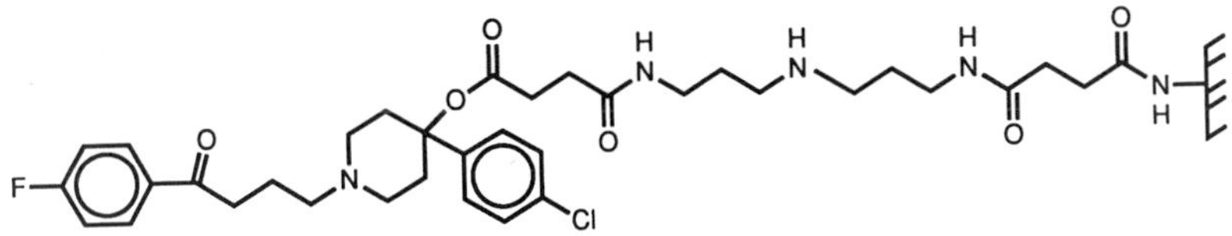

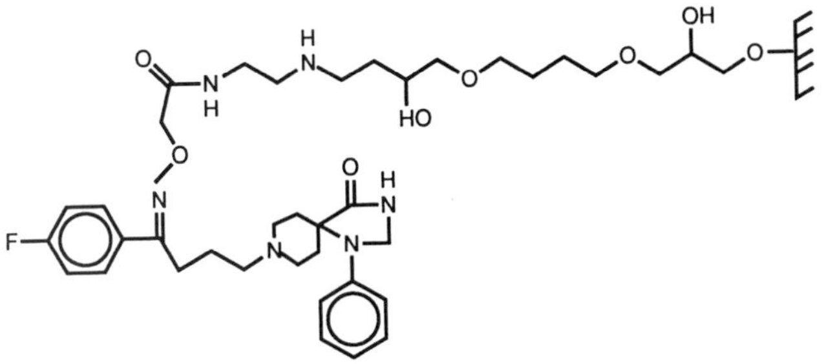

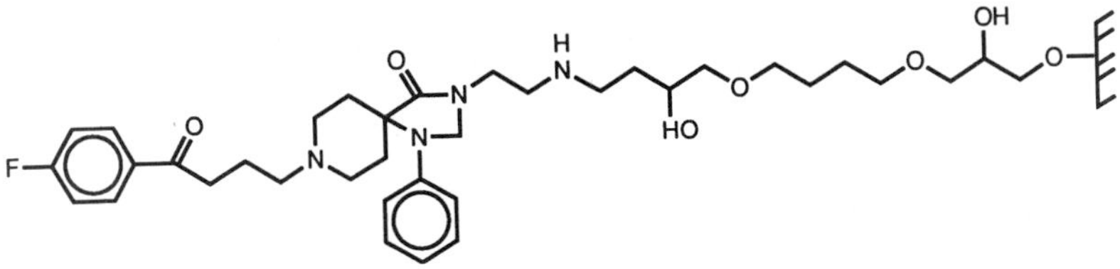

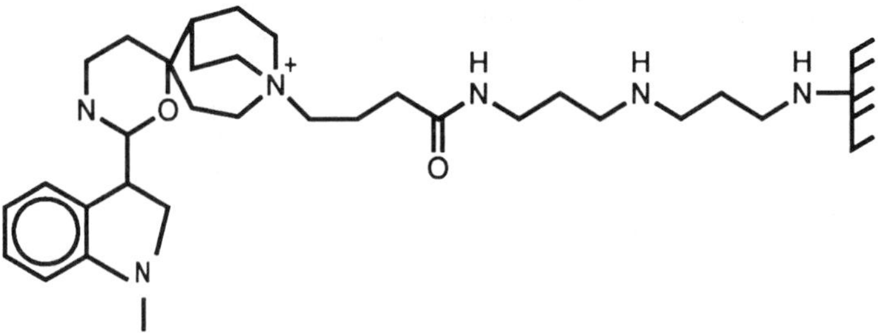

Figure 12 Affinity chromatography matrix for the 5-HT_3 receptor.

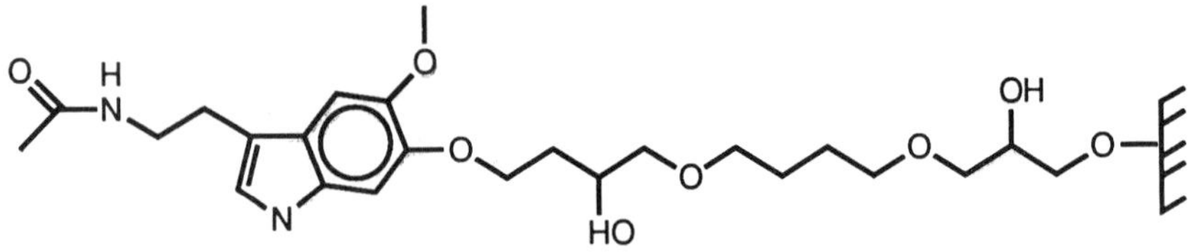

Figure 13 Immobilized melatonin.

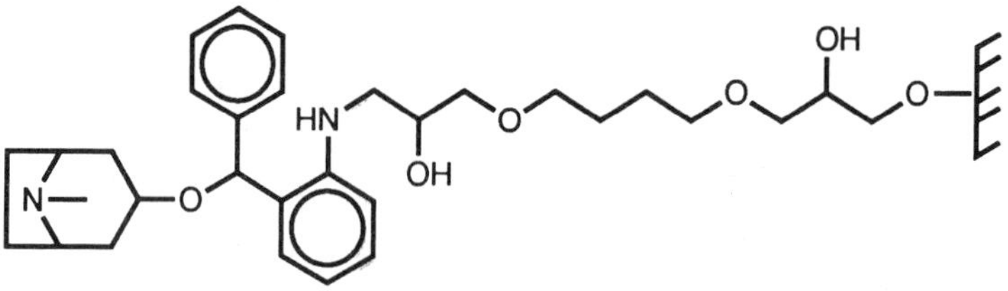

Figure 14 Immobilized antagonist of the muscarinic acetylcholine receptor.

receptor solubilized in CHAPS or digitonin has been purified by affinity chromatography using antagonists such as, respectively, N_6-aminobenzyladenosine (37) linked to agarose or xanthine derivatives (Figure 15) (38–41) linked to Affi-Gel supports. Munshi and colleagues (37) reported that CHAPS-solubilized receptor required additional asolectin for successful affinity chromatography. The A_2 adenosine receptor from mouse mastocytoma cell membranes was solubilized in digitonin and purified using 5′-*N*-ethylcarboxamidoadenosine cross-linked to Sepharose 6B (42). Soluble adenosing-binding proteins, also designated as "P-type receptors," have been isolated from the cytoplasm of brain and other tissues by affinity chromatography using 5′amino-5′-deoxyadenosine (43) linked to agarose and N^6-aminononyladenosine (44) linked to Sepharose.

E. Prostaglandin Receptor

Thromboxane is a very labile molecule and is not suitable as a ligand for affinity chromatography. However, the development of high-affinity antagonists such as that shown in Figure 16 enabled the successful purification of a receptor from platelets (45).

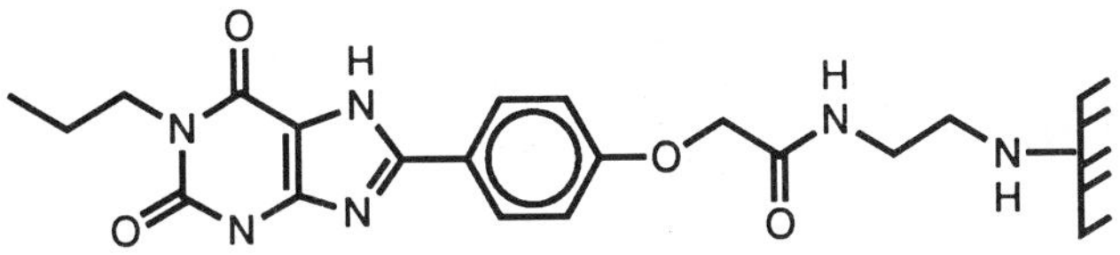

Figure 15 Immobilized xanthine amine congener.

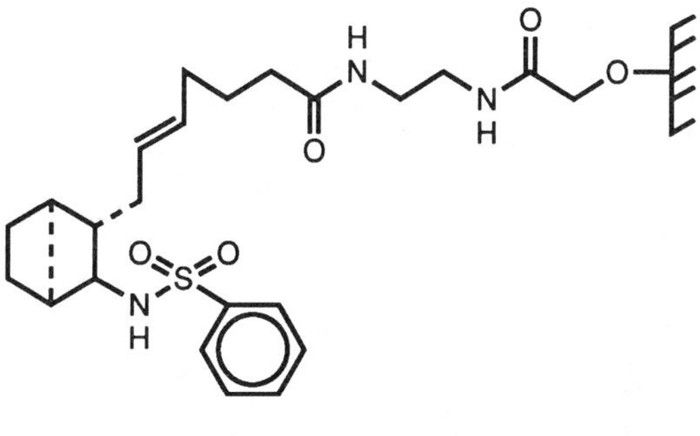

Figure 16 Immobilized prostaglandin antagonist.

F. Polypeptide and Peptide Hormone Receptors

The peptide receptors have offered considerable scope for the development of affinity methods. The ligands have been immobilized on a variety of supports, or biotinylated and then immobilized with avidin, or captured using immobilized antibodies. The specificity of the interactions involved have often resulted in quite spectacular levels of purification, going from crude extract to homogeneity in a single step, as summarized in Table 3.

Table 3 Affinity Purification of Polypeptide Receptors[a]

Receptor	Method	Purification factor	Yield (%)	Refs.
Adrenocorticotrophic hormones	BA	NA	NA	56
Cholecystokinin	IL	182	30	57
Complement C3b binding proteins	IL	NA	NA	58
Fibroblast acidic growth factor	IL	60	39	59
Fibroblast basic growth factor	IL	NA	NA	60
Gastrin-releasing peptide	IL	300	43	61
Gonadotrophin releasing hormone receptors	BA	NA	NA	62, 63
Granulocyte macrophage colony-stimulating factor (GM-CSF)	BA	20,000	11	64
Interleukin-2	IL	NA	NA	65
Interleukin-4	IL	800	33	66
Putative mitochondrial receptor for imported proteins	IL	NA	NA	67
Insulin receptors	BA	NA	60	56
Lactotransferrin	AB	NA	30	68
α-Latroxin	IL	5,300	50	69
Leuteinizing hormone	IL	5,500	20	70, 71
α-Macroglobulin	IL	10,000	NA	72
Neurotensin	IL	15,600	72	73
Opioid receptors	IL	5,611 (19,600 after reconstitution)	8	74–76
Placental lactogen	IL	30,000	32	77
Somastostatin receptors	IL	2	50	78, 79
Thyrotrophin	IL	269	4	80–82
Vasoactive-intestinal peptide	IL	12,500	36	84
Vasopressin	IL	64	16	83

[a]The method used for immobilization of the ligand is indicated by AB for "antiligand antibodies," BA for "biotin–avidin," and IL for "covalent cross-linking." Purification factor and yield are estimates for the affinity purification step only, not overall figures for the complete purification procedure. NA in these columns indicates that an estimate cannot be made.

IV. CYTOSKELETON AND CELL ADHESION RECEPTORS

A. Adhesion Receptors

Adhesion receptors have been characterized as having a high affinity for peptides containing the sequence Arg–Gly–Asp. In platelets, two glycoproteins could be isolated from octylglucoside extracts using an affinity matrix of Lys–Tyr–Gly–Arg–Gly–Asp–Ser coupled to Sepharose 4B (46). The proteins could be specifically eluted with a small soluble peptide (Gly–Arg–Gly–Asp–Ser–Pro). Fibronectin receptors from chicken embryo fibroblasts were isolated using a cell-binding fragment of chicken fibronectin as an affinity matrix (47). The receptor could be specifically eluted using the small peptide Gly–Arg–Gly–Asp–Ser–Pro (48,49).

B. Collagen Receptors

Collagen receptors have been purified using gelatin of collagen cross-linked to Sepharose (50–52). The gelatin–Sepharose was reported to be more stable than immobilized collagen.

C. Lectins

A number of carbohydrate-binding proteins from mammalian cell surfaces, which are apparently involved in cell–cell interactions, have been isolated using immobilized sugars. These include a membrane homing receptor from hemopoietic cells (53), an advanced glycosylation end-product receptor from macrophages (54), and some mammalian hepatic lectins (55).

REFERENCES

1. P. J. Cuatrecasas, *Biol. Chem., 245*:3059 (1970).
2. M. Wilchek and E. A. Bayer (eds.), Avidin-biotin technology, *Methods in Enzymology*, Vol. 184, Academic Press, New York, 1990.
3. A. Helenius and K. Simons, *Biochim. Biophys. Acta, 415*:29 (1975).
4. L. M. Hjelmeland and A. Chrambach, *Methods Enzymol., 104*:305 (1984).
5. M. G. P. Page, J. P. Rosenbusch, and I. Yamato, *J. Biol. Chem., 263*:15897 (1988).
6. J. K. Wright, H. Schwarz, E. Straub, P. Overath, B. Bieseler, and K. Beyreuther, *Eur. J. Biochem., 124*:545 (1982).
7. J. B. Robinson, J. M. Strottman, D. G. Wick, and E. Stellwagen, *Proc. Natl. Acad. Sci. USA, 77*:5847 (1980).
8. D. A. Klaerke, J. Petersen, and P. L. Jorgensen, *FEBS Lett., 216*:211 (1987).
9. T. Zeuthen, P. M. Andersen, K. E. Eskesen, and B. D. Cherksey, *Comp. Biochem. Physiol. (A), 90*:687 (1988).
10. P. W. Feit, E. K. Hoffman, M. Schiodt, P. Kristensen, F. Jessen, and P. B. Dunham, *J. Membr. Biol., 103*:135 (1988).
11. T. Kitlar, A. I. Morrison, R. Kinne, and J. Deutscher, *FEBS Lett., 234*:115 (1988).

12. R. Bolli, K. A. Nalcecz, and A. Azzi, *J. Biol. Chem., 264*:18024 (1989).
13. A. Szewczyk, M. J. Nalcecz, C. Broger, L. Wojtczak, and A. Azzi, *Biochim. Biophys. Acta, 894*:252 (1987).
14. R. S. Kaplan, J. A. Mayor, N. Johnston, and D. L. Oliviera, *J. Biol. Chem., 265*:13379 (1990).
15. M. D. Cunningham and E. K. Michaelis, *J. Biol. Chem., 265*:7768 (1990).
16. D. R. Hampson and R. J. Wenthold, *J. Biol. Chem., 263*:2500 (1988).
17. Y. Ohmori and K. Kuriyama, *Biochem. Biophys. Res. Commun., 172*:22 (1990).
18. E. Sigel, F. A. Stephenson, C. Mamalaki, and E. A. Barnard, *J. Biol. Chem., 258*:6965 (1983).
19. E. Sigel, C. Mamalaki, and E. A. Barnard, *FEBS Lett., 147*:45 (1982).
20. J. W. Lomasney, L. M. F. Leeb-Lundberg, S. Cotecchia, J. W. Regan, J. F. De Bernardis, M. G. Caron, and R. J. Lefkowitz, *J. Biol. Chem., 261*:7710 (1986).
21. M. G. Caron, Y. Srinirasan, J. Pitha, K. Kociolek, and R. J. Lefkowitz, *J. Biol. Chem., 254*:2923 (1979).
22. B. D. Cherksey, S. A. Mendelsohn, J. A. Zadunaisky, and J. Altszuler, *Membr. Biol., 84*:105 (1985).
23. A. Sidhu, *J. Biol. Chem., 265*:10065 (1990).
24. D. M. Clagett, R. Schoenleber, C. Chung, and J. F. McKelvy, *Biochim. Biophys. Acta, 986*:271 (1989).
25. L. Antonian, E. Antonian, R. B. Murphy, and D. I. Schuster, *Life Sci., 38*:1847 (1986).
26. S. E. Senogles, N. Amlaiky, A. L. Johnson, and M. G. Caron, *Biochemistry, 25*:749 (1986).
27. J. Ramwani and R. K. Mishra, *J. Biol. Chem., 261*:8894 (1986).
28. H. Kanety, M. Schreiber, Z. Elazar, and S. Fuchs, *Neuroimmunology, 18*:25 (1988).
29. Z. Elazar, H. Kanety, C. David, and S. Fuchs, *Biochem. Biophys. Res. Commun., 156*:602 (1988).
30. R. M. McKernan, N. P. Gillard, K. Quirk, C. O. Kneen, G. I. Stevenson, C. J. Swain, and C. I. Ragan, *J. Biol. Chem., 265*:13572 (1990).
31. S. A. Rivkees, R. W. Conron, and S. M. Reppert, *Endocrinology, 127*:1206 (1990).
32. E. A. Kapp and C. G. Whiteley, *Biochim. Biophys. Acta, 1034*:29 (1990).
33. R. Olsen, J. C. Meunier, and J.-P. Changeux, *FEBS Lett., 28*:96 (1972).
34. H. Nakayama, M. Shirase, T. Nakashima, Y. Kurogochi, and J. M. Lindstrom, *Brain Res. Mol. Brain Res., 7*:221 (1990).
35. P. G. Waser, D. M. Bodmer, and W. H. Hopff, *Eur. J. Pharmacol., 172*:231 (1989).
36. K. Haga and T. Haga, *J. Biol. Chem., 260*:7927 (1985).
37. R. Munshi and J. J. Linden, *Biol. Chem., 264*:14853 (1989).
38. H. Nakata, *Mol. Pharmacol., 35*:780 (1989).
39. H. Nakata, *J. Biol. Chem., 264*:16545 (1989).
40. M. E. Olah, K. A. Jacobson, and G. L. Stiles, *FEBS Lett., 257*:292 (1989).
41. H. Nakata, *J. Biol. Chem., 265*:671 (1990).
42. H. Nakata, *J. Biochem. (Tokyo), 105*:700 (1989).
43. K. Ravid, R. A. Rosenthal, S. R. Doctrow, and J. M. Lowenstein, *Biochem. J., 258*:653 (1989).
44. M. E. Bembenek, *Biochem. Biophys. Res. Commun., 168*:702 (1990).

45. F. Ushikubi, M. Nakajima, M. Hirata, M. Okuma, M. Fujiwara, and S. J. Narumiya, *Biol. Chem., 264*:16496 (1989).
46. S. C. Lam, E. F. Plow, S. E. Souza, D. A. Cheresh, A. L. Frelinger, and M. H. Ginsberg, *J. Biol. Chem., 264*:3742 (1989).
47. U. Hofer, J. Syfrig, and R. J. Chiquet-Ehrismann, *Biol. Chem., 265*:14561 (1990).
48. A. P. Mould, L. A. Wheldon, A. Komoriya, E. A. Wayner, K. M. Yamada, and M. J. Humphries, *J. Biol. Chem., 265*:4020 (1990).
49. B. A. Bottger, U. Hedin, S. Johansson, and J. Thyberg, *Differentiation, 41*:158 (1989).
50. A. Erdei and K. B. Reid, *Behring Inst. Mitt., 84*(July):216 (1989).
51. M. L. Lu, D. A. Beacham, and B. S. Jacobson, *J. Biol. Chem., 264*:13546 (1989).
52. S. P. Sugrue, *J. Biol. Chem., 262*:3338 (1987).
53. T. Matsuoka and M. Tavassoli, *J. Biol. Chem., 264*:20193 (1989).
54. S. Radoff, A. Cerami, and H. Vlassara, *Diabetes, 39*:1510 (1989).
55. N. Ali and A. Salahuddin, *FEBS Lett., 246*:163 (1989).
56. F. M. Finn and K. Hofman, *Methods Enzymol., 184*:244 (1990).
57. J. Szecowka, G. Hallden, I. D. Goldfine, and J. A. Williams, *Regul. Pept., 24*:215 (1989).
58. M. W. Nickells and J. P. Atkinson, *J. Immunol., 144*:4262 (1990).
59. L. W. Burrus and B. B. Olwin, *J. Biol. Chem., 264*:18647 (1989).
60. A. Mereau, I. Pieri, C. Gamby, J. Courty, and D. Barritault, *Biochimie, 71*:865 (1989).
61. R. I. Feldman, J. M. Wu, J. C. Jenson, and E. Mann, *J. Biol. Chem., 265*:17364 (1990).
62. E. Hazum, *Methods Enzymol., 184*:285 (1990).
63. E. Hazum, *J. Chromatogr., 510*:233 (1990).
64. S. Chiba, K. Shibuya, K. Miyazono, A. Tojo, Y. Oka, K. Miyagawa, and F. Takaku, *J. Biol. Chem., 265*:19777 (1990).
65. J. E. Smart, P. C. Familleti, D. V. Weber, R. F. Keeney, and P. Bailon, *J. Invest. Dermatol., 94*:158S (1990).
66. J. P. Galizzi, B. Castle, O. Djossou, N. Harada, H. Cabrillat, S. A. Yahia, R. Barrett, M. Howard, and J. Banchereau, *J. Biol. Chem., 265*:439 (1990).
67. H. Ono and S. Tuboi, *J. Biochem. (Tokyo), 107*:840 (1990).
68. J. Mazurier, D. Legrand, W. L. Hu, J. Montreuil, and G. Spik, *Eur. J. Biochem., 179*:481 (1989).
69. A. G. Petrenko, V. A. Kovalenko, O. G. Shamotienko, I. N. Surkova, T. A. Tarasyuk, Yu. A. Ushkaryov, and E. V. Grishin, *EMBO J., 9*:2023 (1990).
70. K. Alpaugh, K. Indrapichate, J. A. Abel, R. Rimerman, and J. Wimalasena, *Biochem. Pharmacol., 40*:2093 (1990).
71. P. C. Roche and R. J. Ryan, *J. Biol. Chem., 264*:4636 (1989).
72. S. K. Moestrup and J. Gliemann, *J. Biol. Chem., 264*:15574 (1989).
73. J. Mazella, J. Chabry, N. Zsuger, and J. P. Vincent, *J. Biol. Chem., 264*:5559 (1989).
74. Z. H. Song, D. P. Barbas, P. S. Portoghese, and A. E. Takamori, *Prog. Clin. Biol. Res., 328*:69 (1990).
75. J. Simon, S. Benyhe, J. Hepp, E. Varga, K. Medzihradszky, A. Borsodi, and J. J. Wolleman, *J. Neurosci. Res., 25*:549 (1990).
76. S. Loukas, F. Panetsos, and C. Zioudrou, *Prog. Clin. Biol. Res., 328*:97 (1990).
77. M. Freemark and J. Comer, *J. Clin. Invest., 83*:883 (1989).

78. H. T. He, K. Johnson, K. Thermos, and T. Reisine, *Proc. Natl. Acad. Sci. USA, 86*:1480 (1989).
79. F. Reyl-Desmars, S. Le-Roux, C. Linard, F. Benkouka, and M. J. Lewin, *J. Biol. Chem., 264*:18789 (1989).
80. T. Akamizu, M. Saji, and L. D. Kohn, *Biochem. Biophys. Res. Commun., 170*:351 (1990).
81. L. C. Harrison and P. J. Leedman, *Clin. Biochem., 23*:43 (1990).
82. P. J. Leedman, J. D. Newman, and L. C. Harrison, *J. Clin. Endocrinol. Metab., 69*:134 (1989).
83. Z. Georgoussi, S. J. Taylor, S. B. Bocckino, and J. H. Exon, *Biochim. Biophys. Acta, 1055*:69 (1990).
84. A. Couvineau, T. Voisin, L. Guijarro, and M. Laburthe, *J. Biol. Chem., 265*:13386 (1990).

6

Purification of Nucleic Acid–Binding Proteins by Affinity Chromatography

Vincent Moncollin and Jean M. Egly

Institut de Chimie Biologique, Strasbourg, France

I. INTRODUCTION

In the past few years, progress in genetic engineering has provided a useful tool for studying the mechanisms by which specific genes are expressed in a temporal or tissue-specific manner. Molecular genetics approaches have been used to identify and characterize nucleic acids sequences involved in basic processes such as replication, transcription, splicing, or translation. In each case these nucleic acids sequences have been shown to exert their effects through interaction with proteins (for a review, see Ref. 1). To understand the underlying mechanisms of these reactions, the purification of the various components (repressor, transcription factors, spliceosome, etc.) has been undertaken from various sources (HeLa cells, calf thymus, *Drosophila* cells, etc.) (2–5). However, the efficiency of the genetic regulation implies a low abundance of these proteins in the cell, which renders their purification very difficult to achieve by conventional chromatography techniques. Large amounts of starting material are necessary to obtain microgram quantities of purified protein. By reducing the number of purification steps required and by increasing the yield, affinity chromatography represents a major breakthrough in the purification of those proteins that bind specific nucleic acid sequences. The feasibility of using site-specific DNA affinity chromatography for the purification of sequence specific DNA-binding proteins was demonstrated by Herrick (6), who showed that lac repressor bound more tightly to DNA containing the lac operator than to DNA lacking the operator when the DNA was coupled to either a cellulose or an agarose matrix. Elution of the protein with high salt buffer resulted in a highly purified preparation of the repressor. Following these initial observations, various procedures have been reported to carry out the affinity chromatography. Indeed, this technique is now widely used in molecular biology laboratories to study the mechanism of transcription, replication, splicing, or chromatine structure.

In this chapter we describe protocols for the purification of DNA-binding proteins, which can also be applied to the purification of any RNA-binding protein. RNA affinity chromatography has been used successfully for the purification of the RNA polymerase III transcription factor TFIIIA and the components of the spliceosome complex (7,8). We also discuss the problems specific to the DNA coupling procedure. Crucial parameters for the optimization of the chromatography and the successful purification of the protein will then be presented as well as detailed experimental protocols for DNA affinity chromatography using the cyanogen bromide coupling procedure. Finally, an example of affinity purification of a DNA-binding protein is presented to demonstrate the effectiveness of this method. Structural studies of protein–nucleic acid interaction have been reviewed recently (9) and will not be detailed here.

II. COUPLING OF NUCLEIC ACIDS TO THE MATRIX

The main points in designing a nucleic affinity support concern the choice of the support, the design of the ligand, and the coupling procedure.

A. Problems Specific to DNA Coupling

As stated previously, the affinity matrix should possess all the characteristics of an ideal chromatographic support (as described in Chapter 2, Section II). The supports should be porous while having a certain rigidity to support a high flow rate. In addition, the matrix should be as inert as possible to avoid nonspecific interactions while having chemical groups available for activation. Consequently, hydrophobic supports such as polystyrene or polyvinyl should be avoided. If the purification has to be scaled up, it will be convenient to select rigid supports such as composites which usually possess a silica core coated with hydrophilic polymers such as acrylic, methacrylate, polyvinyl, and dextran. Agarose beads are commonly used at the laboratory scale.

The coupling procedure itself is also important, especially if the ligand–protein interaction is weak. Indeed, coupling procedures using either cyanogen bromide (10) or bisepoxyrane (11) result in the introduction on the matrix of a partial positive charge and a hydrophobic spacer, respectively, which may increase nonspecific interactions. Moreover, due to the heterogeneity of the chemical groups belonging to the nucleic acid (such as amino, keto, or phospho), it is difficult to design an affinity support in which the ligand will be attached at only one point, and consequently the full nucleic acid sequence may not be available for interaction with the sequence-specific binding protein.

B. Choice and Preparation of the DNA Fragment

The recognition sequence of the protein should be precisely identified by methods such as DNAase I footprinting, methidiumpropyl EDTA–Fe(II) footprinting, or dimethyl sulfate methylation protection, whereas the affinity of the protein for the sequence, which will reflect the strength of the interaction for its binding site, can be estimated by gel shift assay or nitrocellulose filter binding assay. A brief description of the various assays is given below. Once the recognition sequence is determined, one should design the DNA ligand that will be coupled on the chromatographic support, taking into account the size of the protein to be purified. The DNA fragment should be long enough to avoid problems of steric hindrance between the protein and the matrix itself, and/or between two proteins if the binding sequence is multimerized. However, additional spacing sequences will increase the possibility of nonspecific binding of contaminating proteins.

The DNA fragment can be prepared by construction of a plasmid containing multiple tandem copies of the recognition sequence in head-to-tail direct orientation to stabilize the insert. This plasmid can be amplified and purified and the excised DNA fragment coupled to the resin. Alternatively, recent progress has allowed the chemical synthesis of single-stranded oligonucleotides long enough to produce (after hybridization with its complementary strand) any desirable ligand which could be multimerized in vitro (provided that restriction sites were previously designed at each end of the oligonucleotides) and in sufficient amounts (mg) to prepare a large-scale affinity resin. Since the coupling of the DNA to the CNBr-activated resin probably occurs by primary amino groups on unpaired bases, a four-base protruding 5′ end (like GATC) is usually included in the synthetic oligonucleotides. This also allows polymerization of the oligonucleotides. In all cases, polymerization of the specific DNA fragment increases the capacity of binding by reducing problems of steric hindrance. As a control, a resin that does not possess any recognition site can be used to identify proteins that bind nonspecifically to DNA–agarose; however, in most cases a mutated DNA fragment can be used to prepare a control affinity resin and to verify that the specific protein binds to the wild-type specific sequence but not to the mutated one.

C. Coupling Methods

Although cyanogen bromide activation of agarose remains the most popular method for synthesis of an affinity resin, other techniques have been developed and these will be discussed briefly. The cyanogen bromide activation technique is now well established (see Figure 1a). However, this coupling procedure involves the formation of cationic charges on the gel, which could be the target for nonspecific interactions (10). Furthermore, it should also be known that partial leakage of ligand may occur at pH < 5 or pH > 10 (12). Another disadvantage of cyanogen bromide is its high toxicity, although this problem can be avoided since CNBr-activated supports are now commercially available.

1. Avidin–Biotin System

The avidin–biotin system takes advantage of the very strong interaction between biotin and either streptavidin or avidin. In this approach a biotinylated DNA fragment is incubated with the protein fraction. The specific protein–DNA complexes formed are then purified from contaminants by chromatography on an avidin matrix (Figure 1b). The DNA fragment can be biotinylated by incorporation of biotinylated deoxynucleotides via an enzymatic reaction, or alternatively, the DNA fragment can be produced by polymerase chain reaction in which two primers, flanking the protein binding site, are chemically synthetized and coupled to biotin. After purification by reversed-phase high-performance liquid chromatography (HPLC), these biotinylated oligonucleotides are used to amplify the DNA fragment (13,15). The latter method can give microgram amounts of DNA fragment with the advantage that every DNA molecule includes a biotin residue.

The avidin–biotin system has some advantages. As the interaction between protein and DNA is made in solution, the reaction parameters can easily be optimized in preliminary tests (e.g., gel shift assays). In addition, since coupling to the matrix is made after the complex formation, the access or the protein to the DNA is not hindered by the matrix, and the efficiency of binding is high. Note that in this system the DNA is bound to the matrix by one of its ends, which is not necessarily the case in CNBr-activated matrices, where multiple attachment sites of the DNA to the matrix can reduce the potential binding sites for the protein. However, practically, this technique remains preferentially used for analytical purpose.

2. Teflon-Linked Oligonucleotides

A method that uses a Teflon fiber support has been developed to produce an affinity matrix of very high capacity (14). In this case the oligonucleotide synthesis takes place in the DNA synthesizer directly onto the Teflon matrix. A soluble synthetic oligonucleotide of complementary base sequence is then

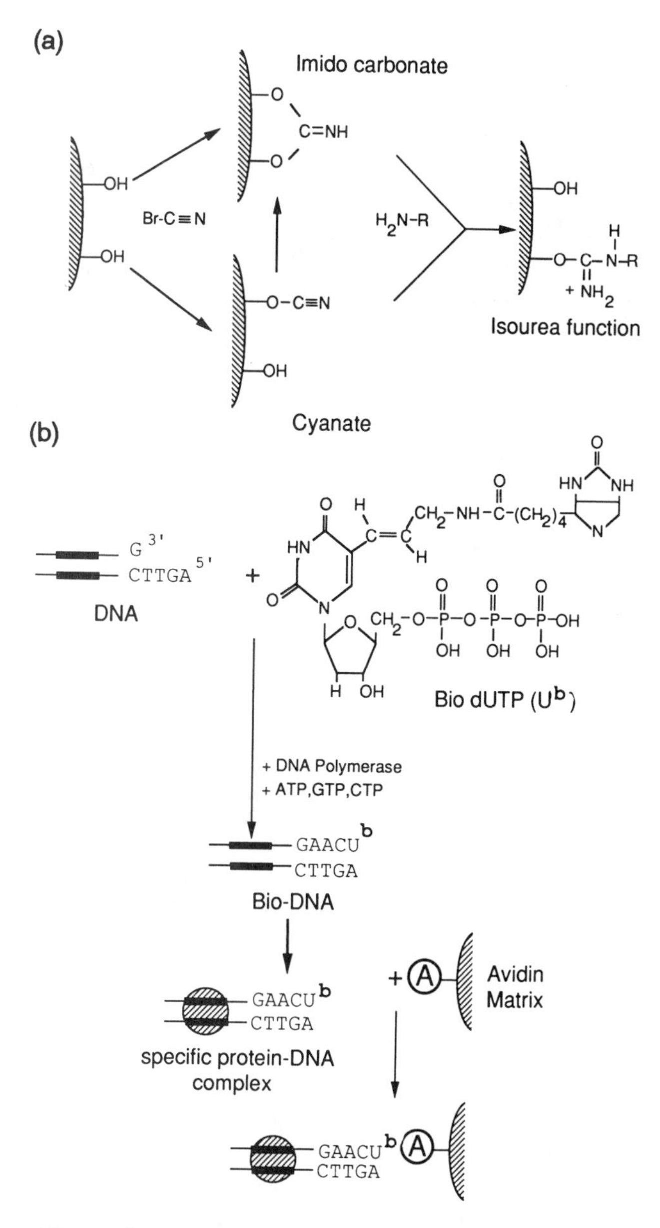

Figure 1 Preparation of DNA affinity support. (a) CNBr coupling: binding of nucleic acids is usually done by the free amino group of the base; however, this does not exclude that binding may also occur through phospho groups and/or sugar primary alcohol. (b) Avidin–biotin system: the byotinylated UTP (U^b) is incorporated in the DNA fragment by enzymatic reaction or by direct chemical synthesis (see the text). The nucleoprotein complex resulting from specific interaction between the biotinylated DNA fragment and the target protein is then specifically retained on an avidin matrix (A).

incubated with this matrix under hybridization conditions. Using this method, a matrix with 0.15 μmol (4 mg) of double-stranded synthetic oligonucleotides in a 70-μL packed volume can be prepared. In addition, this support is inert, presents good flow properties, and is easily handled since Teflon fibers do not shrink or swell. These properties made this technique very attractive for the synthesis of an affinity matrix. However, due to some steric hindrance, the hybridization reaction is not complete, and as a consequence some nonspecific interactions of proteins with the single-stranded DNA occur. It should also be mentioned that some supports having polylinker DNA possessing some restriction enzyme sites have been designed to allow ligation of various specific DNA fragments containing one of these restriction sites. As mentioned previously, in this case the ligation reaction is not total, and as a consequence unsubstituted polylinker remains; the advantage of this technique is that the ligand is attached by one end only with no point of nonspecific interactions.

3. Magnetic DNA Affinity

The magnetic DNA affinity technique combines the power of affinity chromatography with speed. The specific DNA fragment is coupled to Dynabeads, which can be selectively trapped by a magnet (13,15). The DNA fragment is labeled at one end with a biotinylated deoxyribonucleotide and bound to Dynabeads coated with streptavidin. The beads are then incubated with the protein fraction to form specific DNA complexes. A magnet is placed against the wall of the tube to separate the magnetic beads (with the specific protein attached) from contaminant proteins remaining in solution. After washing, the specific protein is eluted by resuspending the beads in a buffer of high ionic strength and the beads are again removed by magnetic separation. The entire procedure take a few minutes.

III. AFFINITY CHROMATOGRAPHY

A. Design of the Affinity System

In addition to the wild-type DNA affinity support, it is strongly recommended that three negative control columns be run in parallel: (a) a nonactivated support, (b) an activated support that has been then deactivated (if a CNBr-activated support is used, part of the support is then deactivated either with lysine or with some inert protein), and (c) a CNBr-activated support containing the mutated DNA or RNA [this nucleic acid mutant should have no (or very few) interactions with the target protein]. Typically, the first support will provide information concerning the ability of the protein to flow through the column; the second and third columns will tell us about nonspecific interactions resulting from the coupling procedure and/or to the mutated ligand, respectively. These three columns will also provide information concerning the washing and/or the elution of the nonspecific proteins.

B. Sample Preparation

Whatever the method chosen, optimization of the chromatography itself relies on several important points which have to be considered carefully. Although not always required, preliminary conventional chromatography steps, such as ion exchange or gel filtration, are usually beneficial for preservation of the affinity column. For example, the crude protein fraction often contains nuclease activity, which has to be removed prior to the affinity chromatography, thus allowing multiple uses of the column without significant loss of capacity. In addition, these preliminary purification steps may significantly reduce the amount of contaminating proteins that could otherwise bind nonspecifically to the DNA and interfere with DNA binding of the specific protein.

Formation of the protein–DNA complex is optimized by varying a number of parameters. Thus the nature of the salts, the ionic strength, the temperature, the pH, and the magnesium concentration have to be defined in each particular case. In addition, the presence of a small amount of detergent (e.g., 0.1% NP 40) often reduces the proportion of specific protein that is found in flow-through from the affinity resin, probably by reducing the binding of nonspecific DNA-binding proteins to specific sites. Furthermore, nonspecific DNA competitor should be included in the incubation step (especially when crude material will be applied) in order to trap these nonspecific DNA-binding proteins. In most cases, synthetic polydeoxynucleotides such as poly(dI-dC), poly(dG-dC), or poly(dA-dT) have been used successfully. However, various nonspecific DNA has to be tried since it has been reported that some of the specific DNA-binding proteins also interact strongly with particular nonspecific DNA. It is necessary to determine experimentally the appropriate amount and type of competitor DNA to be added by preliminary DNAase I footprinting or gel shift assays using the starting material. The amount of competitor DNA used in the affinity chromatography is usually estimated to one-fifth of the quantity that was obtained by adjustment of the small-scale DNA-binding assay to full scale (16). Removal of these nonspecific DNA-binding proteins can also be carried out by preliminary chromatography on a nonspecific DNA resin. This method should be considered when the interaction between the protein and the DNA is very strong and specific.

C. Adsorption and Elution

Once the sample to be applied is well characterized, one should decide which adsorption procedure will be used: batch incubation of the protein fraction with the affinity resin is often preferred to column application, since it reduces the time of equilibration of the load with the matrix. A low-speed centrifugation (to avoid agarose beads breaking) allows fast recovery of the beads. Alternatively, the batch of resin can be packed into a glass column. Direct adsorption of the sample is recommended when small volumes of chromatographic supports are used to avoid lost of resin. In this case low flow rate and/or recycling of the flow-through will favor protein–ligand interactions, especially when the kinetics of interaction are slow. Thus if the protein has a low affinity for the nucleic acid sequence, it can nevertheless be purified since the interaction may be sufficient to delay its progress down the column respective to that of the impurities.

Whatever the adsorption procedure, the resin should be washed thoroughly to remove all remaining protein nonspecifically bound to the agarose matrix. In most cases the specific DNA-bound protein is eluted by a buffer of increasing salt concentration. However, change in pH, temperature, or solvent can be used, provided that such elution buffers do not alter in an irreversible manner either the DNA or the biological activity of the protein to be purified. If the purity required is not achieved, a second or even third cycle of DNA affinity chromatography may be used to remove contaminants. We describe below the protocols related to each crucial step in the design of the DNA affinity chromatography, followed by an example demonstrating how this technique has been used to purify to homogeneity a particular DNA-binding protein.

IV. DNA-BINDING ASSAYS

The efficiency of the DNA affinity chromatography relies on various parameters that have to be optimized using a DNA-binding assay. Several methods are described briefly below; the detailed protocols can be found in the references cited.

A. Gel Shift Assay

The gel shift assay method is based on the electrophoretic separation of a protein–DNA complex from free DNA on a nondenaturing polyacrylamide gel (17–19). This assay is not only very easy to perform but also very sensitive; even a very small amount of complex can be detected since it is separated from the remaining DNA. The specificity of the complex should be checked by competition with wild-type or mutant DNA fragments.

B. DNAase I Footprinting

The DNAase I is an endonuclease that cuts DNA almost independently of the nucleotidique sequence (20). In the assay, a 5′-end-labeled DNA fragment is digested by DNAase I in the presence or absence of the protein fraction. The digestion products are analyzed by electrophoresis on a polyacrylamide–urea gel. The specific binding of a protein will protect the DNA sequence from DNAase I hydrolysis totally or partially. In addition, some hypersensitive sites are usually observed at the limits of the protected area. The amount of DNAase I and the time of digestion should be optimized so that a regular pattern of DNA sequence is obtained. This technique allows several protein-binding sites to be seen simultaneously, but it is less sensitive than the gel shift assay. Furthermore, DNAase I has some specificity and does not cut poly dA-dT sequences efficiently.

C. Dimethyl Sulfate Protection Assay

The principle of the dimethyl sulfate (DMS) protection assay (21) is very similar to that of the DNAase I footprint, but instead of a nuclease action, a chemical (DMS) is used to modify the G residues (methylation) as in the Maxam and Gilbert sequencing protocol. The treatment with piperidine results in cleavage at this position unless the binding protein prevented such methylation. This assay visualizes intimate contacts between the protein and DNA but requires that G residues are present in the binding region.

D. Methydium–propyl-EDTA Footprinting

Methydium–propyl-EDTA (Fe^{2+}) is a complex that breaks the phosphodiester bonds of DNA without any specificity. After gel electrophoresis, a very regular sequence ladder is obtained and protein-protected areas are revealed by the absence of the corresponding bands (22).

E. Nitrocellulose Filter Binding Assay

The double-stranded DNA alone binds very poorly to nitrocellulose while most protein–DNA complexes are retained. It is thus possible to separate free DNA from complexed DNA by filtration on nitrocellulose. The amount of labeled DNA retained is a function of the amount of protein–DNA complex formed (23). This assay is fast and easy to perform, but it does not make a distinction between single or multiple protein–DNA complexes. Furthermore, even strong complexes are not 100% retained. Woodbury and von Hippel (24) have detailed the method to estimate the binding parameters.

V. PROTOCOL

The protocol presented here is based primarily on the original publication of Kadonaga and Tjian (25) and on more recent reports on this method (12) as well as on our own experiment (26).

A. Preparation of the DNA Affinity Resin

1. 5′-Phosphorylation of the Oligonucleotides

To 130 μL of DNA [500 μg of each oligonucleotide in TE buffer (10 m*M* Tris–HCl (pH 8), 1 m*M* EDTA)] add 20 μL of 10X T4 polynucleotide kinase buffer (500 m*M* Tris–HCl (pH 7.6), 100 m*M* $MgCl_2$, 50 m*M* DTT, 1 m*M* spermidine, 1 m*M* EDTA) and incubate at 88°C for 2 min, 65°C for 10 min, 37°C for 10 min, and 25°C for 5 min. (Alternatively, following 2 min at 88°C, the water bath is switched off so that the temperature decreases progressively to 25°C.)

Divide the DNA solution into 2 × 75 μL and for each add 15 μL of 20 m*M* ATP (pH 7) containing 5 μCi of γ^{32}P-ATP and 10 μL of T4 polynucleotide kinase (100 units). Incubate at 37°C for 2 h. Add 50 μL of 10 *M* ammonium acetate and 100 μL of water. Heat at 65°C for 15 min to inactivate kinase. Add 750 μL of ethanol. Spin 15 min to pellet DNA. Discard supernatant and dissolve the pellet by vortexing in 225 μL of TE. Extract once with 250 μL of phenol, then with 250 μL of $CHCl_3$, and transfer the aqueous phase to a new tube. Repeat the precipitation with ethanol and 25 μL of 3 *M* NaOAc (sodium acetate). Wash with 800 μL of 70% ethanol. Dry the pellets.

2. Ligation of the Oligonucleotides

To each tube, add 65 μL of water and 10 μL of 10X linker-kinase buffer (660 m*M* Tris–HCl (pH 7.6), 100 m*M* $MgCl_2$, 150 m*M* DTT, 10 m*M* spermidine). Dissolve DNA by vortexing. Add 20 μL of 20 m*M* ATP (pH 7) and 5 μL of T4 DNA ligase (30 Weiss units). Incubate at 16°C overnight. Depending on the oligonucleotides used, the optimal temperature for ligation will vary from 4 to 30°C. Short oligonucleotides (<15 mers) tend to ligate better at lower temperatures. Monitor the ligation by agarose gel electrophoresis of 0.5 μL of aliquot. The average length of the ligated oligonucleotides should be at least 10 mers.

After ligation is complete, extract once with phenol and once with $CHCl_3$ and transfer the aqueous phase to a new tube. Precipitate with 33 μL of 10 *M* ammonium acetate and 133 μL of isopropanol at –20°C for 20 min. Spin 15 min to pellet DNA. Dissolve the pellet in 225 μL of TE and precipitate with 25 μL of 3 *M* NaOAc and 750 μL of ethanol. Wash DNA with 70% ethanol. Dry the pellets. Dissolve the DNA in 50 μL of distilled water. Store at –20°C.

3. Coupling of the DNA to Sepharose

Wash 10 mL (settled volume) of Sepharose CL-2B (Pharmacia, Uppsala, Sweden) with about 500 mL of bidistilled water in a splintered glass funnel. Transfer the Sepharose to a 25-mL graduated cylinder. Add water to 20 mL of final volume. Transfer this slurry to a 150-mL glass beaker already equilibrated in a water bath to 15°C, over a magnetic stirrer in a fume hood. Dissolve 1.1 g of cyanogen bromide (*Danger*: Toxic) in 2 mL of *N,N*-dimethylformamide. Add this solution dropwise over 1 min to the stirring slurry of Sepharose in the fume hood. Immediately add slowly dropwise 1.8 mL of 5 *M* NaOH to the stirring mixture (at 15°C) over 10 min. The pH of the reaction, which generates HBr as a by-product, should not exceed pH 10. Immediately add 100 mL of ice-cold water and pour the mixture into a sintered glass funnel. It is very important that the activated Sepharose is not suction-filtered into a dry cake. Wash the resin with 4 × 100 mL of ice-cold water and 2 × 100 mL of ice-cold 10 m*M* potassium phosphate, pH 8. Immediately transfer the resin to a 15-mL polypropylene screw-cap tube and add about 4 mL of 10 m*M* potassium phosphate pH 8 to give a thick slurry. Immediately add the DNA (100 μL). Incubate on a rotating wheel overnight at room temperature.

Transfer the resin to a sintered glass funnel and wash with 200 mL of water and 100 mL of 1 *M* ethanolamine–HCl (pH 8). Check for the presence of radioactivity in the first few milliliters of the filtrate. Usually, all detectable radioactivity is present only in the resin. Transfer the resin to a 15-mL polypropylene screw-cap tube and add 1 *M* ethanolamine–HCl (pH 8) to give a final volume of 14 mL. Incubate the tube on a rotating wheel at room temperature for 4 to 6 h (inactivation of the unreacted cyanogen bromide–activated Sepharose). Wash the resin successively with 100 mL of 10 m*M* potassium phosphate (pH 8), 100 mL of 1 *M* potassium phosphate (pH 8), 100 mL of 1 *M* KCl, 100 mL of water, and 100 mL of column storage buffer. Store the resin at 4°C (do not freeze). The resin is stable for at least 1 year at 4°C.

B. Affinity Chromatography

The chromatography is usually performed at 4°C. The DNA affinity resin (1 mL) is equilibrated in buffer B (50 m*M* Tris–HCl (pH 7.9), 17.4% glycerol, 5 m*M* $MgCl_2$, 0.1% NP-40, 1 m*M* DTT) containing 0.05 *M* KCl. The protein fraction (in buffer B + 0.05 *M* KCl) is incubated 15 min at 4°C with the nonspecific competitor DNA. Add the affinity resin and incubate 15 min further with occasional mixing. Pour the resin in a Bio-Rad Econo-Column and wash with 10 column volumes of buffer B + 0.005 *M* KCl. Elute the protein by successive addition of 1-column-volume portions of buffer B containing increment of 0.1 *M* KCl from 0.1 *M* to 0.9 *M* KCl, then with 4 column volumes of buffer B + 1 *M* KCl. Collect 1-mL fractions and dialyze against buffer D (50 m*M* Tris–HCl, pH 7.9, 50 m*M* KCl, 25% glycerol, 0.1 m*M* EDTA, 0.5 m*M* DTT). Freeze the protein fractions in liquid nitrogen and store at –80°C. Wash the affinity resin with 10 column volumes of regeneration buffer [10 m*M* Tris–HCl (pH 7.8), 1 m*M* EDTA, 2.5 *M* NaCl, 1% (v/v) Nonidet P-40] and 10 column volumes of storage buffer [10 m*M* Tris–HCl (pH 7.8), 1 m*M* EDTA, 0.3 *M* NaCl, 0.04% (w/v) NaN_3]. Store the resin at 4°C.

VI. PURIFICATION OF THE UEFy BY AFFINITY

In addition to the TATA-box motif located 25 to 30 base pairs upstream from the initiation site, the adenovirus 2 major late promoter (Ad-2 MLP) has an upstream element (27) also found in several mammalian genes (28,29) and which is required for optimal in vitro and in vivo transcription of this promoter. These two sequences are the target of the BTF1/TFIID (4,30) and UEF (30–32) factors, respectively, two DNA-binding proteins that have been identified in HeLa cells (see Figure 2a). As an analog of the BTF1/TFIID has been identified, purified, and cloned from yeast (33–35), this prompted us to look for the presence of an analog of the human UEF in the yeast *Saccharomyces cerevisiae*. This led to the identification and purification of UEFy (36), a yeast analog of the human UEF. We thus present as an example to illustrate the DNA affinity procedure, the method used to purify this UEFy from a crude yeast extract.

A. Methods

The UEFy was purified as follows. An extract of the yeast *S. cerevisiae* S-100 (5 mL; 37 mg/mL) prepared as described previously (37) was dialyzed 12 h against buffer E (50 m*M* Tris–HCl (pH 7.9), 50 m*M* KCl, 17.4% glycerol, 2 m*M* $MgCl_2$, 0.5 m*M* DTT, 0.1 m*M* PMSF) and incubated 15 min at 4°C with 100 μg poly(dI-dC)(dI-dC)/mL (Pharmacia, Uppsala, Sweden) and 0.1% NP-40 and then 15 min at 4°C with 1 mL of the sequence-specific DNA affinity resin. The resin was made as follows. The two 35-mer synthetic oligonucleotide strands (nucleotides –41 to –71 with respect to the Ad-2 MLP start site with a GATC tetramer added at the 5′ terminus) were hybridized and ligated with T4 DNA ligase. The polymers were fixed onto a Sepharose CL-4B resin (Pharmacia, Uppsala, Sweden) preactivated by CNBr.

The resin was then packed in a Pasteur pipette and washed with 10 column volumes of buffer F [50 m*M* Tris–HCl (pH 7.9), 50 m*M* KCl, 17.4% glycerol, 2 m*M* KCl] and 5 column volumes of buffer F containing 1 *M* KCl. After dialysis against buffer F, the 1 *M* KCl fraction was incubated 15 min at 4°C with 50 μg/mL of poly(dI-dC)(dI-dC) and reapplied on the DNA affinity column. The 1 *M* KCl eluate from this second affinity column (~2 μg/mL) was dialized against buffer D [50 mM Tris–HCl (pH 7.9), 50 m*M* KCl, 25% glycerol, 0.1 m*M* EDTA, 0.5 m*M* DTT] and stored at –80°C. Purified fractions were analyzed on 9% SDS-polyacrylamide gels.

B. Results

To purify the UEFy, a yeast S-100 extract was applied to the DNA affinity matrix as described in Section VI.A. The 1 *M* KCl eluate of the first DNA affinity

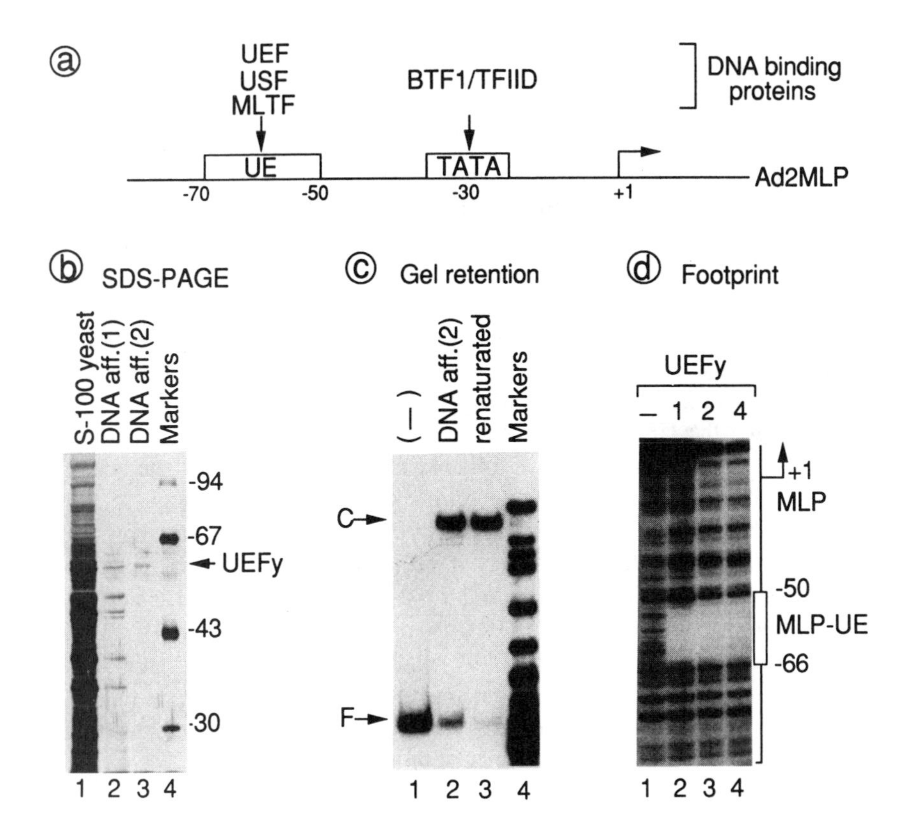

Figure 2 (a) Schematic organization of the Ad-2 MLP. The binding sites of the UEF (UE) and BTF1/TFIID (TATA) as well as the initiation site (+1) are represented. (b) SDS-polyacrylamide gel analysis of each stage of purification of the UEFy. Lane 1:0.5 μL of yeast S-100 extract. Lanes 2 and 3: 100 μL of the 1 *M* KCl eluate of the first and second DNA affinity columns, respectively. Lane 4: molecular weight markers. The arrow indicates the 60-kDa polypeptide (UEFy). (c) Nucleoprotein complexes formation between the UEFy and the ^{32}P-labeled SacII-BamHI fragment (–245 to +33) of pM677. Binding reaction included 0.1 ng of the labeled template and either 0.05 μL of the purified UEFy (lane 2) or 4 μL of the renatured 60-kDa polypeptide eluted from the SDS gel (lane 3) or no protein (lane 1). Lane 4: size markers. C and F indicate the nucleoprotein complexes and the free DNA, respectively. (d) DNAase I footprint on the coding strand of the Ad-2 MLP. Increasing amounts of UEFy, as indicated (in ng), were incubated with ~1 ng of the ^{32}P-labeled DNA fragment. The binding site as well as the initiation site are indicated.

chromatography was dialyzed and reapplied on the DNA affinity column. As shown by SDS-PAGE analysis (Figure 2b) the 1 *M* KCl eluate of the second affinity column contains a single polypeptide of 60 kDa (lane 3). This fraction was able to form a specific complex with the labeled DNA fragment as detected by gel retention (Figure 2c, lane 2). Furthermore, after elution from the SDS-polyacrylamide gel and renaturation, (26), only the fraction containing the 60-kDa polypeptide was able to form the specific nucleoprotein complex (lane 3). In addition, the 1 *M* KCl eluate of the second DNA affinity chromatography protects from digestion by DNAse I, the –50 and –66 region of the MLP sequence (Figure 2d, lanes 2–4), which correspond to the previously identified binding site of the UEF (31). The specificity of binding has been demonstrated both in gel retention and in DNAse I footprint by competition with cold oligonucleotide encompassing the UE (–41 to –71 of the MLP sequence). Thus preincubation of UEFy with the wild-type UE oligonucleotide but not with the mutated oligonucleotide prevents the formation of nucleoprotein complex in gel retention as well as the presence of a protected area in DNAase I footprint (not shown). All these results demonstrate the efficiency of the DNA affinity chromatography since the UEFy has been purified from a crude yeast extract (12,000-fold) in a two-step procedure with a very high yield (53%). This procedure is fast (a few hours) and gives purified protein with both DNA binding and transcription stimulatory activities (36). Thus the affinity chromatography has been, and should remain, a powerful technique for the purification of multiple DNA or RNA binding proteins which are involved in various fundamental processes in the cell.

ACKNOWLEDGMENTS

We are very grateful to P. Chambon for very helpful discussions. We thank N. Burton for critical reading of the manuscript, P. Jalinot for advice on the avidin–biotin system, S. Golla for excellent technical assistance, and C. Werlé and E. Rauscher for preparing the manuscript.

REFERENCES

1. P. J. Mitchell and R. Tjian, *Science, 245*:371 (1989).
2. T. Matsui, J. Segall, P. A. Weil, and R. G. Roeder, *J. Biol. Chem., 255*:11992 (1980).
3. M. Samuels, A. Fire, and P. A. Sharp, *J. Biol. Chem., 257*:14419 (1982).
4. B. L. Davidson, J. M. Egly, E. R. Mulvihill, and P. Chambon, *Nature, 301*:680 (1983).
5. W. S. Dynan and R. Tjian, *Cell, 32*:669 (1983).
6. G. Herrick, *Nucleic Acids Res., 8*:3721 (1980).
7. B. R. Braun, D. L. Riggs, G. A. Kassavetis, and E. P. Geiduschek, *Proc. Natl. Acad. Sci. USA, 86*:2530 (1989).
8. P. J. Grabowski and P. A. Sharp, *Science, 233*:1294 (1986).

9. T. A. Steitz, *Q. Rev. Biophys., 23*:205 (1990).
10. M. Wilchek and T. Miron, *Methods Enzymol., 34*:74 (1974).
11. L. Sundberg and J. Porath, *J. Chromatogr., 90*:87 (1974).
12. I. Parikh, S. March, and P. Cuastrecasas, *Methods Enzymol., 34*:96 (1974).
13. O. S. Gabrielsen, E. Hornes, L. Korsnes, A. Ruet, and T. B. Oyen, *Nucleic Acids Res., 17*:6253 (1989).
14. C. H. Duncan and S. L. Cavalier, *Anal. Biochem., 169*:104 (1988).
15. M. Uhlen, *Nature, 340*:733 (1989).
16. J. T. Kadonaga, *DNA Protein Eng. Tech., 2*:73 (1990).
17. M. G. Fried and D. M. Crothers, *Nucleic Acids Res., 9*:6505 (1981).
18. M. M. Garner and A. Revzin, *Nucleic Acids Res., 9*:3047 (1981).
19. A. Varshavsky, *Methods Enzymol., 151*:551 (1987).
20. D. Galas and A. Schmitz, *Nucleic Acids Res., 5*:3157 (1978).
21. U. Siebenlist, R. T. Simpson, and W. Gilbert, *Cell, 20*:269 (1980).
22. M. W. Van Dyke, R. P., Hertzberg, and P. B. Dervan, *Proc. Natl. Acad. Sci. USA, 79*:5470 (1982).
23. M. Yarus and P. Berg, *Anal. Biochem., 28*:479 (1967).
24. C. P. Woodbury, Jr., and P. H. von Hippel, *Biochemistry, 22*:4730 (1983).
25. J. T. Kadonaga and R. Tjian, *Proc. Natl. Acad. Sci. USA, 83*:5889 (1986).
26. V. Moncollin, M. Gerard, and J. M. Egly, *J. Chromatogr., 510*:243 (1990).
27. R. Hen, P. Sassone Corsi, P. Corden, M. P. Gaub, and P. Chambon, *Proc. Natl. Acad. Sci. USA, 79*:7132 (1982).
28. R. W. Carthew, L. A. Chodosh, and P. A. Sharp, *Genes Dev., 1*:973 (1987).
29. L. A. Chodosh, R. W. Carthew, J. G. Morgan, G. R. Crabtree, and P. A. Sharp, *Science, 238*:684 (1987).
30. M. Sawadogo and R. G. Roeder, *Cell, 43*:165 (1985).
31. N. G. Miyamoto, V. Moncollin, J. M. Egly, and P. Chambon, *EMBO J., 4*:3563 (1985).
32. R. W. Carthew, L. A. Chodosh, and P. A. Sharp, *Cell, 43*:439 (1985).
33. S. Hahn, S. Buratowski, P. A. Sharp, and L. Guarente, *Cell, 58*:1173 (1989).
34. B. Cavallini, I. Faus, H. Matther, J. M. Chipoulet, B. Winsor, J. M. Egly, and P. Chambon, *Proc. Natl. Acad. Sci. USA, 86*:9803 (1989).
35. P. M. Eisenmann, C. Dollard, and F. Winston, *Cell, 58*:1183 (1989).
36. V. Moncollin, R. Stalder, J. M. Verdier, A. Sentenac, and J. M. Egly, *Nucleic Acids Res., 18*:4817 (1990).
37. S. Dézeélée, F. Wyers, A. Sentenac, and P. Fromageot, *Eur. J. Biochem., 65*:543 (1976).

7

Nucleic Acid and Its Derivatives

Herbert Schott

Universität Tübingen, Tübingen, Germany

I. IMMOBILIZATION OF NUCLEIC ACID AND ITS DERIVATIVES

A. Mononucleotides and Nucleosides

The immobilization of mononucleotides and nucleosides (1,2) summarized in Table 1 can be classified into (a) phosphate-linked, (b) ribosyl-linked, and (c) nucleobase-linked. Figure 1, for example, demonstrates the possible positions on adenosine-5′-phosphate that can be functionalized for the immobilization to a polymer support.

B. Oligonucleotides

Three different strategies have been developed to immobilize oligonucleotides: covalent linkage of the oligonucleotides by means of their terminals using (a) one-step or (b) multistep reactions and (c) immobilization by a secondary valence.

1. *Covalent Binding of the Oligonucleotide Terminals*

a. One-Step Reaction

Following oligomerization of mononucleotides, the terminal phosphate groups are reacted with the free hydroxyl groups of cellulose (3). The final product size should thus consist of cellulose containing a mixture (in size) of the bound homopolymer chains connected by phosphodiester at the 5′ or 3′ terminals of the

Table 1 Immobilization of Mononucleotides and Nucleosides Through Phosphate, Ribosyl, and Nucleobase Linkage

Ligand	Spacer	Support	Method (linkage[a])	Refs.
Guanosine-5′-diphosphate, adenosine-5′-diphosphate, adenosine-5′-triphosphate, uridine-5′-diphosphate	6-Amino-1-hexanol phosphate	CNBr–agarose	Activatiaon by CNBr (a)	4
Uridine-5′-phosphate	1,6 Diaminohexane	CNBr–Sepharose	Activation by CNBr (a)	5
Ribonucleoside or ribonucleotide phosphate	Different carbonic acid hydrazide	Hydrazide–Sepharose/ Agarose	1. Periodate oxidation 2. Hydrazone formation (b) (see Scheme 1)	4, 6, 7
Adenosine-5′-phosphate	—	6-Aminohexyl-Sepharose, 2-Aminoethyl-cellulose	1. Periodate oxidation 2. Schiff bases (b)	8
Ribonucleoside or ribonucleoside-phosphate	—	Aminooxybutyl-cellulose	1. Periodate oxidation 2. Schiff bases (b)	9
Inosine-5′-phosphate	2′,3′-*O*-[1-(6-aminohexyl) levulinic acid amide]acetal	CNBr–agarose	Activation by CNBr (b)	10
Guanosine, uridine, cytosine, adenosine, thymidine		Polymethyacrylate	1. Esterification 2. Copolymerisation (b)	15–17
Nucleotides or nucleosides containing primary amino groups	Different alkyl chains	CNBr–Sepharose	Activation by CNBr (c)	1, 2
		Amino derivative of polysaccharide	1. Glutaraldehyde coupling 2. Reductioan with KBH_4 (c)	11
Inosine-5′-phosphate, guanosine-5′-triphosphate, guanosine-5′-diphosphate	8-(6-Aminohexyl) residue	CNBr–agarose	Activation by CNBr (c)	10, 13
Adenosine, cytosine, guanosine, uridine, thymidine, inosine	—	Cellulose	Coupling by epichlorohydrin (c)	14
Poly(9-vinyladenine)	—	Macroporous silica gel		18
Nucleobase derivatives		Polyethyleneimine-coated silica gel		12

[a](a) Phosphate, (b) ribosyl, (c) nucleobase linkage.

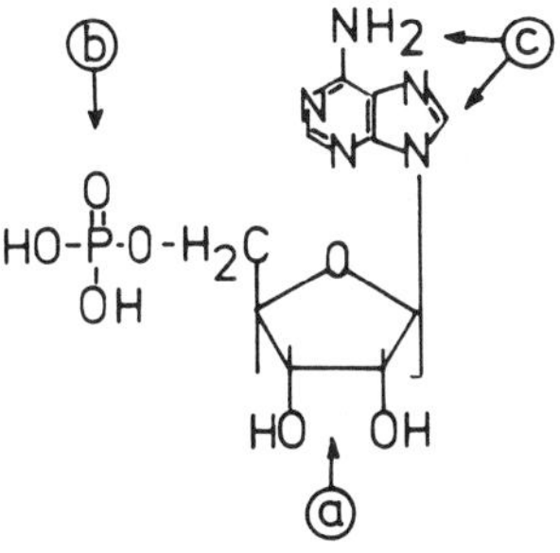

Figure 1 Positions of AMP that can be used for the immobilization: (a) ribosyl linked; (b) phosphate linked; (c) N^6-linked or C-8 linked. (From Ref. 2.)

chains, depending on whether the 5′ or the 3′ nucleotides are used. The oligonucleotides that may be incorporated in this way are limited in base sequence and chain length to those that can be prepared chemically in anhydrous solution. In the case of polycondensation of deoxycytidylic, deoxyadenylic, and deoxyguanylic acid it is necessary to use protecting groups for the nucleobases. The protecting groups are removed after the linkage of the oligonucleotides to the cellulose.

b. Multistep Reaction

A multistep procedure has been reported (3) which is applicable to oligonucleotides of both synthetic and natural origin and has the advantage that polymers may also be incorporated covalently onto cellulose paper strips as well as cellulose powder. In this system the terminal phosphate group or polyphosphate group of a defined oligonucleotide is activated in aqueous solution with a water-soluble reagent. A disadvantage of this method is the occurrence of side reactions at nucleobases and with internucleotide phosphodiesters.

As an alternative to the immobilization of oligo(dT) via their terminal phosphate groups, the preparation of *p*-aminophenyl oligo(dT)–Sepharose has been reported (20). The attachment of 5′-NH_2-terminated oligo(dT) by coupling with CNBr-activated cellulose occurs in high yield and exclusively via the 5′ terminal (21). However, a problem is the leakage of nucleotides from the solid support in basic medium an elevated temperatures, a phenomenon inherent to CNBr activation. A mixed anhydride of oligonucleotides and mesitoic acid was also applied to achieve immobilization of oligonucleotides via their phosphomonoester groups on aminohexyl-Sepharose (22).

The immobilization of three different functionalized 5′-NH_2-terminated oligonucleotides (Figure 2) on 2,4,6-trichloro-*s*-triazine– or 2-amino-4,6-dichloro-*s*-triazine–activated cellulose has been reported (23). The immobilized

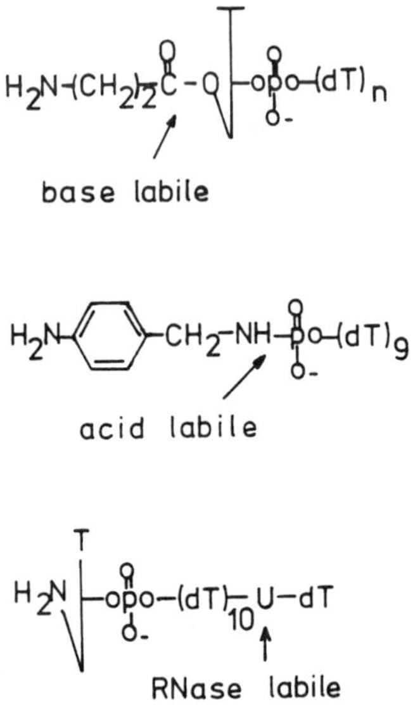

Figure 2 Functionalized oligonucleotides suitable for reversible attachment to activated cellulose. (Adapted from Ref. 23.)

oligonucleotides could be released quantitatively from the solid support by base or acid treatment or RNase digestion.

As an alternative to the polysaccharide matrix, cross-linked polyvinyl alcohol (PVAL) has been used as a support for the attachment of oligo(dG). Base-protected oligo(dG) is synthesized and oligomers with four and more monomeric units are bound to the free hydroxyl groups of an insoluble PVAL-gel via the phosphodiester linkage with their 5′-phosphate groups. The oligo(dG) gel, obtained after cleavage of the protecting groups, is suitable for preparative chromatography (24). Multimerized synthetic oligodeoxyribonucleotides covalently attached to agarose support can be used for the chromatography of sequence-specific binding proteins (27).

2. *Secondary Valence Binding of Polymer-Bound Oligonucleotides*

To minimize the side reactions that attend direct covalent immobilization, defined oligonucleotides are covalently bounds to the free hydroxyl groups of a soluble PVAL by means of their 5′-phosphate terminals. The polymer-bound oligonucleotide thus obtained can be purified and is subsequently irreversibly adsorbed as polyanion on the anion exchanger DEAE–cellulose using a batch procedure. The method allows the preparation of DEAE–cellulose loaded with oligonucleotides of known constitution on any scale required and avoids undesired side reactions during the immobilization (25,26). The oligonucleotide–DEAE–cellulose possesses both high capacity and good flow rate, under low pressure, when used as column packing. The schematical structure of PVAL-p(dC)$_n$–DEAE–cellulose is given as an example for the other oligonucleotide–DEAE–cellulose (see Figure 3).

5′-Tritylated oligonucleotides binding hydrophobically to low-substituted trityl cellulose–Sepharose retain their base-pairing specificities. The salt, dielectric constant, and temperature dependence of these noncovalently anchored oligonucleotides permits the isolation of a variety of RNAs. Medium trityl-Sepharose has a high binding specificity, equivalent to oligo(dT)–cellulose (28).

C. Polynucleotides and Nucleic Acids

Since in affinity chromatography interactions between the components should mimic the conditions of solution, careful consideration has to be given to the nature of the insoluble support, the ligand, and the method of immobilization, in

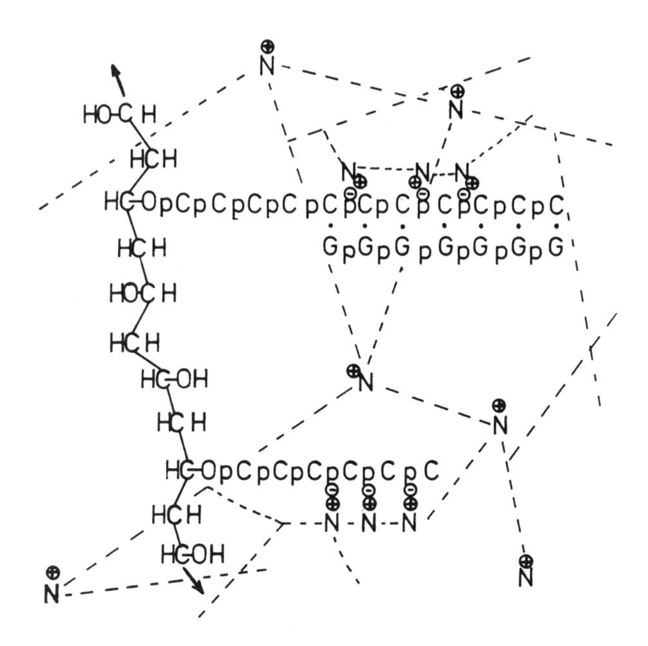

Figure 3 Schematical structure of the PVAL-p(dC)$_n$–DEAE–cellulose that adsorbed oligo(dG) by base pairing. (From Ref. 108.)

order to preserve the native structure of the immobilized polynucleotides. Several strategies for the immobilization of polynucleotides have been developed (3,30,31,44).

1. Immobilization via Physical and Mechanical Forces or Irradiation

a. Adsorption onto Cellulose

The attachment of DNA to cellulose by drying and subsequently lyophilization of a DNA–cellulose paste leads to DNA–cellulose (3). The DNA is bound reversibly and is slowly desorbed. No DNA is bound to the cellulose without prior drying. Single-stranded (ss) DNA–cellulose must be made by a modified procedure. Complexes between denatured DNA and cellulose are quite stable, whereas RNA and sonicated native DNA complexes with cellulose break down more rapidly. Drying tRNA with cellulose does not produce any detectable complex. Due to the mild procedure, damage to the ligand is kept to a minimum. Since DNA–cellulose prepared by adsorption exhibits only limited stability to desorbing conditions such as high temperatures, formamide, and low ionic strengths of the buffers used, it cannot be recommended for hybridization experiments.

b. Ultraviolet Irradiation

Polynucleotides can be immobilized to insoluble supports by irradiating with ultraviolet (UV) light. The UV irradiation is presumably a consequence of the production of intermolecular cross-linkages via the nucleobases of the polynucleotides, and the supports may serve only to immobilize the cross-linked polynucleotides. Both poly(U) and poly(C) have been linked in the presence of UV light to fiberglass filters (3). Attachment of poly(A), poly(U), and poly(C) viral RNA (35) to cellulose by UV irradiation has also been reported. Covalent cross-linking of DNA upon UV irradiation leads to immobilization of DNA onto nitrocellulose or nylon membranes (36).

c. Entrapment Within the Insoluble Support

The entrapment of phage fd DNA and denatured calf thymus DNA in 4% agar during gel formation has been reported (3). The entrapment procedure also allows immobilization of circular double-stranded (ds) DNA (39). Poly(A), poly(G), and poly(I) associate reversibly with agarose gels at high NaCl molarities over the pH range 6 to 10 at 20 to 40°C. Poly(C) and poly(U) could not be immobilized in agarose gels under these conditions. However, poly(C) could be immobilized in agarose without precipitation between pH 3.2 and 4.0 (40). Acrylamide gel beads are used as a matrix to immobilize poly(U) or DNA (3). The gel is made by dissolving DNA in the acrylamide solution before cross-linking polymerization. The resulted gel can be used in the column. DsDNA is irreversible trapped during crossed-field gel electrophoresis (41).

d. Reversible Complexation of Ribonucleotides with Immobilized Dihydroxyboryl Residues

Terminal *cis*-diol groups in ribopolynucleotides are capable of forming specific cyclic complexes with supports containing covalently bound dihydroxyboryl groups (Scheme 4) (76–81). At pH 6, the complex breaks down, allowing recovery of the immobilized ribonucleotides from the support. Dihydroxylboryl-substituted methacrylic acid polymer (borate gel) (74) is suitable for chromatography under pressure, which is advantageous for preparative separations.

2. *Immobilization Using Chemical Activation*

Table 2 summarized different methods for the chemical linkage of polynucleotides and nucleic acids to polymer support. Selected coupling reactions are illustrated in Schemes 2 and 3. Methods commonly used for covalent immobilization of ssDNA have been applied to several solid supports (Sephadex G-25 and Cellex 410) as well as to a number of macroporous materials (Sepharose C1-6B and C1-2B; Sephacryl S-500 and S-1000). Coupling efficiencies and stability of covalently bound DNA were compared for both classes of materials. The immobilized DNAs are lost from the supports in a biphasic manner, with about 10 to 20% loss per day during the first 2 to 3 days at 45°C, followed by only about 1% loss per day thereafter. The accessibility of immobilized DNA depend more crucially on the method of immobilization than on the type of support used for fixation (82,83).

II. CHROMATOGRAPHY OF DNA, OLIGONUCLEOTIDES, MONONUCLEOTIDES, AND NUCLEOSIDES

A. Various Possibilities of DNA Isolation

Cellulose-bound ssDNA complementary to ovalbumin mRNA was used to isolate ssDNA of the ovalbumin gene. Purification resulted in a DNA fraction that was enriched 2300-fold in the ovalbumin sequence (183). Ss nucleic acid probes are immobilized on solid supports. Optionally, the ss nucleic acid may contain several sequences capable of hybridizing with the target nucleic acid. This method increases the sensitivity of detection of the target (84). A method for the rapid isolation of mtDNA from the yeast *Saccharomyces cerevisiae* has been described. The mtDNA is extracted directly from the cell lysate by poly-L-lysine/kieselguhr exchange chromatography (86). RNA complementary to kinetoplasts of *Crithidia luciliae* was labeled with fluorescein and hybridized with Sephadex beads to which kinetoplast DNA or heterologous DNA had been covalently bound as well as to *C. luciliae* preparations. The fluorescein labeled RNA was found to hybridize specifically with homologous DNA both on the beads and in the cells. The sensitivity of the hybrid detection could be increased by applying an indirect

Table 2 Immobilization of Polynucleotides and Nucleic Acids Using Chemical Activation Methods

Ligand	Support	Method	Coupled ligand (μg) per mL support	Refs.
ssDNA, RNA, polyribonucleotides	CNBr-Sepharose	Activation by CNBr	?	44, 49
DNA, RNA	ECD–agarose	Activation by CNBr	20–100	50
ssDNA derivatives	Aminopentane–Sepharose	Azo coupling	75	52
DNA, RNA, poly(I)	Diazobenzyloxymethyl (DBM) paper	Azo coupling	25 (per cm^2)	53–56
Nucleic acids	Arylamine-substituted papers	Azo coupling (see Scheme 2)	?	57
DNA derivatives	Enzacryl AH gel	Azo coupling	50	58
DNA	Sephadex G-200	Condensation by carbodiimide	$6–38 \times 10^3$	152
DNA fragments	Silica carriers	Condensation by carbodiimide	3×10^3	60
$(dT)_{18}$	Macroporous silica	Condensation by N-hydroxysuccinimidyl ester		43
RNA	Phosphocellulose filters	Condensation by 1,1′-carbonyldiimidazole (see Scheme 3)	?	61
DNA, RNA	Cellulose	Activation by cyanuric chloride	20×10^3	62, 63
DNA, RNA	Cellulose, Sepharose	Activation by bisoxirane	3	65, 66
Oligodeoxyribonucleotides having 5′-terminal 2-aminoethylamino groups	Glass beads			70
Polyribonucleotides having 3′-aldehydo groups	Aminopropyl-containing support, hydrazine-containing support	Schiff bases, hydrazone formation	5–40 ?	3, 71–73
Oligodeoxyribonucleotides up to 70 units	Cellulose, Sepharose, and glass derivatives	Condensation		46
Nucleic acid probes containing cytidine residues outside the specific binding region	Amino group–containing supports	Bisulfite-mediated transamination		48
Oligo (dT)	Glyceryl porous glass	Condensation		64
Oligonucleotides synthesized in situ	Glass beads	Condensation		59

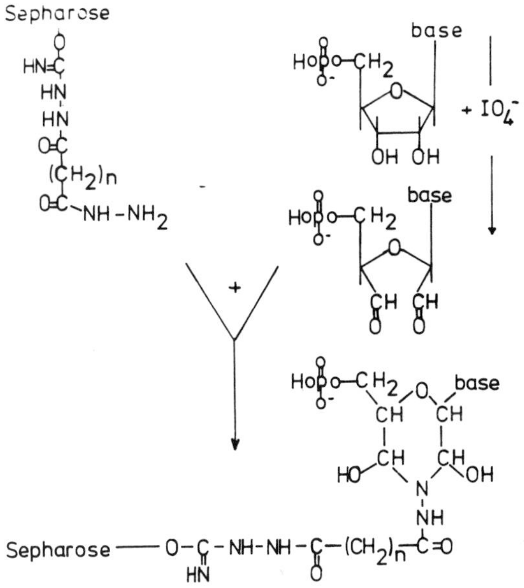

Sepharose
O
HN=C
HN
HN
O=C
$(CH_2)_n$
O=C–NH–NH$_2$
HO–P(O)(O$^-$)–O–CH$_2$
base
OH OH
+ IO_4^-
HO–P(O)(O$^-$)–O–CH$_2$
base
CH CH
O O
+
HO–P(O)(O$^-$)–O–CH$_2$
O
base
CH CH
CH CH
HO OH
N
NH
Sepharose —— O–C–NH–NH–C–$(CH_2)_n$–C=O
HN O

Scheme 1

$CH_2CHCH_2O(CH_2)_4OCH_2CHCH_2$
O O
or
CH_2CHCH_2Br
O
+ cellulose filters
cellulose
O
$CH_2CHCH_2O(CH_2)_4O$
HO
CH_2
$CHCH_2$
O
or
cellulose
O
CH_2CHCH_2
O
+ HS NH$_2$
cellulose — O
$CH_2CHCH_2O(CH_2)_4$
HO
cellulose
or OCH_2CHCH_2S
OH
NH$_2$
"cellulose-R-S– –NH$_2$"
1. HNO_2
2. nucleic acid
cellulose-R-S
N=N–nucleic acid

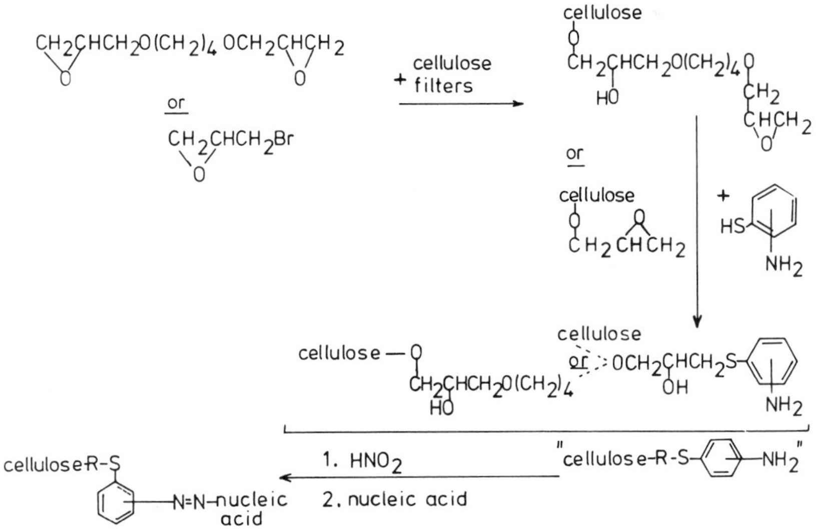

Scheme 2

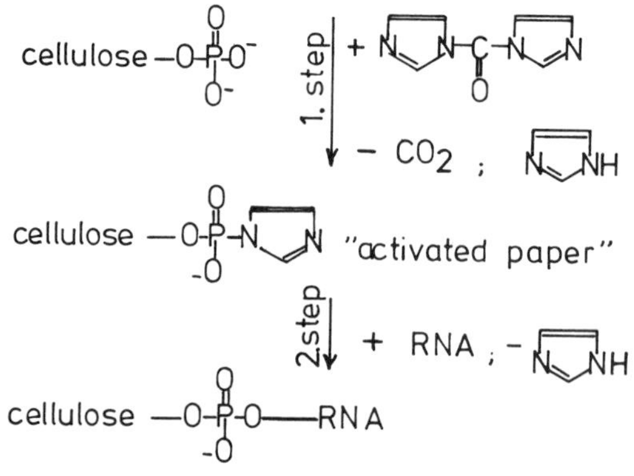

cellulose—O-P(=O)(O⁻)-O⁻ + imidazole–C(=O)–imidazole

1. step ↓ − CO_2 ; imidazole

cellulose —O-P(=O)(-O)-imidazole "activated paper"

2.step ↓ + RNA ; − imidazole

cellulose —O-P(=O)(-O)-O——RNA

Scheme 3

matrix HN-phenyl-B(HO)(OH) + R, B ribose (HO OH) ⇌ (pH >8 / pH <7) matrix HN-phenyl-boronate ester R, B + H_2O

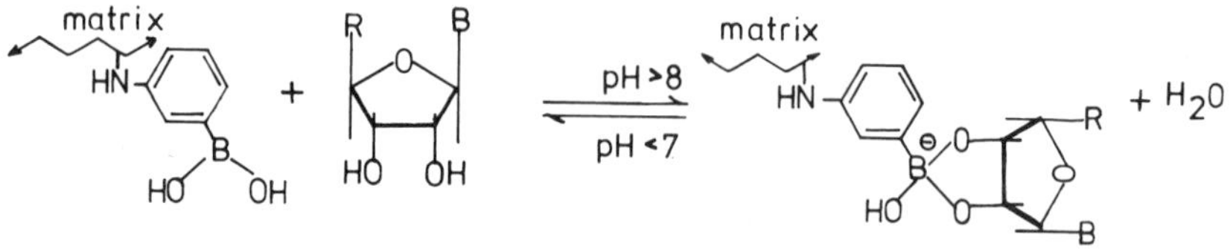

B ≙ nucleo base

R ≙ CH_2OH or oligonucleotide or ribonucleic acid

Scheme 4

immunofluorescence reaction using rabbit antiserum raised against the hapten fluorescein (110). DNA-cellulose competition binding assays were used to measure the ability of cloned DNA fragments of the chicken vitellogenin II gene to displace the estrogen–receptor complex from total chicken DNA coupled to cellulose (155). Separation of native DNA fragments, with structural genes at the end, was carried out by obtaining fragments of native DNA of a definite size, splitting a small section of 3′-terminal chains from these fragments with *Escherichia coli* exonuclease III, hybridizing poly(A)$^+$ mRNA with these sections, and separating the hybridized DNA fragments on poly(U)–Sepharose (91).

A procedure for enriching DNA for specific sequences based on R looping has been developed. R loops are formed with the DNA using mRNAs containing the sequence of interest and then isolated on poly(U)–Sepharose via the poly(A) tail of the mRNA (92). The isolation of picomole quantities of nascent mercurated DNA from a mixture of cellular nucleic acids using thiol-agarose is described (100). A histone–silica column, which binds both DNA and RNA, is used in the separation of nucleic acids by high-performance liquid chromatography (HPLC) (101). The isolation of structural genes whose transcripts do not contain terminal poly(A) sequences has been described (113). Affinity generation of ssDNA for dideoxy sequencing following the polymerase chain reaction (104) and the selective enrichment of a large genomic DNA fragment (137) used the biotin–avidin complexation. Biotin-labeled DNA probes are hybridized to DNA or RNA immobilized on nitrocellulose filters. After removal of residual probe, the filters are incubated with a performed complex made with avidin-DH and biotinylated polymers of intestinal alkaline phosphatase. The filters are then incubated with a mixture of 5-bromo-4-chloro-3-indolyl phosphate and nitro blue tetrazolium, which results in the deposition of a purple precipitate at the sites of hybridization. This procedure will detect target sequences in the range 1 to 10 pg (109). A method for quantification of specific DNA immobilized in microtiter wells based on the hybridization with biotinylated DNA probe (151). DNA from human placenta was digested and incubated on an anti-5-methylcytosine antibody column. Methylated DNA is thus enriched from total genomic DNA (105). Applications of antinucleic acid antibody for the selective isolation of DNA and RNA are reported (138). The isolation of DNA from tissue homogenate is performed by arcidinium chromatography (116). The quantitative recovery of DNA fragments (51) and DNA (47) from agarose gel by affinity chromatography is described. Table 3 summarizes special approaches of the detection and isolation of DNA fragments based on a Watson–Crick base-pairing mechanism.

B. Oligonucleotides, Mononucleotides, and Nucleosides

The specificities of four different lectins were used to separate nucleotide sugars (94). The preparation of immobilized bovine pancreatic ribonuclease by covalent attachment to Sepharose 4B is described. The strength of binding of mononucleotides, at 4°C, follows the order 5′-GMP > 5′-AMP > 3′-UMP > 3′-CMP. When binary mixtures of a 3′-pyrimidine nucleotide and a 5′-purine nucleotide are chromatographed jointly, a cooperative effect is found and elution of either or both ligands is retarded (153). Alkali-stable dinucleotides are found in a low proportion (about 3%) in RNA. Partial purification on the milligram scale was achieved by means of a combination of charge-transfer chromatography on acriflavin–Sephadex G-25 and chromatography on Sepharose 4B–RNAase (98).

Thermal chromatography on oligonucleotide cellulose has been used to resolve mixtures of homooligonucleotides (102). Oligo(U) were found to form hybrid

Table 3 Selected Examples of DNA Isolation Using Affinity Chromatography

Example	Method	Refs.
Diagnosis of potato spindle tuber viroid disease by nucleic acid hybridization	Hybridiation of DNA to immobilized viroid RNA	90
Isolation of a ss fragment of the tick-borne encephalitis virus DNA	Oligonucleotide (34-base) column	99
Isolation of fragments of pigeon DNA containing sequences complementary to globin mRNA	Poly(U)–Sepharose	93
Purification of large DNA fragments enriched in globin gene sequences	Poly(U)–Sepharose	97
Use of immobilized plasmids to purify rabbit globin complementary DNAs	Plasmid pCR1 immobilized to Sepharose	96
Isolation of duplex DNA fragments containing (dG-dC) clusters	Poly(C)Sephadex	87
Analysis of (dA-dT) clusters in yeast mtDNA	Poly(U)–Sephadex	88
Preparation of a oligo(dG)-tailed plasmid vector	Oligo(dC)–cellulose	89
Preparation of oligo(dG)-tailed DNA fragments	Oligo(dC)–cellulose	154
Isolation of a fragment of the enterotoxigenic gene of *E. coli*	Oligo(dT)10–magnetic bead support	95

structures with oligo(dA)–cellulose columns having "melted temperatures" (T_m) lower than those observed for oligo(dT) of equal length. Comparison of oligo(dA) and oligo(A) using oligo(dT)–cellulose again showed interactions with oligo(A) to be less stable (Table 4). In the case of oligo(dA), the replacement of an internal dA by dC/dG or dT causes destabilization. In the case of oligo(A), replacement of an internal A residues by C or U resulted in a similar destabilization. However, replacement of an internal A residue by G resulted in significantly less destabilization. These false monomer units in the complementary oligonucleotide lower its melting temperature. Oligonucleotides with a low content of false sequences are thus eluted before the completely complementary oligonucleotides during the temperature gradient. The order of elution may be explained by the base-pairing mechanisms as follows. The degree of hybridization increases with the number of complementary paired bases within the nucleotide chain. An increase in temperature reduces the degree of base pairing. At temperature above the "melting point," base pairing of short complementary oligonucleotides disappears, whereas longer oligomers remain hybridized. The melting points of complementary

Table 4 Thermal Elution of Oligo$(dA)_n$, Oligo$(A)_n$, and Oligo$(dT)_n$ from Cellulose–p$(dT)_9$ and Cellulose–p$(dA)_9$ Columns

Oligonucleotide cellulose	Deoxyribo-nucleotide	T_m (°C)	Ribonucleotide	T_m (°C)
Cellulose–p$(dT)_9$	p$(dA)_6$	8.5	A$(pA)_6$	12.5
	p$(dA)_7$	18.0	A$(pA)_7$	18.0
	p$(dA)_8$	26.0	A$(pA)_8$	25.0
	p$(dA)_9$	32.0	A$(pA)_9$	25.5
	p$(dA)_{10}$	35.0	A$(pA)_{10}$	26.0
	p$(dA)_{11}$	37.0	A$(pA)_{11}$	28.5
Cellulose–p$(dA)_9$	p$(dT)_7$	14.0		
	p$(dT)_8$	21.0		
	p$(dT)_9$	26.0		
	p$(dT)_{10}$	28.5		
	p$(dT)_{11}$	31.0		
	p$(dT)_{12}$	33.5		

Source: Ref. 108.

oligonucleotides shift to higher temperatures with an increasing number of paired bases, but at the same time the difference in melting point for each homolog unit decreases. As a result, a mixture of the longer-chain homologous oligonucleotides cannot be fully resolved. In addition, the melting range widens, apparently because the immobilized oligonucleotides of the stationary phase are not uniform in their molecular structure or are not accessible to the same degree. Thus partial hybridization between mobile and immobilized complementary oligonucleotides is favored. Partially hybridized oligonucleotides "melt" at lower temperatures than do fully hybridized ones. Their partial hybridization can lead to simultaneous elution of partial hybridized oligonucleotides of higher chain length with lower homologs which undergo complete hybridization. Even if the stationary phase contains only uniform oligonucleotide molecules, partial hybridization caused by steric hindrance cannot be excluded.

Oligonucleotide–DEAE–cellulose (25,26) is particularly suitable for use in preparative affinity chromatography. DNA can be partially hydrolyzed to oligomers of adenylic, thymidylic, or guanylic acid by chemical methods. Isolation of homologs with six or more monomer units can be achieved with affinity chromatography. For example, the alkaline hydrolysis of oxidized DNA from herring sperm yields a complex mixture of oligo(dA) (103). After removal of fragments containing one to three monomer units, approximately 10% of the partial hydrolysate remains. This remaining fraction is separated into two fractions by the base-pairing mechanism on a PVAL(pT)$_n$–DEAE–cellulose column with a two-step temperature gradient (Figure 4). In peak I of Figure 4, all compounds (~95%) are eluted which do not form base pairs with the immobilized oligo(dT). The compounds (~5%) adsorbed from the mobile phase at –4°C, however, are desorbed only after raising the temperature to 30°C and are eluted in peak II of Figure 4. The products in peak II are desalted, enzymatically dephosphorylated, and fractionated according to the number of monomer units on QAE–Sephadex (Figure 5).

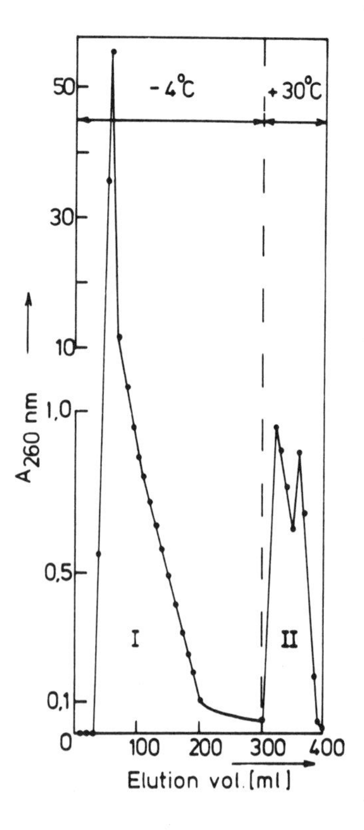

Figure 4 Elution profile of about 1200 A_{260} units of the fraction, which contained high-molecular-weight oligo(dA) isolated from the partial hydrolysate of oxidized herring sperm DNA. The separation is carried out on a PVAL-p(dT)$_n$–cellulose column (20 × 2 cm) in a two-step temperature gradient. Flow rate: 40 ml/hr. The products of peak II are combined, desalted, lyophilized, enzymatically dephosphorylated, and rechromatographed (see Fig. 5). (From Ref. 103.)

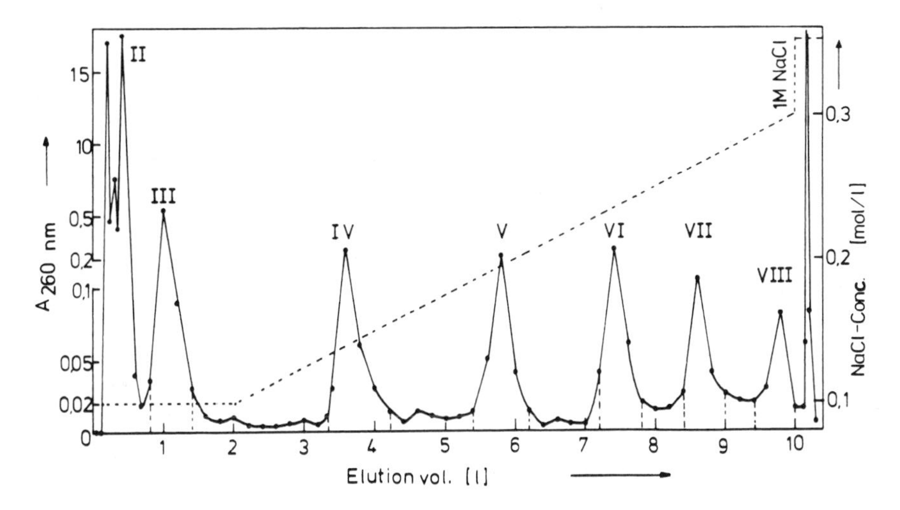

Figure 5 Elution profile of the rechromatography of the products that are eluted in peak II of Fig. 4. About 940 A_{260} units are rechromatographed at 20°C on a QAE–Sephadex A25 column (47 × 2 cm) using a NaCl gradient. Peaks II to VIII contain homologs of the $(dA)_{2\text{-}8}$ series. (From Ref. 103.)

In this manner oligo(dA) with up to eight monomer units is obtained in preparative amounts. In another approach, herring sperm DNA is chemically degraded to oligo(dT), which is subsequently fractionated on a QAE–Sephadex column according to the increasing charges of their phosphate groups. The final fraction contains a mixture of long-chain oligo(dT). Chromatographically pure, single substances of the $(dT)_{3\text{-}12}$ series are separated from the mixture using affinity chromatography and rechromatography on QAE–Sephadex. Oligo(dG), which is isolated from the partial hydrolyzate of herring sperm DNA, is separated on two differently substituted matricizes of PVAL-$(dC)_n$–DEAE–cellulose. Purine oligonucleotides are adsorbed at 0°C on PVAL-p$(dC)_n$–DEAE–cellulose, and pyrmidine oligonucleotides on oligo(guanylic acid) gel according to the base-pairing mechanism if their sequences contain at least three or more homologs, consecutive guanylic or cytidylic moieties (106).

III. ISOLATION OF DIFFERENT RNA

Affinity chromatography is an important method for the isolation of specific RNA sequences (107,108).

A. Various Possibilities of the RNA Isolation

A procedure is described that permits a one-step separation of total plant nucleic acids into DNA, rRNA, sRNA, oligonucleotides, two poly(A) RNA fractions, and residual polysaccharides by chromatography on columns of Sephadex G-75, Sepharose 4B, and poly(U)–Sepharose 4B, coupled serially (199). A method for the selection of RNA–DNA hybrids has been developed (200). The hybridization mixture was fractionated on a Sepharose 2B column. The intact probe DNA as well as the RNA–DNA hybrids are eluted with the void volume. Nonhybridized RNA, in contrast, is included in the gel matrix. The RNA–DNA hybrids were denatured in 90% formamide. The RNA selected was separated from the DNA by poly(U)–Sepharose. Restriction endonuclease fragments of DNA of large enough size to exclude them from the column were also used for hybridization. In these experiments hybridizations were carried out under conditions that would allow R-loop formations, and the hybridized RNA was separated from unhybridized RNA by Sepharose column.

A method for the determination of the electrophoretic profile of the various poly(A) RNA species in a RNA sample consists of the following steps (112). The molecules in a RNA sample are first separated according to their molecular weight by electrophoresis in agarose at low ionic strength. The molecules thus separated are then submitted to second electrophoresis in "binding buffer" in a direction perpendicular to the first. In the course of this electrophoresis, the poly(A) RNA species are separated from other RNA species as they bind to a poly(U) glass fiber filter that is placed across the electrophoresis path. Poly(A) RNA can be eluted from the poly(U) filter with formamide and subjected to electrophoresis without subsequent precipitation in ethanol. No measurable quantities of rRNA or tRNA are retained on the poly(U) filters. The hybridization technique enables a quantitative retention of poly(A) molecules representing a wide range of chain lengths.

Antibodies specific to dsRNAs were coupled to CNBr-activated cellulose (114). Synthetic and naturally occurring dsRNAs were bound specifically by the antibody-linked cellulose an were eluted with alkali solution (pH 10.5) or 10% dimethyl sulfoxide (Figure 6).

m-Aminophenylboronate-substituted agarose binds specifically RNA chains carrying a mature 5′ cap (115). The binding occurs at pH $\geq$ 8. The positive charge introduced by the m^7G methylation is necessary for efficient binding, although two closely spaced *cis*-diol groups alone are sufficient for binding. Immobilized boronic acid potentially could be used to separate aminoacyl-tRNA from unacylated species (Figure 7) (74). Highly purified isoaccepting species of tRNA were prepared by use of a polyacrylamide support substituted with nitrobenzene boronic acid as functional groups (81). It is shown that yeast $tRNA^{phe}$, chemically coupled by its oxidized 3′-CpCpA end, behaves exactly as

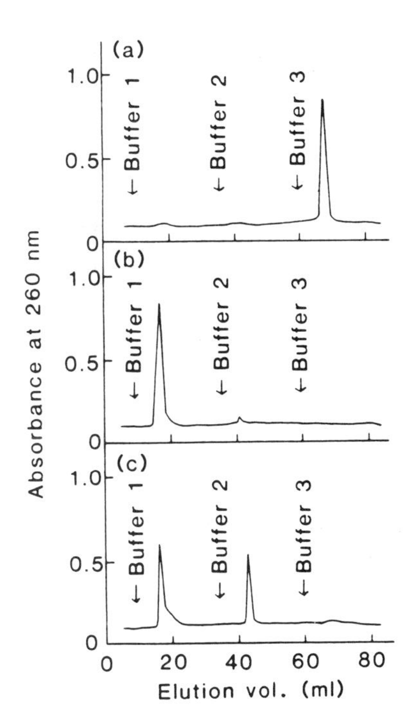

Figure 6 Chromatography of nucleic acids on anti-poly(I)·poly(C) antibody-linked cellulose. The column (1.0 × 3.2 cm), which contained 2.8 mg of antibodies coupled to 0.6 g of cellulose, was equilibrated with 0.01 *M* Tris–HCl (pH 7.6) containing 0.15 *M* NaCl (buffer 1). Various nucleic acids (100 μg) were eluted stepwise from the column at a flow rate of 40 mL/h by 0.1 *M* Tris–HCl (pH 7.6)/0.15 *M* NaCl (buffer 1); 0.01 *M* Tris–HCl (pH 7.6) containing 1 *M* NaCl (buffer 2); and 0.02 *M* sodium bicarbonate (pH 10.5) (buffer 3). (a) Poly(I)·poly(C); (b) baker's yeast tRNA; (c) *E. coli* rRNA. (From Ref. 114.)

free $tRNA^{Phe}$ in its ability to form a specific complex with *E. coli* $tRNA_2Glu$ having a complementary anticodon (148).

Elongation factor Tu coupled to CNBr–Sepharose 4B retained its ligand-binding properties (150). It specifically binds GDP and GTP but does not interact with other nucleotide 5′-phosphates. A conversion of immobilized EF-Tu·GDP to EF-Tu·GTP can be achieved by simple equilibration with GTP. The immobilized EF-Tu·GTP make possible easy purification of aminoacyl-tRNA species from bulk tRNA (149).

Table 5 summarizes selected examples of the isolation of RNA from different cell sources using affinity chromatography. Table 6 summarizes different

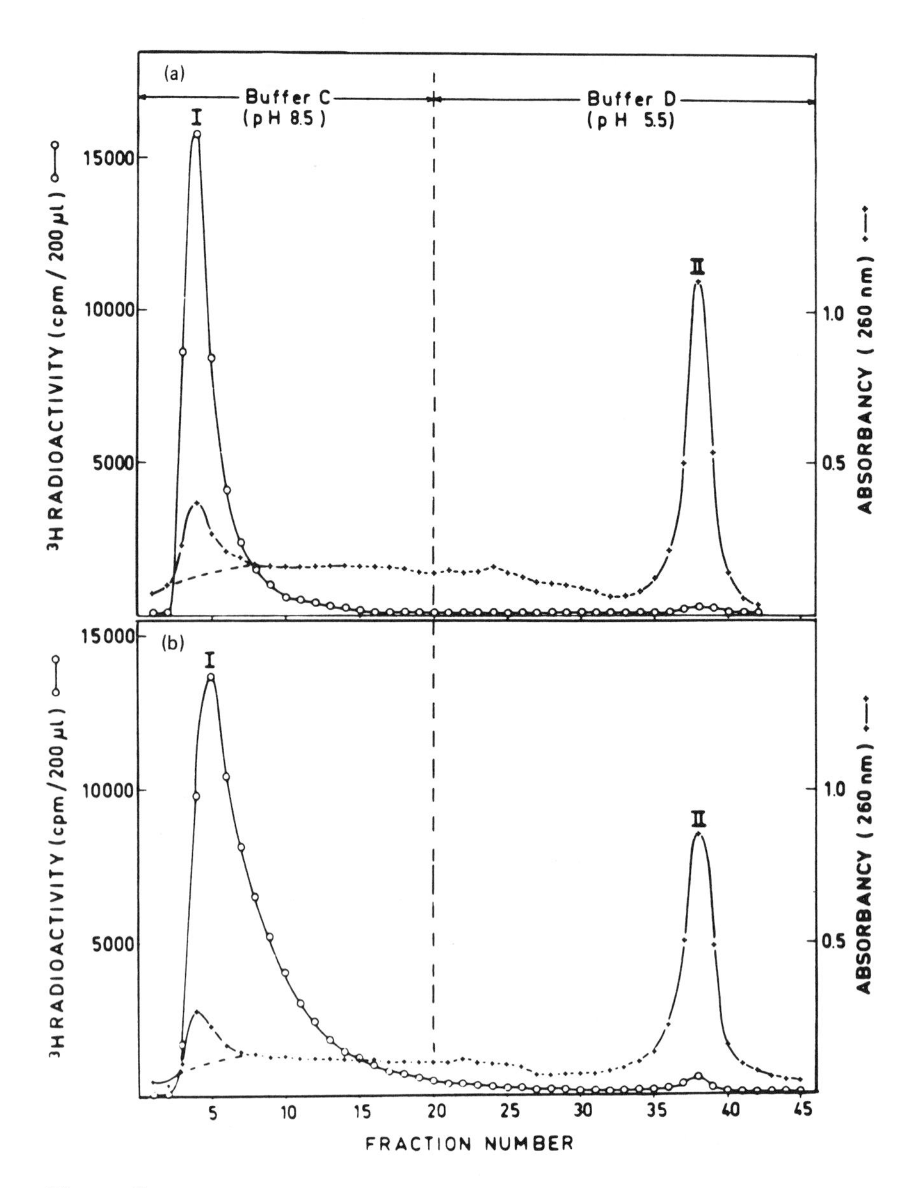

Figure 7 Purification of *E. coli* aminoacyl-tRNA by column chromatography on borate gels. (a) Separation of a tRNA mixture containing 674 pmol of [^{3}H]glutamyl-tRNA (1.6%) in a total of 26.3 A_{260} units of tRNA; (b) separation of a tRNA mixture containing 1030 pmol of [^{3}H]lysyl-tRNA (2.4%) in a total of 27.9 A_{260} units of tRNA. Elution buffers were applied as indicated by the arrows. Fractions of 4.6 mL volume were collected every 5 min in tubes containing 200 µL of 1 *M* sodium, acetate pH 4.5, in order to neutralize the eluant immediately. Aliquots of 200 µL were assayed for acid-insoluble radioactivity (o). Absorbancy of the individual fractions was followed at 260 nm (+); the absorbancy background of the elution buffer alone (- - -) is not subtracted in the figure. (From Ref. 74.)

Table 5 Selected Examples of the Isolation of RNA from Different Sources

RNA	Source	Method[a]	Refs.
SV 40 virus	Transformed mouse cells	B	117
Sendai virus	Enders strain of Sendai virus	D	118
Tabacco etch virus	Infected tissue	A	119
Tomato spotted wilt virus	Infected leaves	A	121
Newcastle disease virus	Chick embryonic cells	B	120
Mouse mammary tumor virus	GR tumor cell line	A	122
Mouse 615 hepatomas	Mouse hepatoma 615	A	67
Poly(A) RNA	Tissue		
Type D retrovirus	Human cell line	B	123
Infectious flacherie virus	Silkworm larvae	B	157
Barley stripe mosaic virus	BSMV strains propagated in wheat plants	A	158
4.5S RNA	Various cell lines	B	124
Pulse-labeled RNA	*E. coli, B. subtilis, B. brevis*	A	126–129
Plant	Wounded potato tuber tissue; crown gall tumor	A	130
	Potato bud tissue	B	135
	Maise	A	132
Petroselium hortense	Parsley	A	131
Vacia faba	Root	B	133
Yellow lupin	Root nodules	A	134
Heterogeneous nuclear RNA	HeLa cells	A	136
	Baby hamster fibroblasts cells	C	139
	Friend erythroleukemic cells	C	140
	Chick embryonic cells	A	141
Mitochondrial RNA of *Trichoderma viride*	Germinating conidia	A	142
Yeast	*S. cerevisiae*	B	144
Mycobacteria	*Mycobacterium phlei* and *smegmatis*	B	68
Drosophila melanogaster	Embryos	A	160
Larvae of the common cattle grub	Esophagi of cattle	A	125
Fish	Fish pituitary	A	45
Rat	Rat calvaria	E	111
	Liver	A	145
	Brain	A	146
Rabbit	Brain	A	147
Bovine mammary glands	Lactating mammary glands	A	33
Articular cartilage	Adult articular cartilage	A	37
Human mitochondrial RNA	HeLa cells	A	161
Sea urchin embryos	Eggs of *Hemicentrotus pulcherrimus*	B	159
	Reticulocyte polysomes	A	156

[a]Using A, oligo(dT)–cellulose; B, poly(U)—Sepharose; C, poly(A)–Sepharose; D, poly(U)–cellulose; E, oligo(dA)–cellulose.

Table 6 Investigation of Synthesis and Degradation of Poly(A) RNA

Investigation	Method[a]	Refs.
Storage and metabolism of mRNA in germinating cotton seeds	C	162
Stimulation of synthesis and translational activity of mRNA in wounded potato tubers by 2,4-dichlorophenoxyacetic acid	A	163
Effect of 2,4-dichlorophenoxyacetic acid on polysomal profiles and poly(A) RNA in excised tuber tissue of Jerusalem artichoke	A	176
Action of cordycepin on nascent nuclear RNA and poly(A) synthesis in regenerating liver	B	164
Synthesis of poly(A) RNA during the germination of *Neurospora crassa* conidia	A	165
Synthesis of poly(A) RNA during zoospore differentiation and germination of *Blastocladiella emersonii*	A	166
Effect of estrogen on gene expression in the chick oviduct	A	167
Synthesis of RNA in oocytes of the grass frog	A	179
Milk-protein mRNA and poly(A) in the involuting rat mammary gland return to the levels found during late pregnancy	A	178
Isolation and characterization of poly(A) mRNA polyoma "early" and "late"	A	168
Kinetics and efficiency of poly(A) of "late" polyoma nuclear RNA	A	177
Poly(A) RNA from rat brain synaptosomes	A	169
Metabolism of RNA in the developing rat brain (see also Figure 8)	A	170
Age changes in the metabolism of mRNA in liver and brain cortex cells of the rat	B	171
Functional assays for mRNA detect many new messages after male meiosis in mice	A	172
Turnover of mRNA in fission yeast	A	173
Effect of age on the properties of mRNA in *Physarum polycephalum*	B	174
Changes in polysomes and mRNA in aging soybean cotyledons (see also Figure 9)	A	175
Poly(A) RNA during the development of *Ceratitis capitata*	A	180
Population-kinetic approach to RNA formation and degradation in growing and resting cells	A	181
Quantitative aspects of RNA synthesis in one-cell and two-cell mouse embryos	B	182
Poly(A) RNAs as error sources in ribosomal RNA turnover analyses	A	143

[a]Using A, oligo(dT)–cellulose; B, poly(U)–Sepharose; C, poly(U)–cellulose.

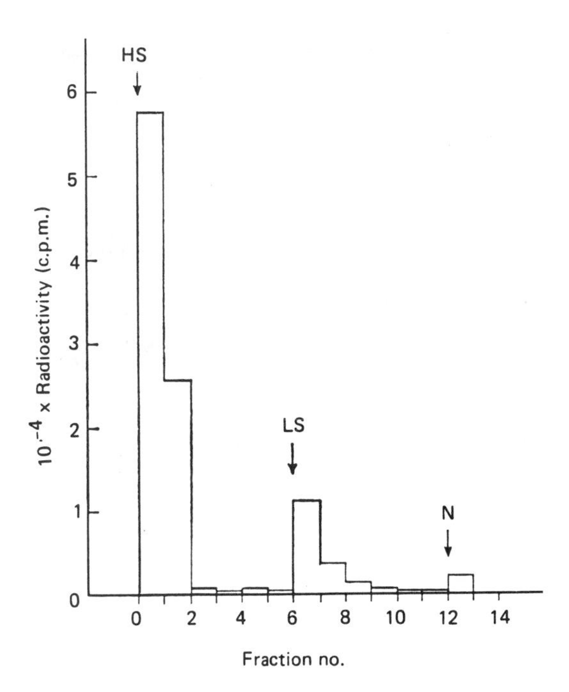

Figure 8 Isolation and characterization of [^{32}P]-labeled poly(A) RNA. High-molecular-weight RNA was isolated from forebrains of 8-day-old rats 5 h after the intracranial injection of 1 mCi of [^{32}P]P_i per gram of brain. The RNA (80 × 10^3 cpm in 6 A_{260} units) was applied in 0.4 *M* NaCl, 10 m*M* Tris–HCl (pH 7.4), 0.2% sodium dodecyl sulfate, onto a column of 200 mg of oligo(dT)–cellulose. The RNA not containing poly(A) was eluted in 2-mL fractions with high-salt buffer (HS). Poly(A) RNA was then eluted with 10 mM Tris–HCl (pH 7.4), containing 0.2% sodium dodecyl sulfate (low-salt buffer; LS). Residual radioactivity in the column was removed with 0.1 *M* NaOH (N). (From Ref. 170.)

investigations of synthesis and degradation of poly(A) RNA using affinity chromatography.

B. Isolation of Messenger RNA Using Various Poly(A) Adsorption Methods

An important objective for many studies on gene regulation is to isolate specific mRNAs for structure–function analysis, and to prepare specific radioactively labeled cDNA to be used as a hybridization probe for the quantitation of mRNA.

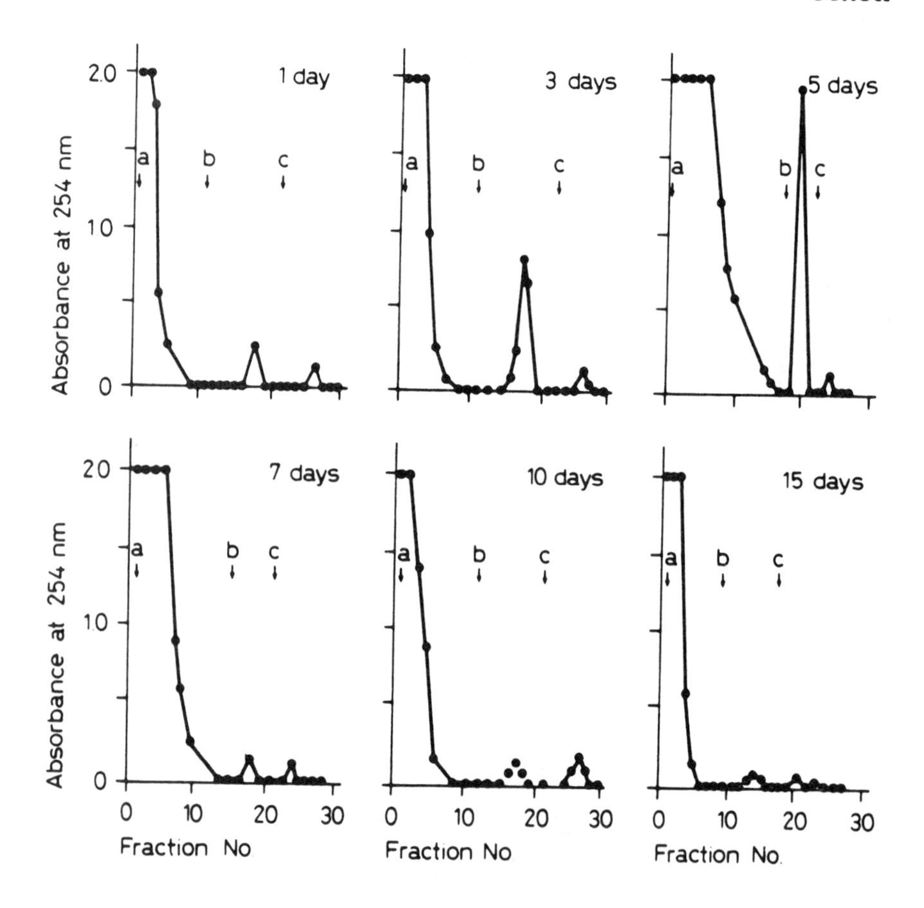

Figure 9 Fractionation of RNA from polysome pellets of soybean cotyledons aged 1, 3, 5, 7, 10, and 15 days by oligo(dT)–cellulose. 160 A_{254} units were loaded onto the column. The buffers were changed as indicated by the arrows to elute the various fractions. a, 0.5 *M* KCl–Tris buffer; b, 0.01 *M* KCl–Tris buffer; c, 0.1 *M* Tris–HCl (pH 7.5). (From Ref. 175.)

The cDNA can also be used as an affinity probe for the isolation of specific gene segments and RNA transcription products, and for the determination of nucleotide sequences.

mRNA in animal cells appears to be generated by cleavage of high-molecular-weight precursors in the nucleus. These mRNA precursors are contained in the metabolically active pool of RNA molecules (HnRNA). Polysomal mRNA as well as a substantial part of the HnRNA have been shown to contain segments of poly(A) at the 3′ terminus. In addition to this there is a nonpolysomal pool of ribonuclear protein particles in the cytoplasm, which also contains poly(A) RNA. This pool is thought to contain mRNA en route to translation.

Various techniques are commonly employed in the isolation of specific mRNAs (108,183). However, to be applicable to a particular system, each of these methods may require modification. Poly(A) mRNAs can be separated from other cellular RNAs by adsorption to nitrocellulose filters or by chromatography on unmodified cellulose, tritylcellulose–Sepharose, oligo(dT)–cellulose, poly(U)–agarose/Sepharose, or DNA–cellulose. Some examples of the isolation of mRNA are described in the following section.

Although nitrocellulose filters are effective in the isolation and detection of nucleic acid RNA (186), they also nonspecifically adsorb rRNA and small amounts of ssDNA, even after a readsorption step. Lower recoveries of mRNAs from nitrocellulose filtration techniques are also observed, apparently due to a requirement for longer poly(A) segments (183). However, in 12.2 *M* NaI and at 25°C or below, mRNA bound to nitrocellulose, whereas DNA and RNA did not. Neither the poly(A) tract nor the cap were required for binding. The immobilized RNA could be translated, reverse transcribed, hybridized with radioactive probes, or released for further manipulation. mRNA was efficiently transferred from polyacrylamide to nitrocellulose in NaI. When cells dissolved in 12.2 *M* NaI were filtered through nitrocellulose, mRNA became selectively bound (quickblot). The quickblot system utilizing protease and detergents to prepare cells for NaI solubilization was especially suitable in quantitative, rapid screening of cells for expression of specific genes (184).

Unmodified cellulose has been shown to bind poly(A) RNA (183,187,188) at neutral pH in high-salt solution and poly(A) RNA manipulation is released in low-salt solution. The cellulose–poly(A) interaction is consistent with association by π–π interaction between aromatic constituents of lignin of the cellulose and the purine bases of the poly(A). ssDNA behaves like poly(A) in these respects, but pyrimidine homopolymers do not. Similarly, poly(A)–poly(U) hybrids have no affinity for cellulose preparations. In high-ionic-strength solution, excess poly(U) removes most of the bound poly(A) from the poly(A)–cellulose complex but does not affect the bound ss DNA. Poly(U) can be used to obtain substantial separation of poly(A) RNA from nucleic acid mixtures that contain ssDNA by selectively eluting the poly(A) RNA–containing material from the cellulose-bound mixture of the two. Fractional elution of poly(A) RNA from cellulose can be accomplished by gradually decreasing the salt content of the elution buffer. The poly(A) RNA oligomers that disengage from cellulose at rather high salt concentrations seem to be shorter than those that are released only at lower salt concentrations. This discrimination may serve as a basis for the fractionation of poly(A) RNAs of which the poly(A) suffixes differ in length. However, mRNA prepared by this technique is usually contaminated with substantial amounts of rRNA.

Several modified celluloses were tested for their ability to bind mRNAs (Figure 10) (190,191). Conditions that enhance polynucleotide hybridization, such as high-ionic-strength solutions and low temperatures, also enhance poly(A) RNA binding, and conditions that destabilize polynucleotide hybrids, such as low-ionic-strength solutions, high temperature, and formamide, also disrupt the binding of poly(A) RNA to the matrix-bound complementary polynucleotide and cause elution of poly(A) RNA from the column. An incomplete binding of poly(A) RNA to oligo(dT)–cellulose was found to be caused by flow rates that can readily be achieved when the column is run by gravity elution. To avoid this lack of binding, during the equilibration of the column the level of the exit tubing is

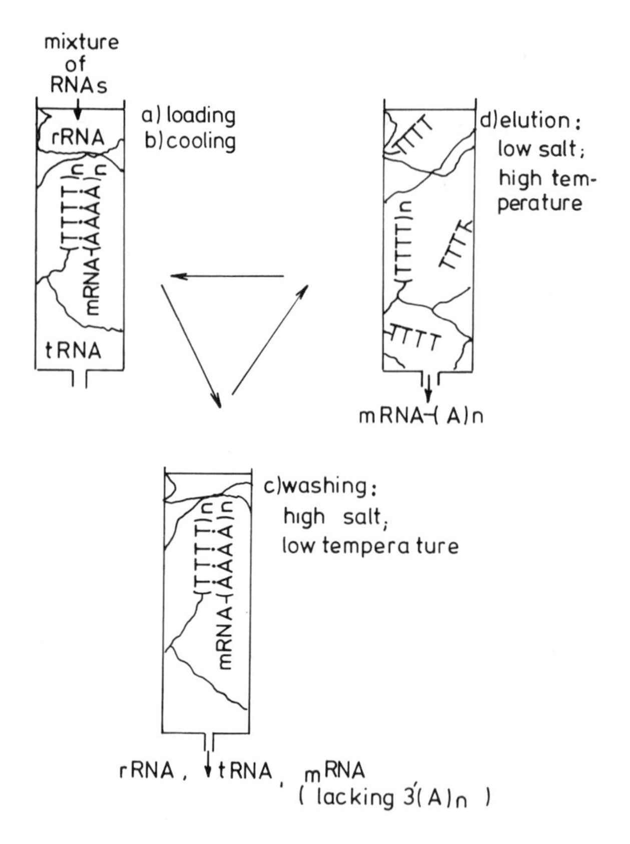

Figure 10 Separation of mRNA from a mixture of RNAs using oligo(dT)–cellulose column. A poly(U)–Sephadex column can be used instead of the oligo(dT)–cellulose. (From Ref. 108.)

adjusted to give a flow rate of <2 mL/min, and this setting is maintained during the application of the sample to the column. Under these conditions, one passage of up to 40 mg of mammary RNA over a 1-g oligo(dT)–cellulose column will give a good yield of poly(A) RNA (192). It was shown that poly(A) interacts nonspecifically with matrices such as acriflavin or DNA–Sepharose as well as with oligo(dT)–cellulose or poly(U)–Sepharose. It was shown that in addition to specific binding of poly(A) sequences with poly(U), the chromatography of poly(A) RNAs on poly(U)–Sepharose is accompanied by nonspecific irreversible adsorption of polynucleotide by Sepharose gel (193). The efficient separation of poly(A)$^+$ RNA preparations from poly(A) RNAs is achieved only after double chromatography of RNA on poly(U)–Sepharose.

The poly(U) ligand is usually somewhat longer than the oligo(dT) ligand and therefore requires stronger elution conditions, such as 70 to 90% formamide or elevated temperature (194). Since the poly(U) chain is approximately 100 monomers long and is attached to the matrix at several points, breaks in the chain lead to the loss of only small amounts of polymeric material. Several weeks of storage in suspension at elevated temperatures results in the loss of only 1 to 2% of the poly(U) and no detectable loss of binding activity from RNA. The disadvantages of chromatography on poly(U)–Sepharose are the nonquantitative recovery of poly(A) mRNA and insufficient adsorbent capacity. A simple method is described for the purification of poly(A) mRNA by chromatography on oligo(dT)–cellulose or poly(U)–Sepharose under high-salt conditions. RNA containing a poly(A) tract as short as 20 residues will bind to the support and is eluted with a low-salt buffer, or a buffer containing formamide in the case of poly(U)–Sepharose, whereas rRNA and tRNA do not bind to the support (185). The purification of poly(A) RNA by oligo(dT)–latex particles is described as an alternative to the oligo(dT)–cellulose method (215).

High variability and low recoveries are obtained when conventional ethanol precipitation is used to recover RNA from the 70–90% formamide-containing solutions that are used to elute poly(U)–Sepharose columns (195). Precipitation of RNA at different formamide concentrations shows that the variability of the recovery increasing formamide concentration and that the recovery decreases by 7% for each 10% increase in the formamide concentration. The formamide concentrations curve shows that these factors can be optimized by diluting the formamide-containing solutions to at least 30% formamide prior to precipitation.

A method for the isolation of mRNA from polysomes (see Figure 11) has been described (183,197,198,232). Polysomes are dissolved in a solution containing 0.5 *M* NaCl and sodium dodecyl sulfate and applied to an oligo(dT)–cellulose column. Poly(A) RNAs are retained by the column, whereas ribosomal proteins and other RNA species are washed off. The column is then eluted with a buffer not containing NaCl.

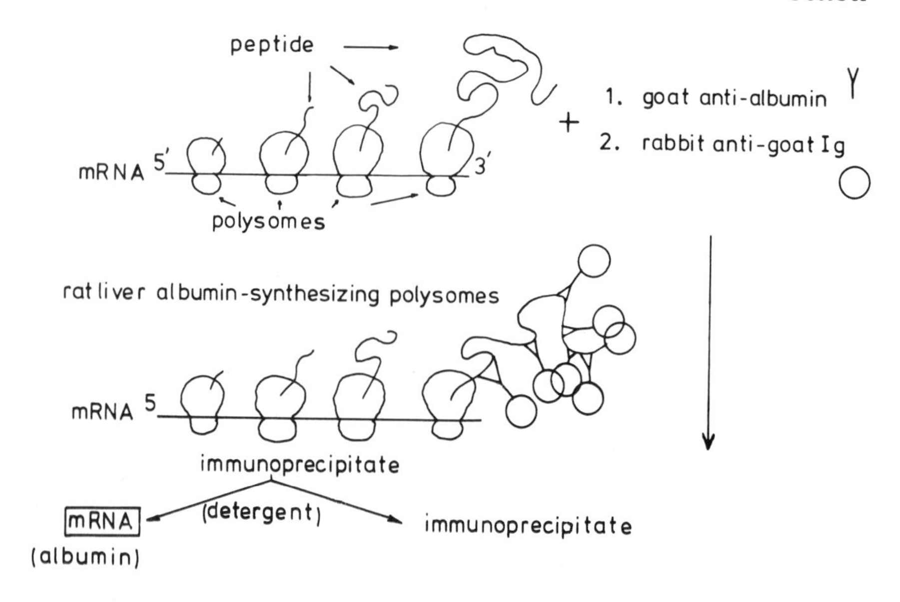

Figure 11 Isolation of specific polysomes with their mRNA. (From Ref. 108.)

DscDNA was made to partially purified mRNA for the small seminal vesicle secretory protein IV and V (201). This cDNA was then inserted into the Pst 1 site of pBR 322 by the (G-C) homopolymer tailing technique. Bacterial transformants harboring plasmids with specific inserts were identified by translation of mRNA that was hybridized to plasmid DNA immobilized on nitrocellulose.

A method for eluting mRNA from agarose gels has been described. Chick ovalbumin mRNA and globin mRNA were electrophoresed on 2% agarose–6.0 *M* urea cylindrical gels at pH 5.5. The position of the RNA after electrophoresis was determined in stained gels, and corresponding sections of unstained gels were cut out. These slices were pulverized and layered on top of an oligo(dT)–cellulose column to which diffusing RNA was bound. The recovery of poly(A) RNA varied between 30 and 50% (206). Efficient recovery of mRNA from agarose gels via transfer to an ion-exchange membrane has been described (207). RNA species are separated by gel electrophoresis and then blotted onto a paper sheet to which poly(U) has been covalently bound. This mRNA affinity paper ("mAP") specifically binds, in a reversible manner, poly(A)$^+$-containing molecules. A replica picture of the agarose gel is thus obtained on the mAP, from which bound mRNA molecules can be eluted by heating in water (210).

An expedient procedure for preparing mRNA from mammalian cells has been developed. It is totally devoid of phenol-based organic solvents and utilizes the

Table 7 Isolation of mRNAs from Various Organisms

mRNA from organism	Cell source	Methods[a]	Refs.
Procaryote	*Rhodospirillum rubrum*	D	240
Yeast	*S. cerevisiae*	A, D	32, 211, 212
Protozoa	*Tetrahymena pyriformis*	B, D	213, 214
Slime mold	*Dictyostelium discoideum*	B	216
Plant	Leaves of light-grown pea seedlings	B	217
	Tobacco	D	237
	Lemma gibba G-3	D	218
	Carrot root disks	F	219
	Developing endosperms from ears of field-grown wheat	D	245
Insect	Wing imaginal disks of *Pieris brassicae*	B	243
Oocysts	*Eimeria tenella*	D	220
Artemia salina embryos	Cytoplasmic fractions	D	221
Mouse	Kidney	D	224
	AKR-2B mouse embryo	D	225
	Liver	D	226
Rat	Liver	D	238
Guinea pig	Spleens or lymph nodes	A	241
Rabbit	Erythroid-enriched bone marrow cells	B	242
HeLa cell	Polysomes	D, C	228–231
?	Polysomes	B	232
Mouse plasmacytoma cells	Myeloma cells (5563), MPC-11	D	233
Navikoff hepatoma	Polysomes	E	234
Friend leukemia cells	Polysomes, cytoplasmic RNA	C, D	235, 236
Physarum polycephalum	Microplasmodia	B, E	38, 202
Herpes simplex virus type 1		E	204
Polyoma virus	Infected mouse cell	E	205
Human malaria parasite	*Plasmodium falciparum*	D	239
Human	Term placenta	D	244
Brain capillaries, endothelial cells	Tissue culture ECL cells	D	29

[a]Using A, poly(U)–cellulose; B, poly(U)–Sepharose; C, poly(A)–Sepharose; D, oligo(dT)–cellulose; E, DNA–cellulose; F, oligo(dG)–cellulose.

Table 8 Selection of Isolated mRNAs Coding for Various Proteins

Coded protein	Cell source	Methods[a]	Refs.
Major structural protein of the adenovirus type 2 (Ad-2) capsid	Ad-2–infected HeLa cells	A	246
Structural proteins of simian adenovirus SA7	Infected African green monkey kidney cells	A	247
N protein of vesicular stomatis virus (VSV)	Indiana serotype of VSV	A	248
Preputial gland β-glucuronidase	Preputial glands of sexually mature female rats	A	249
Antifreeze peptides	Winter flounder livers	A	250
Preprochymosin	Mucosal layer of the fourth stomach of a freshly slaughtered suckling calf	A	251
Prorennin	Mucus layer of two abomasum of newborn calves	B	257
Neuron-specific enolase 14-3-2	Rat brain	B	252
Precursor of ornithine transcarbamoylase; precursor of β subunit of propionyl-CoA carboxylase and cystationine β-synthetase	Rat liver	A	253
Precursors of proline-rich protein	Human submandibular gland	A	254
Subunits of thyrotropin	Mouse thyrotropic pituitary tumors	A	256
Variant surface glycoproteins	*Trypanosoma brucei*	G	258, 259
Photoreaction center and light-harvesting antenna polypeptides	*Rhodospirillum rubrum*	A	260
Phytochrome	Seedlings of *Avena sativa*	A	255
Corticotropin	Neurointermediate lobe of bovine pituitary β-lipotropin precursor	A, B	261
Globin	Erythroblasts from bone marrow of anemic chicken	A	262
	Pigeon	F	263
	Friend cells	D	183
	Immature red blood cells of duck	A	265
	Erythropoietic spleen cells of mice	A	266
Leghemoglobin	Soybean nodules	A	267
β Subunit of follicle-stimulating hormone	Pituitary glands of castrate male sheep	A	268
Transferrin, conalbumin	Chick oviduct	A	269

Table 8 (Continued)

Coded protein	Cell source	Methods[a]	Refs.
α-Fetoprotein	Male rats	A	270
Dopamine β-hydroxylase	Rat pheochromocytoma cell line	A	271
Apo VLDL-II(apoprotein in avian very low density lipoproteins)	Liver of 4-week-old cockerels	A	272
Ceruloplasmin	Rat liver	B	273
Uteroglobin	Endometrium of rabbits	A	274
Immunoglobulin	Human B lymphocytic cell line	A	278
L-chain	MOPC-321 mouse myeloma	A	3
H-chain	IF-2 cell line derived from mouse myeloma MOPC 21	A	276, 277
K-chain	Mouse myelomas	C	279
L- and H-chain	Mouse plasmacytomas	A	280
αH-chain	Mouse myelomas	G	281
β_2-Microglobulin	Mouse liver	A, C	282
α_{2U}-Globulin	Mature male rats	A	283
Myosin	12-Day-old embryonic chick muscle	A	284
Collagen	Calvaria of 15-day-old chick embryos	A	183
	Lung and dermis of neonatal rats	A	286
	Fetal calf skin	A	287
α-Crystalline,	Calf lens	A	3, 289
noncrystalline	Rat lens	A	183
Fibroin	Silk worm	E	28
Nonfibroin	*Bombyx mori* larvae	A	290
Elastin	Thoracic aortas of 16-day-old chick embryos	A	292
Keratin	Epidermis from newborn and adult mouse	A	293
Casein	Lactating rat mammary gland	A	183
Ovalbumin	Oviduct	A	183
		B	183
		C	297
Albumin	Rat liver	B	183
	Mouse hepatoma cell line	A	299
	Liver of male frogs	A	300
	Guinea pig liver	A	301
α-Lactalbumin	Three- to 5-day lactating rat mammary glands	B	302
Zein	Developing maize kernels	A	303
Storage protein	Developing cotyledons of *P. sativum*	A	304
X. laevis vitellogenin	Livers of estradiol 17β–treated *X. leavis*	B	305

Table 8 (Continued)

Coded protein	Cell source	Methods[a]	Refs.
S-100 protein	Rabbit cerebral hemispheres	A	306
Fibronectin		A	307
Immune interferon		A	308
Mouse	Spleen cell cultures	A	309
Human	Suspension cultures of total and lymphocyte-enriched peripheral white blood cells	A	310
	Human fibroblasts strains 17/1 and FS-4		
	Human splenocytes	A	311
Interleukin-2	Human peripheral blood leukocytes	A	312
	Murine T lymphoma EL4	A	313
	Gibbon lymphosarcoma cell line	A	314
Phosphoenolpyruvate carboxykinase	Liver of starved rats	A	315
	Liver of cyclic AMPtreated rats	A	316
Large subunit ribulose–bisphosphate carboxylase	Wild-type *Euglena aplastidic* mutant cells	A	317
Amidating enzyme	Gamma-irradiated potato buds	B	318
Yeast iso-1-cytochrome c		C	288
Swine pepsinogen	Swine gastric mucosa	A	75
Aldolose B subunit	Rat liver	A	291
Aminoacyl-tRNA synthetases	Yeast	A	294
Thyroglobulin	Membrane-bound thyrosid polysomes	H	295
Preproinsulin	Islet tumor cell	A	296
Prolactin and growth hormone	Bovine pituitaries	A, C	264, 275, 285, 298
Protamine	Trout testes	A	183
Histone	Whole cell and polysomal RNA extracts	A, J	223, 227
Tyrosine aminotransferase	Polysomes	I	203
Pro-opiomelanocortin	Porcine pituitary neuerointer-mediate lobes	A	196
Glutamine synthetese	Embryonic chick retina	A	189
Elastin	Ligmentum nuchae of fetal sheep	A	69
Firefly luciferase	Lanterns of adult fireflies	A	85
Glycine methyltransferase	Rat liver	A	34
(Guanine-7-) methyltransferase	Yeast *S. cerevisiae*	C	19
Enkephalin		K	42

[a]Using A, oligo(dT)–cellulose; B, poly(U)–Sepharose; C, DNA support; D, poly(I)–Sephadex; E, oligo(dC)–cellulose; F, poly(A)–Sepharose; G, DNA-DBM paper; H, poly(U)–Sephadex; I, poly(U)–cellulose; J, cDNA–cellulose; K, defined oligonucleotide–cellulose.

deproteinizing ability of the chaotropic agents lithium chloride and guanidinium chloride. Special considerations were given to preventing RNase action during the preparation (208).

To demonstrate the affinity of RNA-containing polyribosomal components (isolated from L5178y cells) to microtubules, microtubule protein was attached to an insoluble matrix. In contrast to ribosomes, poly(A)$^+$ mRNA and poly(A)$^-$ RNP were found to bind to the matrix. Using synthetic polyribonucleotides, no significant differences in the binding properties of ss and ds polynucleotides of different base composition to microtubule protein were observed. However, binding is dependent on the size of the nucleotides. A minimal chain length of 12 nucleotide units is required (209). Coat protein covalently bound to a solid support selectively binds RNAs containing a hairpin that forms complexes with the protein (222). Table 7 summarizes examples of the isolation of mRNAs from various organisms. Table 8 contains a selection of various isolated mRNAs coding for proteins.

REFERENCES

1. P. V. Sundaram, *Nucleic Acids Res., 1*:1587 (1974).
2. I. P. Trayer and M. A. Winstanley, *J. Biochem., 9*:449 (1978).
3. P. T. Gilham, in *Immobilized Biochemicals and Affinity Chromatography* (R. B. Dunlap, ed.), Plenum, New York, 1974, p. 173.
4. R. Barker, I. P. Trayer, and R. L. Hill, *Methods Enzymol., 34*:479 (1974).
5. V. N. Shibaev, Yu. Yu. Kusov, N. A. Kalinchuk, and N. K. Kochetkov, *Bioorg. Khim., 3*:120 (1977).
6. M. Wilcheck and R. Lamed, *Methods Enzymol., 34*:475 (1974).
7. E. Lanka, C. Edelbluth, M. Schlicht, and H. Schuster, *J. Biol. Chem., 253*:5847 (1978).
8. T. Uchida and Y. Shibata, *J. Biochem., 90*:463 (1981).
9. A. A. Nedospasov and R. M. Kohomutov, *Bioorg. Khim., 4*:645 (1978).
10. Y. D. Clonis and C. R. Lowe, *Eur. J. Biochem., 110*:279 (1980).
11. P. Gacesa and W. J. D. Whish, *Biochem. J., 175*:349 (1978).
12. S. Nagae, T. Miyamoto, Y. Inaki, and K. Takemoto, *Anal. Sci., 4*:575 (1988).
13. H. Kapmeyer, D. A. Lappi, and N. O. Kaplan, *Anal. Biochem., 99*:189 (1979).
14. S. J. Morris, *J. Chromatogr., 82*:349 (1973).
15. H. Schott, G. Greber, and L. Bucsis, *Makromol. Chem., 136*:303 (1970).
16. H. Schott and G. Greber, *Makromol. Chem., 136*:307 (1970).
17. H. Schott and G. Greber, *Makromol. Chem., 144*:333 (1971).
18. M. Akashi, M. Yamaguchi, H. Miyata, M. Hayashi, E. Yashima, and N. Miyauchi, *Chem. Lett., 7*:1093 (1988).
19. C. Locht, J.-L. Beaudart, and J. Delcour, *Eur. J. Biochem., 134*:117 (1983).
20. J. C. Smith and M. A. W. Eaton, *Nucleic Acids Res., 1*:1763 (1974).
21. L. Clerici, F. Campagnari, J. F. M., de Rooij, and J. H. van Boom, *Nucleic Acids Res., 6*:247 (1979).

22. V. V. Shumyantzeva, N. I. Sokolova, and Z. A. Shabarova, *Nucleic Acids Res., 3*:903 (1976).
23. J. F. M. de Rooij, G. Wille-Hazeleger, A. B. J. Vink, and J. H. van Boom, *Tetrahedron, 35*:2913 (1979).
24. H. Schott and H. Watzlawick, *Makromol. Chem., 182*:825 (1981).
25. H. Schott and H. Watzlawick, *Makromol. Chem., 182*:119 (1981).
26. H. Schott, *J. Chromatogr., 115*:461 (1975).
27. J. T. Kadonaga, *Methods Enzymol., 208*:10 (1991).
28. P. Cashion, G. Sathe, A. Javed, and J. Kuster, *Nucelic Acids Res., 8*:1167 (1980).
29. R. J. Boado and W. M. Pardridge, *J. Neurochem., 57*:2136 (1991).
30. H. Potuzak and P. D. G. Dean, *FEBS Lett., 88*:161 (1978).
31. W. H. Scouten, *Affinity Chromatography. Bioselective Adsorption on Inet Matrices*, Wiley, New York, 1981, 348 pp.
32. T. Chow, C. Juby, and L. Yuen, *Anal. Biochem., 175*:63 (1988).
33. A. E. Rachkov and N. F. Starodub, *Mol. Biol., 37*:29 (1984).
34. T. Gomi, H. Ogawa, and M. Fujioka, *Biochem. Int., 9*:25 (1984).
35. G. G. Carmichael, *J. Biol. Chem., 250*:6160 (1975).
36. S. M. Kalachikov, V. A. Adarichev, and G. M. Dymshits, *Bioorg. Khim., 18*:52 (1992).
37. M. E. Adams, D. Q. Huang, L. Y. Yao, and L. J. Sandell, *Anal. Biochim., 202*:89 (1992).
38. R. A. Cox and N. J. Smulian, *Chem. Abstr., 102*:109396j (1985).
39. M. Fuke and C. A. Thomas, *J. Mol. Biol., 52*:395 (1970).
40. S. L. Petrovic, M. B. Novakovic, and J. S. Petrovic, *Biopolymer, 14*: 1905 (1975).
41. J. L. Viovy, F. Miomandre, M. C. Miguel, F. Caron, and F. Sor, *Electrophoresis, 13*:1 (1992).
42. H. J. Wolter, *Chem. Abstr., 112*:51770q (1990).
43. L. R. Massom and H. W. Jarrett, *J. Chromatogr., 600*:221 (1992).
44. A. Weissbach and M. Poonian, *Methods Enzymol., 34*:463 (1974).
45. D. Song and B. Guo, *Shengwu Huaxue Zazhi, 7*:21 (1991).
46. P. Westerman and G. Herrmann, *Chem. Abstr., 113*:191856t (1990).
47. M. Grey and M. Brendel, *Curr. Genet., 22*:83 (1992).
48. K. A. Cruickshank, *Chem. Abstr., 116*:145452x (1992).
49. B. A. Klyaschchitsky, V. Kh. Mitina, G. E. Morozevich, and R. I. Yakubovskaya, *J. Chromatogr., 210*:67 (1981).
50. J. Kempf, N. Pfleger, and J. M. Egly, *J. Chromatogr., 147*:195 (1978).
51. M. W. McEnery, C. W. Angus and J. Moss, *Anal. Biochem., 156*:72 (1986).
52. H. W. Dickerman, T. J. Ryan, A. I. Bass, and N. K. Chatterjee, *Arch. Biochem. Biophys., 186*:218 (1978).
53. M. Goldberg, R. P. Lifton, G. R. Stark, and J. G. Williams, *Methods Enzymol., 68*:206 (1979).
54. J. C. Alwine, D. J. Kemp, B. A. Parker, J. Reiser, J. Renart, G. R. Stark, and G. M. Wahl, *Methods Enzymol., 68*:220 (1979).
55. J. Reiser, J. Renart, and G. R. Stark, *Biochem. Biophys. Res. Commun., 85*:1104 (1978).
56. G. M. Wahl, M. Stern, and G. R. Stark, *Proc. Natl. Acad. Sci. USA, 76*:3683 (1979).
57. B. Seed, *Nucleic Acids Res., 10*:1799 (1982).
58. A. J. Macdougall, J. R. Brown, and T. W. Plumbridge, *Biochem. J., 191*:855 (1980).

59. U. Maskos and E. M. Southern, *Nucleic Acids Res., 20*:1679 (1992).
60. V. I. Latich, V. P. Varlamov, N. N. Semenova, and S. V. Rogozhin, *Biokhimiya, 45*:1597 (1980).
61. T. Y. Shih and M. A. Martin, *Biochemistry, 13*:3411 (1974).
62. S. Biagioni, R. Sisto, A. Ferraro, P. Caiafa, and C. Turano, *Anal. Biochem., 89*:616 (1978).
63. H.-D. Hunger, H. Grutzmann, and C. Coutelle, *Biochim. Biophys. Acta, 653*:344 (1981).
64. T. Mizutani and Y. Tachibana, *J. Chromatogr., 356*:202 (1986).
65. H. Potuzak and P. D. G. Dean, *Nucleic Acids Res., 5*:297 (1978).
66. L. G. Moss, J. P. Moore, and L. Chan, *J. Biol. Chem., 256*:12655 (1981).
67. C. Zou, T. Lan, and Z. Sun, *Huaxi Yike Daxue Xuebao, 19*:116 (1988).
68. V. M. Katoch and R. A. Cox, *Int. J. Lepr. Other Mycobact. Dis., 54*:409 (1986).
69. K.-G. Yoon, M. May, N. Goldstein, Z. K. Indik, L. Oliver, C. Boyd, and J. Rosenbloom, *Biochem. Biophys. Res. Commun., 118*:261 (1984).
70. H. Yamagishi and Y. Mitoma, *Chem. Abstr., 115*:136660r (1991).
71. H. R. Burrell and J. Horowitz, *Eur. J. Biochem., 75*:533 (1977).
72. M. B. Ustav, J. L. Remme, A. J. Lind, and R. L.-E. Villems, *Bioorg. Khim, 5*:365 (1979).
73. K. G. Skryabin, V. P. Varlamov, V. M. Zakhariev, S. V. Rogozhin, and A. A. Bayev, *Bioorg. Khim., 2*:1416 (1976).
74. H. Schott, E. Rudloff, P. Schmidt, R. Roychoudhury, and H. Kössel, *Biochemistry, 12*:932 (1973).
75. K. Sogawa, Y. Ichihara, K. Takahashi, Y. Fujii-Kuriyama, and M. Muramatsu, *J. Biol. Chem., 256*:12561 (1981).
76. E. Schlimme, K. S. Boos, E. Hagemeier, K. Kemper, U. Meyers, H. Hobler, T. Schnelle, and M. Weise, *J. Chromatogr., 378*:349 (1986).
77. S. Hecht, *Tetrahedron, 33*:1671 (1977).
78. M. Akashi, T. Tokiyoshi, N. Miyauchi, and K. Mosbach, *Nucleic Acids Symp. Ser., 16*:41 (1985).
79. R. E. Duncan and P. T. Gilham, *Anal. Biochem., 66*:532 (1975).
80. R. P. Singhal, R. K. Bajaj, C. M. Buess, D. B. Smoll, and V. N. Vakharia, *Anal. Biochem., 109*:1 (1980).
81. B. J. B. Johnson, *Biochemistry, 20*:6103 (1981).
82. H. Bünemann, P. Westhoff, and R. G. Herrmann, *Nucleic Acids Res., 10*:7163 (1982).
83. H. Bünemann, *Nucleic Acids Res., 10*:7181 (1982).
84. A. Yamane, *Chem. Abstr., 116*:122631r (1992).
85. K. V. Wood, J. R. de Wet, N. Dewji, and M. DeLuca, *Biochem. Biophys. Res. Commun., 124*:592 (1984).
86. L. Hwang-Lee, J. Blamire, and S. F. Cottrell, *Anal. Biochem., 128*:47 (1983).
87. D. Zuidema, F. M. Van den Berg, and R.-A. Flavell, *Nucleic Acids Res., 5*:2471 (1978).
88. J. P. M. Sanders and P. Borst, *Mol. Gen. Genet., 157*:263 (1977).
89. A. Leriche, D. Christophe, H. Brocas, and G. Vassart, *Anal. Biochem., 129*:249 (1983).
90. R. A. Owens and T. O. Diener, *Science, 213*:670 (1981).

91. A. P. Ryskov, G. N. Enikolopov, T. V. Vygodina, N. N. Dobbert, and G. P. Georgiev, *Dokl. Akad. Nauk SSSR, 232*:706 (1977).
92. B. M. Tyler and J. M. Adams, *Gene, 10*:147 (1980).
93. V. Z. Tarantul, V. A. Lipasova, and K. G. Gazaryan, *Dokl. Akad. Nauk SSSR, 224*:719 (1976).
94. M. Tokuda, M. Kamei, S. Yui, and F. Koyama, *J. Chromatogr., 323*:434 (1985).
95. M. L. Collins, D. N. Halbert, N. Donald, W. King, and J. M. Lawrie, *Chem. Abstr., 113*:94353u (1990).
96. A. Coutelle, P. Ioannaou, and R. Williamson, *Gene, 3*:113 (1978).
97. A. P. Ryskov, N. E. Maleeva, and S. A. Limborska, *Gene, 3*:81 (1978).
98. C. Arus, M. V. Nogues, and C. M. Cuchillo, *J. Chromatogr., 237*:500 (1982).
99. V. F. Zarytova and I. G. Shishkina, *Anal. Biochem., 188*:214 (1990).
100. G. Banfalvi, S. Bhattacharya, and N. Sarkar, *Anal. Biochem., 146*:64 (1985).
101. C. P. Holstege, M. J. Pickaart, and L. L. Louters, *J. Chromatogr., 455*:401 (1988).
102. S. Gillam, K. Waterman, and M. Smith, *Nucleic Acids Res., 2*:265 (1975).
103. H. Schott, *J. Chromatogr., 187*:119 (1980).
104. L. G. Mitchell and C. R. Merril, *Anal. Biochem., 178*:239 (1989).
105. R. R. Kumar and D. D. Deobagkar, *Biotechnol. Tech., 5*:469 (1991).
106. H. Schott, *Sep. Sci. Technol., 22*:2061 (1987).
107. J. M. Taylor and R. W. Hamilton, in *Affinity Chromatography and Molecular Interactions* (J. M. Egli, ed.), INSERM Symposia Series, Vol. 86, Paris, 1979, p. 265.
108. H. Schott, in *Affinity Chromatography* (J. Cazez, ed.), *Chromatogr. Sci., 27* (1984).
109. J. J. Leary, D. J. Brigati, and D. C. Ward, *Proc. Natl. Acad. Sci. USA, 80*:4045 (1983).
110. J. G. J. Bauman, J. Wiegant, and P. van Duijn, *Histochemistry, 73*:181 (1981).
111. M. Zeichner and D. Breitkreutz, *Arch. Biochem. Biophys., 188*:410 (1978).
112. E. Egyhazi and A. Ossoinak, *Anal. Biochem., 92*:280 (1979).
113. S. G. Arsenyan, T. A. Avdonina, A. Laving, M. Saarma, and L. L. Kisselev, *Gene, 11*:97 (1981).
114. Y. Kitagawa and E. Okuhara, *Anal. Biochem., 115*:102 (1981).
115. H. E. Wilk, N. Kecskemethy, and K. P. Schäfer, *Nucleic Acids Res., 10*:7621 (1982).
116. D. E. Sok, C. H. Jung, Y. B. Kim, and Y. S. Chung, *Han'guk Saenghwa Hakhoechi, 23*:422 (1990).
117. N. S. Grigoryan, *Mol. Biol., 14*:640 (1980).
118. R. J. Colonno and H. O. Stone, *J. Virol., 17*:737 (1976).
119. T. Otal and V. Hari, *Virology, 125*:118 (1983).
120. N. L. Varich, I. S. Lukashevich, and N. V. Kaverin, *J. Virol., 18*:111 (1976).
121. F. N. Verkleij, P. de Vries, and D. Peters, *J. Gen. Virol., 58*:329 (1982).
122. R. Klemenz, M. Reinhardt, and H. Diggelmann, *Mol. Biol. Rep., 7*:123 (1981).
123. E. Mothes, *Arch. Geschwulstforsch, 52*:191 (1982).
124. F. Harada, N. Kato, and H. Hoshino, *Nucleic Acids Res., 7*:909 (1979).
125. K. B. Temeyer and J. H. Pruett, *Ann. Entomol. Soc. Am., 83*:55 (1990).
126. Y. Gopalakrishna, D. Langley, and N. Sarkar, *Nucleic Acids Res., 9*:3545 (1981).
127. Y. Gopalakrishna and N. Sarkar, *Biochemistry, 21*:2724 (1982).
128. I. Hussain, N. Tsukagoshi, and S. Udaka, *J. Bacteriol., 151*:1162 (1982).
129. M. Altmann, N. Kaeufer, and H. von Doehren, *FEMS Microbiol. Lett., 18*:245 (1983).

130. R. K. Tripathi and G. Kahl, *Plant Cell Physiol., 23*:1101 (1982).
131. H. Ragg, J. Schroeder, and K. Hahlbrock, *Mol. Biol. Rep., 2*:119 (1975).
132. J. L. Nichols and L. Welder, *Biochim. Biophys. Acta, 652*:99 (1981).
133. G. Trapy and R. Esnault, *Phytochemistry, 17*:1859 (1978).
134. A. Konieczny and A. B. Legocki, *Acta Biochim. Pol., 28*:83 (1981).
135. K. K. Ussuf and P. M. Nair, *Indian J. Biochem. Biophys., 18*:276 (1981).
136. V. M. Kish and T. Pederson, *Proc. Natl. Acad. Sci. USA, 74*:1426 (1977).
137. R. P. Kandpal, D. C. Ward, and S. M. Weissman, *Nucleic Acids Res., 18*:1789 (1990).
138. Y. Kitagawa, *Tanpakushitsu Kakusan Koso, 31*:1099 (1986).
139. R. H. Burdon, A. Shenkin, J. T. Douglas, and E. J. Smillie, *Biochim. Biophys. Acta, 474*:254 (1977).
140. G. R. Molloy and S. Johnson, *Prep. Biochem., 12*:77 (1982).
141. Z. Lassota, J. Michalik, M. Szyszko, and A. Krówczynska, *Acta Biochim. Pol., 26*:83 (1979).
142. D. Rosen and M. Edelman, *Eur. J. Biochem., 63*:525 (1976).
143. F. N. Onyezili, *Acta Biochim. Biophys. Hung., 25*:37 (1990).
144. G. Padmanaban, F. Hendler, J. Patzer, R. Rayan, and M. Rabinowitz, *Proc. Natl. Acad. Sci. USA, 72*:4293 (1975).
145. P. Cantatore, C. De Giorgi, and C. Saccone, *Biochem. Biophys. Res. Commun., 70*:431 (1976).
146. J. DeLarco, A. Abramowitz, K. Bromwell, and G. Guroff, *J. Neurochem., 24*:215 (1975).
147. J. B. Mahony and I. R. Brown, *J. Neurochem., 25*:503 (1975).
148. H. Grosjean, C. Takada, and J. Petre, *Biochem. Biophys. Res. Commun., 53*:882 (1973).
149. M. Sprinzl and K. H. Derwenskus, *J. Chromatogr. Libr., 45A*:143 (1990).
150. A. Louie, E. Masuda, M. Yoder, and F. Jurnak, *Anal. Biochem., 141*:402 (1984).
151. Y. Nagata, H. Yokota, O. Kosuda, K. Yokoo, K. Takemura, and T. Kikuchi, *FEBS Lett., 183*:379 (1985).
152. M. G. Mykoniatis, *J. Biochem. Biophys. Methods, 10*:321 (1985).
153. M. V. Nogués, A. Guasch, J. Alonso, and C. M. Cuchillo, *J. Chromatogr., 268*:255 (1983).
154. D. R. Forsdyke, *Anal. Biochem., 137*:143 (1984).
155. J.-P. Jost, M. Seldran, and M. Geiser, *Proc. Natl. Acad. Sci. USA, 81*:429 (1984).
156. N. Shaun, B. Thomas, and H. R. V. Arnstein, *Eur. J. Biochem., 143*:27 (1984).
157. Y. Hashimoto, A. Watanabe, and S. Kawase, *Microbiologica, 7*:91 (1984).
158. A. A. Agranovsky, V. V. Dolja, and J. G. Atabekov, *Virology, 129*:344 (1983).
159. K. Takahashi and T. Yanagisawa, *Zool. Mag., 92*:186 (1983).
160. F. Berthier, S. Alziari, M. Renaud, and R. Durand, *C. R. Soc. Biol., 178*:64 (1984).
161. J. Montoya, G. L. Gaines, and G. Attardi, *Cell, 34*:151 (1983).
162. J. R. Hammett and F. R. Katterman, *Biochemistry, 14*:4375 (1975).
163. R. K. Tripathi and G. Kahl, *Biochem. Biophys. Res. Commun., 106*:1218 (1982).
164. R. J. Glazer, *Biochim. Biophys. Acta, 418*:160 (1976).
165. P. E. Mirkes and B. McCalley, *J. Bacteriol., 125*:174 (1976).
166. A. J. Jaworski, *Arch. Biochem. Biophys., 173*:201 (1976).

167. J. J. Monahan, S. E. Harris, and B. W. O'Malley, *J. Biol. Chem., 251*:3738 (1976).
168. L. J. Rosenthal, *Nucleic Acids Res., 3*:661 (1976).
169. J. DeLarco, S. Nakagawa, A. Abramowitz, K. Bromwell, and G. Guroff, *J. Neurochem., 25*:131 (1975).
170. W. Berthold and L. Lim, *Biochem. J., 154*:517 (1976).
171. P. Z. Khasigov, V. F. Glazkov, A. A. Del'vig, D. A. Kuznetsov, and A. Ya. Nikolaev, *Biokhimiya, 48*:179 (1983).
172. H. Fujimoto and R. P. Erickson, *Biochem. Biophys. Res. Commun., 108*:1369 (1982).
173. R. S. S. Fraser, *Eur. J. Biochem., 60*:477 (1975).
174. A. J. P. Brown and N. Hardman, *J. Gen. Microbiol., 122*:143 (1981).
175. L. Barakett and D. T. N. Pillay, *Gerontology, 28*:1 (1982).
176. S. C. Minocha, R. Minocha, and G. Kahl, *Physiol. Plant, 61*:189 (1984).
177. N. A. Acheson, *Mol. Cell. Biol., 4*:722 (1984).
178. V. A. Mezl and S. Nadin-Davis, *Biosci. Rep., 4*:359 (1984).
179. S. M. Elizarov, P. E. Fel'gengauer, and A. S. Stepanov, *Biochem. Acad. Sci. USSR, 49*:69 (1984).
180. J. M. Fominaya, J. M. Garcia-Segura, and J. G. Gavilanes, *Insect Biochem., 14*:307 (1984).
181. F. Jauker and A. R. Rinaldy, *Exp. Cell Res., 143*:163 (1983).
182. K. B. Clegg and L. Pikó, *J. Embryol. Exp. Morphol., 74*:169 (1983).
183. J. M. Taylor, *Annu. Rev. Biochem., 48*:681 (1979).
184. J. Bresser, H. R. Hubbell, and D. Gillespie, *Proc. Natl. Acad. Sci. USA, 80*:6523 (1983).
185. R. J. Slater, *Methods Mol. Biol., 2*:117 (1984).
186. J. Meinkoth and G. Wahl, *Anal. Biochem., 138*:267 (1984).
187. Yu. P. Zerov, *Biokhimiya, 41*:35 (1976).
188. A. M. Kotin and N. P. Teryukova, *Vopr. Med. Khim., 22*:712 (1976).
189. P. K. Sarkar and S. Chaudhury, *Mol. Cell. Biochem., 53*:233 (1983).
190. J.-L. Ochao, J. Kempf, and J. M. Egly, *Int. J. Biol. Macromol., 2*:33 (1980).
191. P. Cashion, A. Javed, V. Lentini, D. Harrison, J. Seeley, and G. Sathe, *Enzyme Eng., 6*:219 (1982).
192. S. Nadin-Davis and V. A. Mezl, *J. Biochem. Biophys. Methods, 11*:185 (1985).
193. A. E. Berman, N. P. Gornaeva, and V. I. Mazurov, *Biokhimiya, 43*:1830 (1978).
194. U. Lindberg and T. Persson, *Methods Enzymol., 34*:496 (1974).
195. S. Nadin-Davis and V. A. Mezl, *Prep. Biochem., 12*:49 (1982).
196. G. Boileau, C. Barbeau, L. Jeannotte, M. Chrétien, and J. Drouin, *Nucleic Acids Res., 11*:8063 (1983).
197. S. H. Kidson and B. C. Fabian, *Biochim. Biophys. Acta, 824*:40 (1985).
198. L. Trombley and C.-S. Wang, *Microbiol. Immunol., 23*:629 (1979).
199. R. Grotha, *Biochem. Physiol. Pflanz., 170*:273 (1976).
200. H. Persson, M. Perricaudet, A. Tolun, L. Philipson, and U. Pettersson, *J. Biol. Chem., 254*:7999 (1979).
201. M. K. Kistler, R. E. Taylor, Jr., J. C. Kandala, and W. S. Kistler, *Biochem. Biophys. Res. Commun., 99*:1161 (1981).
202. R. A. Cox and N. J. Smulian, *FEBS Lett., 155*:73 (1983).

203. V. V. Adler, V. L. Mechitov, and V. S. Shapot, *Dokl. Akad. Nauk SSSR, 233*:719 (1977).
204. E. K. Wagner, R. J. Frink, and K. P. Anderson, *J. Virol., 39*:559 (1981).
205. S. G. Siddell, *Eur. J. Biochem., 92*:621 (1978).
206. L. T. Auger and G. F. Saunders, *Anal. Biochem., 79*:338 (1977).
207. L. J. Holland and L. J. Wangh, *Nucleic Acids Res., 11*:3283 (1983).
208. S. Ohi and J. Short, *J. Appl. Biochem., 2*:398 (1980).
209. H. C. Schröder, A. Bernd, R. K. Zahn, and W. E. G. Müller, *Mol. Biol. Rep., 8*:233 (1982).
210. D. H. Wreschner and M. Herzberg, *Nucleic Acids Res., 12*:1349 (1984).
211. M. J. Holland, G. L. Hager, and W. J. Rutter, *Biochemistry, 16*:8 (1977).
212. R. W. Haylock and E. A. Bevan, *Curr. Genet., 4*:181 (1981).
213. A. Ron, O. Horovitz, and I. Sarov, *J. Mol. Evol., 8*:137 (1976).
214. I. Barahona, L. Galego, and C. Rodrigues-Pousada, *Eur. J. Biochem., 131*:171 (1983).
215. A. Kakizuka, *Jikken Igaku, 7*:2065 (1989).
216. C. M. Palatnik, R. V. Storti, and A. Jacobson, *J. Mol. Biol., 150*:389 (1981).
217. R. E. Gray and A. R. Cashmore, *J. Biol., 108*:595 (1976).
218. E. M. Tobin and A. D. Klein, *Plant Physiol., 56*:88 (1975).
219. D. A. Stuart, T. J. Mozer, and J. E. Varner, *Biochem. Biophys. Res. Commun., 105*:582 (1982).
220. J. Pasternak, R. J. Winkfein, and M. A. Fernando, *Mol. Biochem. Parasitol., 3*:133 (1981).
221. J. M. Sierra, W. Filipowicz, and S. Ochoa, *Biochem. Biophys. Res. Commun., 69*:181 (1976).
222. V. J. Bardwell and M. Wickens, *Nucleic Acids Res., 18*:6587 (1990).
223. G. Childs, S. Levy, and L. H. Kedes, *Biochemistry, 18*:208 (1979).
224. A. J. Ouellette and C. P. Ordahl, *J. Biol. Chem., 256*:5104 (1981).
225. G. P. Siegal, C. P. Hodgson, P. K. Elder, L. S. Stoddard, and M. J. Getz, *J. Cell. Physiol., 103*:417 (1980).
226. R. B. Moffett and D. Doyle, *Biochim. Biophys. Acta, 652*:177 (1981).
227. R. G. Levenson and K. B. Marcu, *Cell, 9*:311 (1976).
228. E. L. Norwek, H. Nakazato, S. Venkatesan, and M. Edmonds, *Biochemistry, 15*:4643 (1976).
229. W. I. Murphy and G. Attardi, *Biochem. Biophys. Res. Commun., 74*:291 (1977).
230. G. R. Molloy, *J. Biol. Chem., 255*:10375 (1980).
231. M. Edmonds and W. M. Wood, *Biochemistry, 20*:5359 (1981).
232. M. Hirama, A. Takeda, and K. J. McKune, *Anal. Biochem., 155*:385 (1986).
233. K. A. Abraham, T. S. Eikhom, R. M. Dowben, and O. Garatun-Tjeldsto, *Eur. J. Biochem., 65*:79 (1976).
234. F. W. Hirsch, H. N. Nall, W. H. Spohn, and H. Busch, *Proc. Natl. Acad. Sci. USA, 75*:1736 (1978).
235. P. K. Katinakis and R. H. Burdon, *Biochim. Biophys. Acta, 653*:27 (1981).
236. A. J. Minty and F. Gros, *J. Mol. Biol., 139*:61 (1980).
237. B. Teyssendier de la Serve, J.-P. Jouanneau, and C. Péaud-Lenoel, *Plant Physiol., 74*:669 (1984).

238. W. Northemann, T. Andus, V. Gross, and P. C. Heinrich, *Eur. J. Biochem., 137*:257 (1983).
239. J. E. Hyde, M. Goman, R. Hall, A. Osland, I. A. Hope, G. Langsley, J. W. Zolg, and J. G. Scaife, *Mol. Biochem. Parasitol., 10*:269 (1984).
240. P. K. Majumdar and B. A. McFadden, *J. Bacteriol., 157*:795 (1984).
241. R. E. Paque, *Cell Immunol., 30*:332 (1977).
242. B. J. Thiele, Ch. Coutelle, H. D. Hunger, and A. P. Ryskov, *Acta Biol. Med. Ger., 37*:1331 (1978).
243. Ph. Tarroux, *Biochimie, 57*:757 (1975).
244. L. A. Actis, A. Flury, and L. C. Patrito, *Mol. Biol. Rep., 9*:203 (1983).
245. D. Bartels and R. D. Thompson, *Nucleic Acids Res., 11*:2961 (1983).
246. G. Akusjarvi and U. Petterson, *Proc. Natl. Acad. Sci. USA, 75*:5822 (1978).
247. O. V. Chaika, S. N. Khilko, N. A. Grodnitskaya, R. S. Dreizin, and T. I. Tikhomenko, *Vopr. Virusol., 4*:422 (1982).
248. D. J. McGeoch and N. T. Turnbull, *Nucleic Acids Res., 5*:4007 (1978).
249. V. C. Hieber, *Biochem. Biophys. Res. Commun., 104*:1271 (1982).
250. Y. Lin and D. J. Long, *Biochemistry, 19*:1111 (1980).
251. T. J. R. Harris, P. A. Lowe, A. Lyons, P. G. Thomas, M. A. W. Eaton, T. A. Millican, T. P. Patel, C. C. Bose, N. H. Carey, and M. T. Doel, *Nucleic Acids Res., 10*:2177 (1982).
252. K. Sakimura, K. Araki, E. Kushiya, and Y. Takahashi, *J. Neurochem., 39*:366 (1982).
253. J. P. Kraus and L. E. Rosenberg, *Proc. Natl. Acad. Sci. USA, 79*:4015 (1982).
254. R. F. Troxler, H. S. Belford, and F. G. Oppenheim, *Biochem. Biophys. Res. Commun., 111*:239 (1983).
255. J. T. Colbert, P. H. Hershey, and P. H. Quail, *Proc. Natl. Acad. Sci. USA, 80*:2248 (1983).
256. J. A. Gurr, J. F. Catterall, and I. A. Kourides, *Proc. Natl. Acad. Sci. USA, 80*:2122 (1983).
257. K. Nishimori, Y. Kawaguchi, M. Hidaka, T. Uozumi, and T. Beppu, *J. Biochem., 90*:901 (1981).
258. J. H. J. Hoeijmakers, P. Borst, J. van den Burg, C. Weissmann, and G. A. M. Cross, *Gene, 8*:391 (1980).
259. S. Z. Shapiro and J. R. Young, *J. Biol. Chem., 256*:1495 (1981).
260. P. K. Majumdar and V. A. Vipparti, *FEBS Lett., 109*:31 (1980).
261. S. Nakanishi, A. Inoue, T. Kita, S. Numa, A. C. Y. Chang, S. N. Cohen, J. Nunberg, and R. T. Schimke, *Proc. Natl. Acad. Sci. USA, 75*:6021 (1978).
262. R. J. Crawford and J. R. E. Wells, *Biochemistry, 17*:1591 (1978).
263. Yu. N. Baranov, T. N. Zabojkina, V. I. Dubovaya, V. Z. Tarantul, and K. G. Gazaryan, *Dokl. Akad. Nauk SSR, 234*:955 (1977).
264. J. H. Nilson, A. R. Thomason, S. Horowitz, N. L. Sasavage, J. Blenis, R. Albers, W. Salser, and F. M. Rottmann, *Nucleic Acids Res., 8*:1561 (1980).
265. R. K. Strair, A. I. Skoultchi, and D. A. Shafritz, *Cell, 12*:133 (1977).
266. T. Cheng and H. H. Kazazian, Jr., *J. Biol. Chem., 252*:1758 (1977).
267. E. Truelsen, K. Gausing, B. Jochimsen, P. Jorgensen, and K. A. Marcker, *Nucleic Acids Res., 6*:3061 (1979).

268. W. L. Miller and D. C. Alexande, *J. Biol. Chem., 256*:12628 (1981).
269. D. C. Lee, G. S. McKnight, and R. D. Palmiter, *J. Biol. Chem., 253*:3494 (1978).
270. M. A. Innis and D. L. Miller, *J. Biol. Chem., 252*:8469 (1977).
271. K. L. O'Malley, A. Mauron, J. Raese, J. D. Barchas, and L. Kedes, *Proc. Natl. Acad. Sci. USA, 80*:2161 (1983).
272. L. Chan, W. A. Bradley, and A. R. Means, *J. Biol. Chem., 255*:10060 (1980).
273. S. A. Neifkah, V. S. Gaitskhoki, N. A. Klimov, L. V. Puckova, M. M. Shavlovski, and A. L. Schwartzman, *Mol. Biol. Rep., 3*:235 (1977).
274. M. Beato and A. Nieto, *Eur. J. Biochem., 64*:15 (1976).
275. W. L. Miller, J. A. Martial, and J. D. Baxter, *J. Biol. Chem., 225*:7521 (1980).
276. N. J. Cowan, D. S. Secher, and C. Milstein, *Eur. J. Biochem., 61*:355 (1976).
277. K. B. Marcu, O. Valbuena, and R. P. Perry, *Biochemistry, 17*:1723 (1978).
278. C. K. Klukas, F. Cramer, and H. Gould, *Nature, 269*:262 (1977).
279. S. M. Deyev, R. S. Mukhamedov, N. K. Sakharova, O. L. Polyanovsky, V. Viklicky, F. Franek, and J. Hradec, *Immunol. Lett., 7*:315 (1984).
280. C. Auffray and F. Rougeon, *Eur. J. Biochem., 107*:303 (1980).
281. C. Auffray, R. Nageotte, J.-L. Sikorav, O. Heidmann, and F. Rougeon, *Gene, 13*:365 (1981).
282. J. R. Parnes, B. Velan, A. Felsenfeld, L. Ramanathan, U. Ferrini, E. Appella, and J. G. Seidman, *Proc. Natl. Acad. Sci. USA, 78*:2253 (1981).
283. A. K. Deshpande, B. Chatterjee, and A. K. Roy, *J. Biol. Chem., 254*:8937 (1979).
284. S. M. Heywood, D. S. Kennedy, and A. J. Bester, *FEBS Lett., 53*:69 (1975).
285. W. L. Miller, J.-P. Thirion, and J. A. Martial, *Endocrinology, 107*:851 (1980).
286. R. J. Rokowski, J. Sheehy, and K. R. Cutroneo, *Arch. Biochem. Biophys., 210*:74 (1981).
287. R. Kaufmann, A. Belayew, B. Nusgens, C. M. Lapiere, and J. E. Gielen, *Eur. J. Biochem., 106*:593 (1980).
288. J. M. Boss, S. Gillam, R. S. Zitomer, and M. Smith, *J. Biol. Chem., 256*:12958 (1981).
289. A. J. M. Vermorken, J. M. C. H. Hilderink, W. J. M. Van De Ven, and H. Bloemendal, *Biochim. Biophys. Acta, 454*:447 (1976).
290. P. Couble, A. Moine, A. Garel, and J.-C. Prudhomme, *Dev. Biol., 97*:398 (1983).
291. K. Tsutsumi and K. Ishikawa, *Biochem. Biophys. Res. Commun., 100*:407 (1981).
292. W. Burnett, K. Yoon, and J. Rosenbloom, *Biochem. Biophys. Res. Commun., 99*:364 (1981).
293. J. Schweizer and K. Goerttler, *Eur. J. Biochem., 112*:243 (1980).
294. M. Sellami, B. Rether, J. Gangloff, J.-P. Ebel, and J. Bonnet, *Nucleic Acids Res., 11*:3269 (1983).
295. G. Vassart, S. Refetoff, H. Brocas, C. Dinsart, and J. E. Dumont, *Proc. Natl. Acad. Sci. USA, 72*:3839 (1975).
296. J. R. Dugiud, D. F. Steiner, and W. L. Chick, *Proc. Natl. Acad. Sci. USA, 73*:3539 (1976).
297. R. E. Rhoads and G. M. Hellmann, *J. Biol. Chem., 253*:1687 (1978).
298. M. A. Shupnik and J. Gorski, *Mol. Cell. Endocrinol., 17*:181 (1980).
299. P. C. Brown and J. Papaconstantinou, *J. Biol. Chem., 254*:5177 (1979).
300. G. J. Dimitriadis, *Eur. J. Biochem., 118*:255 (1981).

301. J. Guozhong, C. Yan, C. Jianwen, and L. Yaozhen, *Acta Genet. Sin., 8*:241 (1981).
302. P. K. Chakrabartty and P. K. Qasba, *Nucleic Acids Res., 4*:2065 (1977).
303. B. A. Larkins, R. A. Jones, and C. Y. Tsai, *Biochemistry, 15*:5506 (1976).
304. I. M. Evans, R. R. D. Croy, P. Brown, and D. Boulter, *Biochem. Biophys. Acta, 610*:81 (1980).
305. D. J. Shapiro, H. J. Baker, and D. T. Stitt, *J. Biol. Chem., 251*:3105 (1976).
306. J. Mahony, I. Brown, G. Labourdette, and A. Marks, *Eur. J. Biochem., 67*:203 (1976).
307. J. B. Fagan, K. M. Yamada, B. de Crombrugghe, and I. Pastan, *Nucleic Acids Res., 6*:3471 (1979).
308. A. Fuse, H. Heremans, G. Opdenakker, and A. Billiau, *Biochem. Biophys. Res. Commun., 105*:1309 (1982).
309. H. Weening, A. Fuse, G. Opdenakker, J. Van Damme, M. De Lay, and A. Billiau, *Biochem. Biophys. Res. Commun., 104*:6 (1982).
310. M. Houghton, A. G. Stewart, S. M. Doel, J. S. Emtage, M. A. W. Eaton, J. C. Smith, T. P. Patel, H. M. Lewis, A. G. Porter, J. R. Birch, T. Cartwright, and N. H. Carey, *Nucleic Acids Res., 8*:1913 (1980).
311. L. A. Lyakh, S. N. Khil'ko, R. D. Aspetov, D. N. Nosik, A. C. Novokhatskii, and T. I. Tikhonenko, *Biochem. Acad. Sci. USSR, 48*:1503 (1984).
312. S. Hinuma, H. Onda, K. Naruo, Y. Ichimori, M. Koyana, and K. Tsukamoto, *Biochem. Biophys. Res. Commun., 109*:363 (1982).
313. R. C. Bleackley, B. Caplan, C. Havele, R. G. Ritzel, T. R. Mosmann, J. J. Farrar, and V. Paetkau, *J. Immunol., 127*:2432 (1981).
314. Y. Lin, B. M. Stadler, and H. Rabin, *J. Biol. Chem., 257*:1587 (1982).
315. S. M. Tilghman, L. M. Fisher, L. Reshef, F. J. Ballard, and R. W. Hanson, *Biochem. J., 156*:619 (1976).
316. P. B. Iynedjian and R. W. Hanson, *Eur. J. Biochem., 90*:123 (1978).
317. D. Sagher, H. Grosfeld, and M. Edelman, *Proc. Natl. Acad. Sci. USA, 73*:722 (1976).
318. P. Madhusudanan Nair and K. K. Ussuf, *Indian J. Exp. Biol., 17*:460 (1979).

III

RESEARCH ON BIORECOGNITION

8

Affinity Chromatography in Biology and Biotechnology

Probing Macromolecular Interactions Using Immobilized Ligands

Irwin Chaiken

SmithKline Beecham, King of Prussia, Pennsylvania

I. FROM PREPARATIVE TO ANALYTICAL AFFINITY CHROMATOGRAPHY: PROOFS OF TWO PRINCIPLES

Development of polymeric support carriers for immobilized ligands and linkage chemistries to attach the ligands has provided a multitude of tools for the preparative use of affinity chromatography (Chapters 1 to 7). At the historical roots of this tie between affinity support chemistry and preparative applications are the groups of Porath and Anfinsen. In the 1960s, the Porath school provided hydrophilic supports friendly for the separation of biological marcomolecules and chemistries for the attachment of ligands to these supports (1) (see Ref. 2 for review). The Anfinsen school then took the plunge and, with the agarose support and cyanogen bromide linkage chemistry reported by Porath and colleagues, immobilized thymidine diphosphate and used the resulting affinity matrix to purify staphylococcal nuclease (3). Their demonstration of an impressive, one-step enzyme purification by an elution sequence of binding to the affinity support under interaction buffer conditions followed by chaotropic dissociation of bound enzyme from support (see Figure 1B) proved the principle of preparative affinity chromatography. Their success established a generalized purification scheme used since then for many proteins and other macromolecules (4,5).

The success of affinity chromatography as a preparative tool led eventually to another principle—that it could be used analytically to characterize macromolecular recognition quantitatively. Two key factors that make affinity

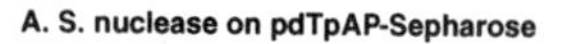

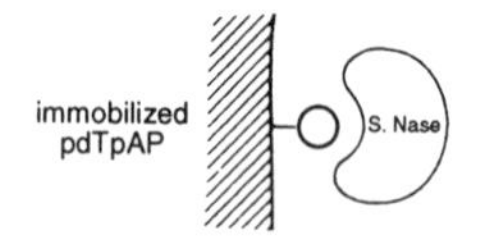

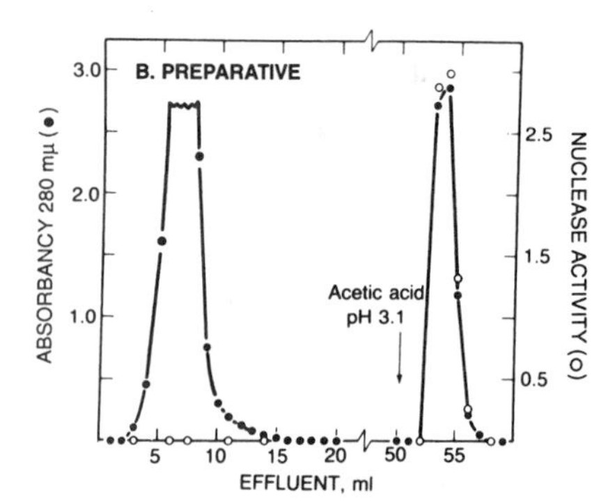

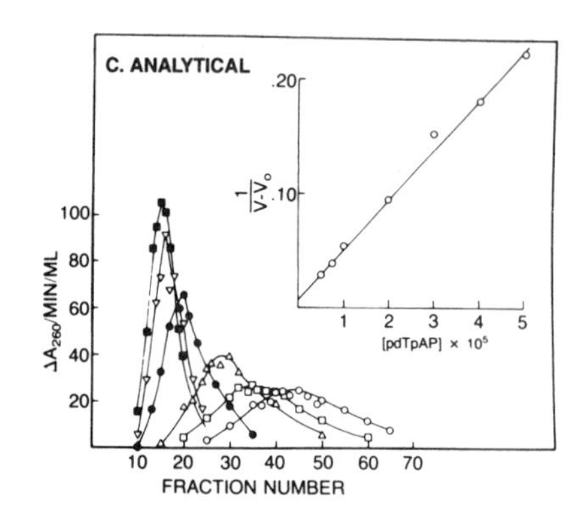

Figure 1 Proofs of two principles in affinity chromatography using staphylococcal nuclease elution on deoxythymidine 3′-phosphate, 5′-aminophenylphosphate–Sepharose. (A) Schematic diagram of elution of S. nuclease on pdTpAP–Sepharose. (B) Demonstration of preparative affinity chromatography by the purification of nuclease (adapted from Ref. 4). Crude fraction from *S. aureus* broth was eluted, with most protein eluted without retardation and nuclease, retained under the initial binding conditions, eluted upon chaotropic elution with dilute acetic acid. (C) Demonstration of analytical affinity chromatography by zonal elutions of nuclease, with experimental values of elution volume V determined from positions of retarded peaks in elution profiles obtained at different concentrations ($[L_T]$) of competitor pdTpAP. Data replotted as $1/(V - V_0)$ versus $[L_T]$, where V_0 is void or unretarded volume, allow quantitation of dissociation constants for nuclease interactions with both immobilized and soluble (competing) ligand. (Adapted from Ref. 7.)

chromatography such a successful and powerful preparative method seem apparent: selectivity (of formation of complexes between accessible immobilized ligands and specific mobile molecules) and reversibility (of the association process to permit elution of selectively bound molecules). It is just these features, selectivity and reversibility, which led to the prediction that affinity chromatography could be configured as an analytical sensor for recognition properties of mobile macromolecules for immobilized ligands. Selectivity implies the likelihood that the interaction process reflects the same biologically relevant binding-site recognition that occurs in solution. Reversibility implies that the elution of selectively bound molecules can be achieved. Hence if elution of macromolecules on an affinity matrix is performed isocratically under binding conditions, chromatographic retardation should be proportional to affinity for immobilized ligand.

Based on such logic, we (6) took another experimental leap of faith with staphylococcal nuclease elution on agarose-immobilized thymidine diphosphate, in this case with conditions (including low-enough immobilized nucleotide concentration) to allow isocratic elution under binding (nonchaotropic) conditions (Figure 1C). Two types of experiments were done with zonal elution: simple retardation with binding buffer and competitive elution with varying concentrations of soluble ligand in the buffer. From the magnitude of retardation with buffer elution and dependence of retardation on soluble ligand concentration, quantitative interaction properties of staphylococcal nuclease with both immobilized and soluble ligands were measured, as $K_{M/P}$ and $K_{L/P}$, respectively. The values of these parameters were found to be consistent with K_d values determined by other methods. The nuclease results, taken with similar results reported in other laboratories at virtually the same time (7,8), provided a proof of principle for analytical affinity chromatography (AAC). Since then, many related studies have been reported. In general, quantitative parameters determined chromatographically for eluting macromolecules with immobilized ligands have been found to agree well with corresponding interaction parameters measured fully in solution by other biophysical as well as functional assays. Nonetheless, the AAC method also offers the opportunity to probe the impact on recognition of solid-phase immobilization of one of the interactors. Since presence of an interactor in a solid phase is a condition that often occurs in biological systems (e.g., membranes), the AAC method can offer a chance to evaluate biomolecular interactions in a more biologically relevant condition than the dilute aqueous solutions often used with other methods.

II. AAC AND CHARACTERIZATION OF MACROMOLECULAR RECOGNITION IN BIOLOGY

AAC by now has been used to characterize interactions of many biologically occurring molecules, including macromolecules (9–12) (see Chapter 10). A dominant virtue of analytical affinity chromatography, and of analytical uses of immobilized ligands generally, is the versatility to measure interactions across a wide range of affinity, including low and modest affinity, to analyze interactions of molecules with a wide range of size, from small to very large, and to do so on a micro scale. Thus AAC can be seen as a generic method to characterize interaction properties of molecules in biological pathways, even when only relatively small amounts of interacting components can be obtained.

This consideration led us some time ago (see Ref. 13 for review) to attempt to define molecular events in the biosynthetic pathway of neurophysins and neurohypophyseal peptide hormones (Figure 2B) using both peptide and protein affinity supports (Figure 2A). The major molecular species along the pathway from biosynthetic precursor to mature peptides and proteins could be analyzed

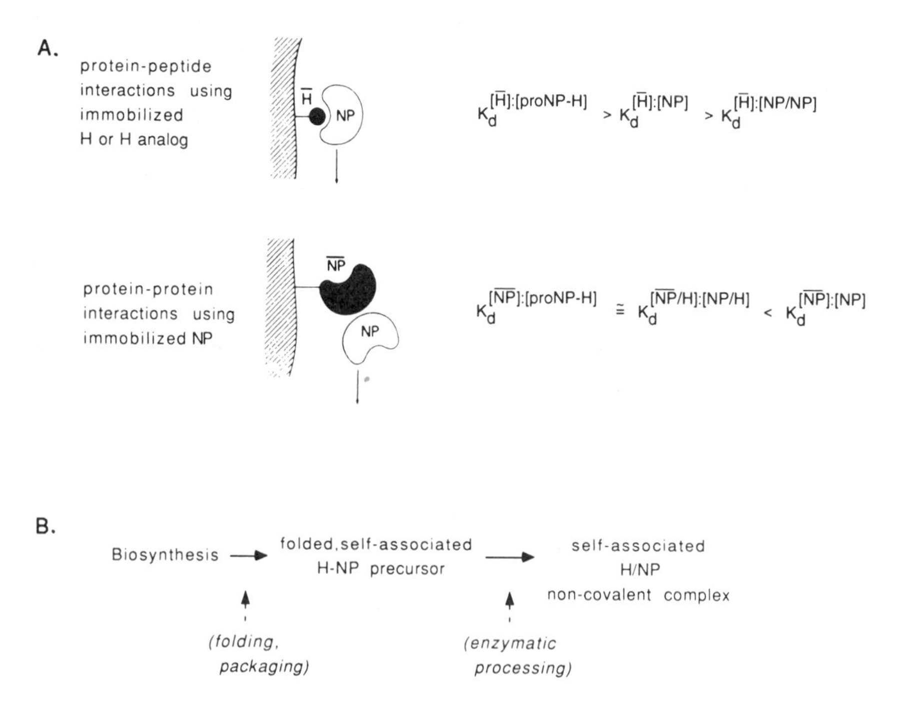

Figure 2 Use of analytical affinity chromatography to characterize molecular events in the biosynthetic pathway of vasopressin, oxytocin, and their associated protein carriers, the neurophysins. (A) Schemes showing the elutions of neurophysin (NP) on affinity matrices containing immobilized NP ($\overline{NP}$) and immobilized hormone or hormone analog ($\overline{H}$) homologous to vasopressin and oxytocin. Part (A) also shows the relationship of dissociation constants determined for the various elutions. In the notation used, the superscripts denote the interacting complex of the matrix ligands ($\overline{H}$, immobilized hormone or hormone analog; $\overline{NP}$, immobilized NP; NP/H, immobilized NP with noncovalently associated H) with eluting precursor (proNP-H), unliganded neurophysin (NP), liganded neurophysin (NP/H), and neurophysin dimer (NP/NP). (B) Molecular events occurring in the biosynthetic pathway in neuroendocrine cells, including the assembly of biosynthetic precursor conformation as deduced by AAC. These events include the folding of precursor into complexes, with hormone and neurophysin domains interacting intramolecularly and, after proteolytic processing, the assembly of neurophysin and hormone into cooperative noncovalent complexes. (Adapted from Ref. 13.)

with AAC, which thus provided a single unifying method to compare the different interactions. Among other findings, the AAC data obtained (Figure 2A) showed that neurophysin binding to peptide ligands potentiates NP self-association, by the higher affinity of NP/H (noncovalent complex of NP and H) for immobilized NP/H than of NP for immobilized NP. Precursor was shown to have the potential to self-associate, since precursor interacted with immobilized NP. The results showed that the peptide and protein domains in the biosynthetic precursor interact with each other, as inferred from the observation that the immobilized NP association affinity of precursor is greater than that of mature neurophysin alone and closer to that of mature NP when complexed noncovalently with peptide hormone (NP/H) and also from the observation that the precursor does not interact with immobilized peptide ligand. Thus measurements with the chromatographic method allow the conclusion that the precursor can self-assemble into a conformationally ordered state. This state may well affect the proteolytic processing that leads to the biologically active peptide hormones oxytocin and vasopressin and their carrier proteins, the neurophysins.

III. AFFINITY SCREENING, MOLECULAR MIMICRY, AND *DE NOVO* DESIGN

Beyond the analysis of biological molecules and their interaction mechanisms, AAC can provide a versatile screening method to identify and optimize molecules as recognition agents. Such recognition molecules can be sought as (a) direct binders to an affinity support containing an immobilized target or (b) effectors of the interaction between soluble and immobilized interactors.

A case in point for the former is the characterization of CD4 mimetics for binding to immobilized HIV gp120. CD4 is the T-cell receptor for HIV, and CD4 mimetics may be useful to antagonize the HIV interaction with T cells by blocking the gp120 binding sites of HIV (Figure 3A). Here a convenient affinity screen has been devised using an affinity column of immobilized gp120 (Figure 3B). This column has been used to measure the interaction of immobilized gp120 with intact sCD4 (Figure 3C). Currently, we are using this column as a direct and rapid assay tool to evaluate mutants and other redesigned forms of sCD4 (D. Myszka, unpublished). Interestingly, a similar assay of CD4-gp120 binding has been devised using the surface plasmon resonance biosensor (Figure 3D). This type of biosensor, which is described in Chapter 9, embodies some of the same basic methodological features originally established with AAC, in particular the use of immobilized ligands to measure interactions of soluble macromolecules flowing over the affinity surface (see Ref. 12 for a previous comparative of these two methods). The biosensor assay promises to contribute to affinity screening efforts with CD4 mimetics and other macromolecules. Of note, the biosensor is

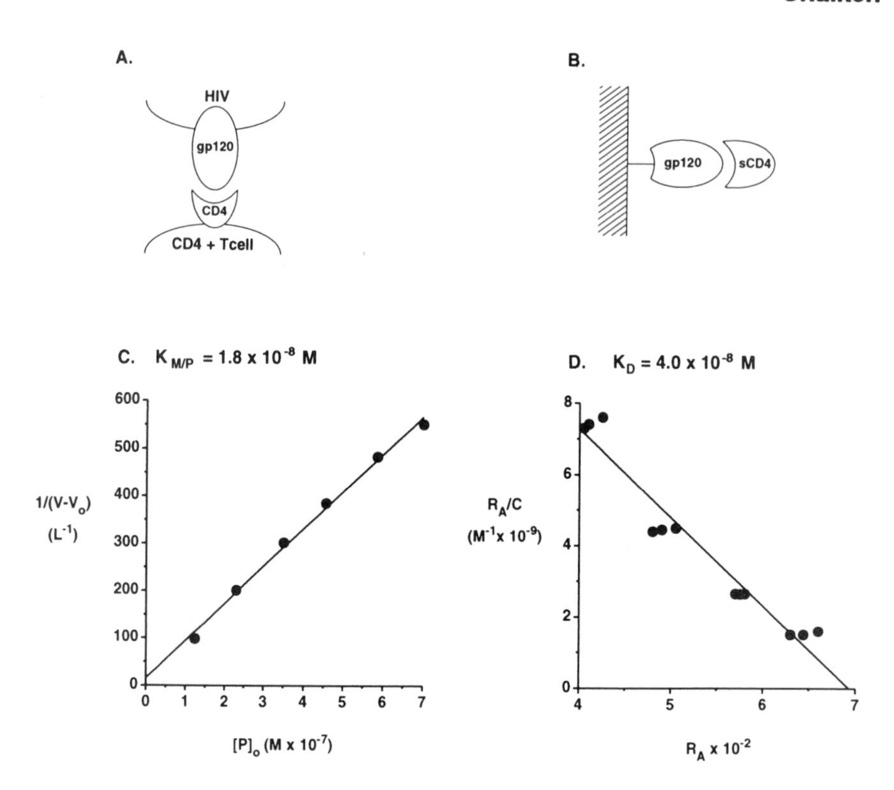

Figure 3 Analytical solid-phase measurements of interaction of soluble CD4 (sCD4) with human immunodeficiency virus envelope protein gp120. (A) Schematic diagram of HIV-1 docking onto CD4-bearing T cell through CD4-gp120 recognition. (B) Scheme of solid-phase affinity systems designed with immobilized gp120. (C) Frontal elution AAC (data from M. Brigham-Burke, D. O'Shannessy, and I. Chaiken, unpublished). Elutions were made of sCD4, from 10 to 100 μL, in the concentration range 1.2 to 7×10^{-7} *M*. Values of elution volume *V* were determined as first derivatives of the fronts of elution in continuous elution chromatography. Experimental conditions were flow rate 360 μL/min, 20 m*M* potassium phosphate–150 m*M* sodium chloride, (pH 7.2). Data were used to obtain the dissociation constant of 1.8×10^{-8} *M* for the complex of sCD4 (P) with immobilized gp120 (M). (D) Surface plasmon resonance biosensor analysis (data from Ref. 17.) (Adapted from Ref. 12.)

configured to measure rate constants of association and dissociation of immobilized ligand/soluble interactor complex. In contrast, AAC measurement of rate constants, which has been suggested theoretically, has been found experimentally to be complicated by mass transfer limitations (10). One notable limitation of the biosensor method is that it requires that the mobile interactor be a macromolecule, while AAC has no restriction to the size of the mobile interactor other than the need to detect its elution.

De novo–designed recognition molecules can also be screened by direct binding to affinity supports. For example, we have used AAC for antisense peptide recognition and design studies. Antisense peptides are sequences encoded in antisense DNA. Chemically synthesized antisense peptides have been found to bind selectively to the corresponding sense peptides and proteins, the native sequences encoded in the corresponding sense DNA. AS peptide binding to sense peptides and proteins has been measured by chromatographic elution of AS peptides on immobilized sense peptide (15,16) (see Ref. 17 for review). Thus it can be predicted that redesigned AS peptide sequences with optimized binding affinity to sense peptide can be sought by affinity screening of AS variants on the same sense peptide supports.

A somewhat different use of affinity screening is to identify effectors of macromolecular interactions by observing the shift in retardation of a macromolecule on an affinity support when eluted with buffers containing potential effectors. Recently, we established an assembly assay for HIV gag p24 with this idea of assembly effectors in mind (18). The p24 protein is a major nucleocapsid protein of HIV. Its self-association and/or that of its biosynthetic precursor p55 gag is believed to be a major step of viral core assembly. Thus effectors of p24 self-association could be therapeutically useful as antiviral agents by antagonizing either virus internalization (by affecting capsid disassembly) or virus maturation (by inhibiting cellular capsid assembly). As an experimental approach to find such effectors, we immobilized p24 monomers on an affinity support and used both frontal and zonal elutions of soluble p24 to visualize self-association (Figure 4). The dimer model assumed in the AAC process (Figure 4A) was confirmed by analytical ultracentrifugation. The K_d values measured, 3.7×10^{-5} and 3.0×10^{-5} M at pH 5.4 for frontal and zonal elutions, respectively (Figure 4B and C), are about the same as that measured by the analytical ultracentrifuge. Thus the AAC system can be considered a sensor of p24 self-association. Small molecule effectors of self-assembly (Figure 4A and D) can be detected by their ability to shift the p24 retardation (either increase or decrease).

IV. CONCLUDING COMMENTS

In affinity chromatography, we have a tool for both preparative and analytical applications. These two wings of the affinity chromatographic field are being developed together. The overall utility of affinity chromatography as a preparative tool is generally appreciated. Analytical utility is less widely recognized. Yet it is likely that the joint development of both aspects of affinity chromatography will

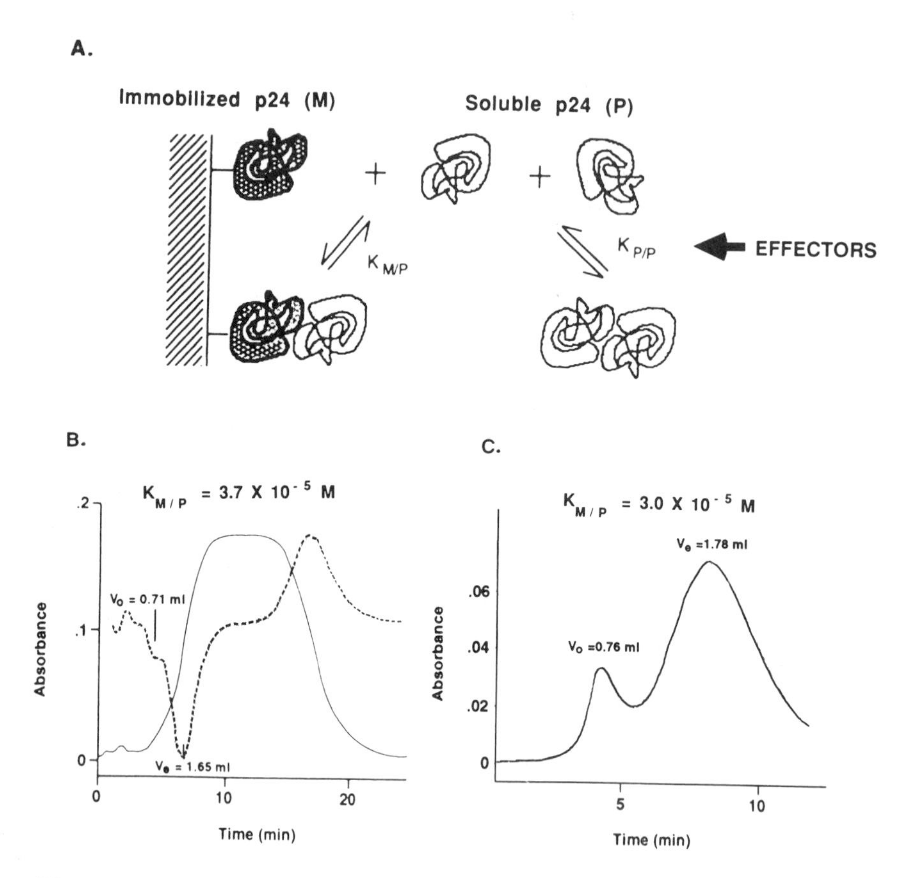

Figure 4 Analysis of HIV-1 p24 self-association and screening for effectors of self-association by elution of soluble p24 (P) on immobilized p24 (M). (A) Scheme showing the dimerization interactions of p24 in solution (defined by the dissociation constant $K_{P/P}$) and on the solid phase (defined by dissociation constant $K_{M/P}$). (B, C) Typical analytical chromatographic elution profiles for the p24 system described in (A). The $K_{M/P}$ values shown were determined from elution data obtained at a range of p24 concentrations. (B) Frontal chromatography profile for p24 at a concentration of 80 μg/mL (continuous line); the minimum of the first derivative of the elution profile (dashed line) defines the experimental elution volume V_e. Experimental conditions were: buffer PBS (pH 7.0), flow rate 0.25 mL/min, temperature 20°C, detection by absorbance at 280 nm. (C) Zonal chromatography profile for p24 at the same experimental conditions as in (B). V_0 is the nonretarded elution volume. Screening for effectors of p24 self-association can be accomplished by determining changes in p24 retardation with compounds added to elution buffers. (Adapted from Refs. 12 and 18.)

make the overall field increasingly impactful in the study of molecular mechanisms in biology and the pursuit of molecular design in biotechnology.

REFERENCES

1. R. Axen, J. Porath, and S. Ernback, *Nature*, 1302 (1967).
2. J. Porath and T. Kristiansen, in *The Proteins* (H. Neurath and R. L. Hill, eds.), Vol. 1, Academic Press, New York, 1975, pp. 94–178.
3. P. Cuatrcasas, M. Wilchek, and C. B. Anfinsen, *Proc. Natl. Acad. Sci. USA, 61*:636 (1968).
4. W. B. Jakoby and M. Wilchek, eds., *Affinity Techniques: Enzyme Purification Part B, Methods in Enzymology*, Vol. 36, Academic Press, New York, 1974.
5. P. D. G. Dean, S. W. Johnson, and F. A. Middle, eds., *Affinity Chromatography: A Practical Approach*, IRL Press, Oxford, 1985.
6. B. M. Dunn and I. M. Chaiken, *Proc. Natl. Acad. Sci. USA, 71*:2382 (1974).
7. L. W. Nichol, A. G. Ogston, D. J. Winzor, and W. H. Sawyer, *Biochem. J., 143*:435 (1974).
8. K.-I. Kasai and S.-I. Ishii, *J. Biochem., 77*:261 (1975).
9. I. M. Chaiken, *Anal. Biochem., 97*:1 (1979).
10. I. M. Chaiken, ed. *Analytical Affinity Chromatography*, CRC Press, Boca Raton, Fla., 1987.
11. D. J. Winzor and J. De Jersey, *J. Chromatogr. Biomed. Appl., 492*:377 (1989).
12. I. M. Chaiken, S. Rose, and R. Karlsson, *Anal. Biochem., 201*:197 (1992).
13. H. E. Swaisgood and I. M. Chaiken, in *Analytical Affinity Chromatography* (I. M. Chaiken, ed.), CRC Press, Boca Raton, Fla., 1987.
14. R. Granzow, T. Van Cott, and L. Mattsson, *Abstracts of the Protein Society Meeting*, Baltimore, June 1991, p. 128.
15. Y. Shai, M. Flashner, and I. M. Chaiken, *Biochemistry, 26*:669 (1987).
16. G. Fassina, M. Zamai, M. Brigham-Burke, and I. M. Chaiken, *Biochemistry, 28*:8811 (1989).
17. I. M. Chaiken, *J. Chromatogr., 597*:29 (1992).
18. S. Rose, P. Hensley, D. J. O'Shannessy, J. Culp, C. Debouck, and I. M. Chaiken, *Proteins—Structure, Function, Genetics, 13*:112 (1992).

9

Surface Plasmon Resonance Detection in Affinity Technologies

BIAcore

Lars G. Fägerstam

Pharmacia Biosensor AB, Uppsala, Sweden

Daniel J. O'Shannessy

SmithKline Beecham, King of Prussia, Pennsylvania

I. INTRODUCTION

The study of macromolecular interactions is central to gaining an understanding of the molecular mechanisms involved in biological processes. Few techniques are available to study such interactions in a label-free model. The analytical affinity chromatographic techniques described in this volume address this need using classical chromatographic approaches. Another technology that has received much attention in recent years, particularly as potential immunosensor devices, is *surface plasmon resonance* (SPR) (1–10). In contrast to most other techniques used to detect and quantitate macromolecular interactions, SPR detection does not require labeled interactants. In addition, SPR detection is a *direct optical sensing technique* that allows one to visualize macromolecular interactions in *real time*. Since detection of interactions is in real time, SPR allows investigations of the *kinetics* of interactions rather than just a determination of the dissociation constant, or K_d, as described for analytical affinity chromatography, immunoprecipitation, ELISA, or other techniques. The following will therefore be a brief description of the principles of SPR detection and examples of the applications of the technology, particularly as it relates to affinity technologies. It should be noted at this point that the SPR technology to be described is similar to analytical affinity chromatography, for example, in that one of the interacting species is in an immobilized form (i.e., one observes the interaction between immobilized ligand and soluble ligate). However, since many biological processes occur at or on surfaces, this should not invoke limitations of the technology.

II. TECHNOLOGY AND SYSTEM DESIGN

A. Principles of Surface Plasmon Resonance

SPR is an optical phenomenon that occurs as a result of total internal reflection (TIR) of light at a metal film–liquid interface. Total internal reflection is observed in situations where light travels through an optically dense medium such as glass and is reflected back through that medium at the interface with a less optically dense medium such as buffer, provided that the angle of incidence is greater than the critical angle required for the pair of optical media. Although the light is totally reflected, a component of the incident light momentum, termed the *evanescent wave*, penetrates a distance on the order of one wavelength into the less dense medium, in this case buffer. The *evanescent wave* phenomenon has been exploited to excite molecules in close proximity to a glass–liquid interface in a process termed *total internal reflection fluorescence* (1).

If, however, the incident light is monochromatic and p-polarized, and the interface between the media is coated with a thin (a fraction of the light wavelength) metal film, the *evanescent wave* will interact with free oscillating electrons, or *plasmons*, in the metal film surface. Therefore, when *surface plasmon resonance* occurs, energy from the incident light is lost to the metal film, resulting in a decrease in the reflected light intensity (see Figure 1). The *resonance* phenomenon occurs only at an acutely defined angle of the incident light. This angle is dependent on the refractive index of the medium close to the metal–film surface. Changes in the refractive index of the buffer solution, to a distance of about 300 nm from the metal–film surface, will therefore alter the resonance angle. Continuous monitoring of the resonance angle allows quantitation of changes in the refractive index of the buffer solution close to the metal–film surface. As discussed in more detail below, the interaction of macromolecules in the buffer solution causes a change in refractive index in close proximity to the metal–film surface, which translates into a change in the resonance angle which is detected and quantitated by the instrument. It is worth noting that light does not pass through the detection volume (defined by the size of the illuminated area at the interface and the penetration death of the *evanescent wave*) and that the optical device is on one side of the metal–film, detecting changes in the refractive index in the buffer on the opposite side. An extensive treatment of the theory of SPR may be found in Ref. 4.

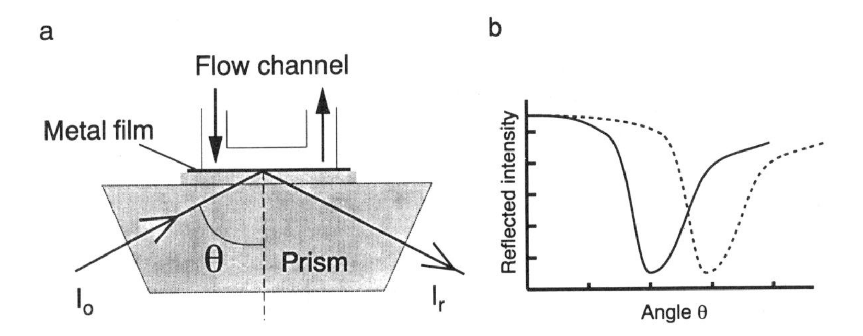

Figure 1 Physics of surface plasmon resonance: (a) optical configuration for excitation of *surface plasmons* at a metal–liquid interface; (b) intensity of reflected light (I_r) as a function of the incident angle, Q.

B. Instrumentation

Pharmacia Biosensor AB (Uppsala, Sweden) has commercialized SPR technology with an instrument called BIAcore. The basic components of the instrument are shown schematically in Figure 2. The instrument consists of a processing unit containing the SPR detector and an integrated microfluidics cartridge that together with the auto-sampler, controls delivery of sample plugs into a buffer stream passing continuously across the sensor chip surface (see below). The entire system is controlled by a personal computer which also acts as a data acquisition and analysis system.

The sensor chip (8,11), depicted in Figure 3, consists of a glass substrate onto which a thin (50-nm) gold film is deposited. The gold film is derivatized with a monolayer of a long chain hydroxyalkyl thiol which serves both as a barrier to proteins and other ligands from coming into direct contact with the metal surface, as well as a functionalized layer for further derivitization. Onto the hydroxyalkyl thiol is covalently attached a 100-nm-thick layer of carboxymethyl dextran. This dextran layer serves to produce a hydrophilic layer suitable for macromolecular interaction studies as well as containing functional groups available for ligand immobilization chemistries (see below).

The microfluidic cartridge (12) contains two identical sets of pneumatic valves and channels, each with two sample loops for injection volumes of between 5 and 45 μL. Located at the surface of the cartridge is a flow cell block with four channels. When the cartridge is "docked" with the sensor chip, four parallel flow cells each with a volume of 60 nL are formed. The four flow cells that are

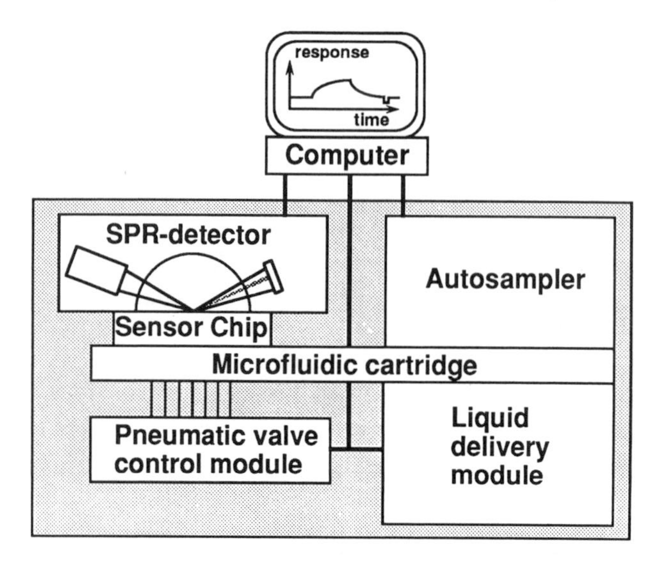

Figure 2 Basic instrument components. The processing unit contains the SPR detector and an integrated microfluidics cartridge that together with an autosampler controls delivery of sample plugs into a transport buffer that passes continuously over the sensor chip surface. Pneumatic valves within the microfluidic cartridge direct the flow. A personal computer serves as a system controller as well as a data acquisition and analysis unit.

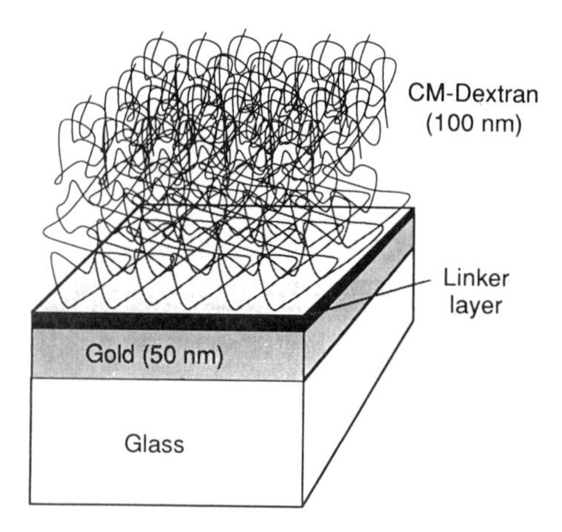

Figure 3 Sensor chip composition. The sensor chip consists of a 50-nm gold film deposited onto a glass slide. The gold film is derivatized with a monolayer of a long chain hydroxyalkyl thiol to which dextran is covalently attached. The dextran extends about 100 nm into the solution phase.

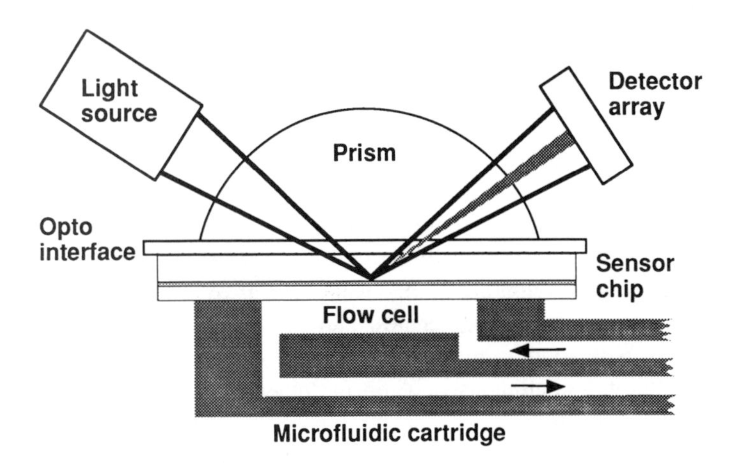

Figure 4 SPR detector in relation to the sensing surface. This schematic represents the SPR detector, sensor chip, and microfluidics cartridge docked together. The position of the reflectance minimum is monitored continuously by the photodiode array. When a molecule introduced via the flow cell is captured onto the sensor chip surface, the position of the reflectance minimum will change. In the instrument, a wedge-shaped light beam and a two-dimensional detector array enables monitoring of four flow cells.

illuminated simultaneously by the transverse wedge of light can be connected to any of the sample loops through the pneumatic valves.

Figure 4 depicts the configuration of the SPR detector with respect to the integrated microfluidics cartridge docked with the sensor chip. The sensor chip is held in contact with the prism of the optical system by the microfluidics cartridge. Plane-polarized light from a high-efficiency, near-infrared LED is focused into a transverse wedge through the prism, onto the side of the sensor chip *opposite* the gold-film. Reflected light is monitored by a fixed, two-dimensional array of light-sensitive diodes positioned such that the resolution between diodes corresponds to a difference in the angle of reflection of 0.1°. Computer interpolation routines process the data from the diode array to determine the resonance angle to an accuracy of $10^{-4\circ}$. Averaged readings are obtained at a frequency of 5 Hz. The use of a fixed-diode-array detector eliminates moving parts from the optical system, increasing the reliability of the instrument and allowing changes in the resonance angle to be detected in *real time.* A layer of silicone polymer, matched in refractive index with the glass of the sensor chip, ensures a precise contact between the removable sensor chip and the fixed optical system.

C. Data Output: The Sensorgram

By continuously monitoring the refractive index (RI), detected as a change in the *resonance angle*, in the detected volume and plotting this value as a function of time, a *sensorgram* is obtained. The *y*-axis of the sensorgram is denoted the *resonance signal* and is indicated in *resonance units* (RU) or *response units.* A change in signal of 1000 RU corresponds to a 0.1° shift in the surface plasmon resonance angle and for proteins is equivalent to a surface concentration of 1 ng/mm^2 (13). The total range covered by the SPR detector is 3°, or 30,000 RU. The resonance signal at any given point in time is the sum of contributions from the sensor chip surface, interacting molecules, and the bulk solution. Under conditions of constant bulk solution refractive index, the amount of interacting molecule can be monitored continuously. If, on the other hand, the refractive index of the ligate solution differs from that of the continuous buffer flow, the amount of interacting ligate may be quantitated from readings taken between sample injections where constant RI buffer flow is operating. For kinetic measurements where the progress of the binding curve rather than the absolute response values are used, correction for sample bulk refractive index is not necessary (see below).

D. Immobilization of Ligands

The methodology relies on immobilization of ligands onto the carboxymethylated dextran of the sensor chip. The immobilization procedure is performed with the sensor chip in place in the instrument and is monitored continuously by the SPR detector. Figure 5 represents a typical sensorgram obtained on immobilization of proteins, in this case the HIV-1 envelope glycoprotein, gp120, using standard *N*-hydroxsuccinimide ester chemistry. In this procedure the carboxyl groups of the carboxy methyl dextran are activated by injection of a mixture of *N*-ethyl-*N'*-(dimethylaminopropyl)carbodiimide hydrochloride (EDC) and *N*-hydroxysuccinimide (NHS), both prepared in water. In a second step, the ligand (protein) is injected over the surface in a low ionic strength solution at a pH value *below* the isoelectric point of the protein. Since only a fraction of the carboxyl groups are activated by the ECD–NHS mixture, the positively charged protein concentrates onto the surface of the sensor chip via electrostatic attraction, and simultaneously, the amines of the protein react with the NHS esters, resulting in the formation of amide links between the protein and the dextran surface (8). It should be noted that protein (ligand) concentrations used for immobilization are usually only in the range 20 to 30 μg/mL, and since only 100 μL is required, a total of only 2 to 3 μg of ligand is needed. Residual NHS, esters remaining after ligand immobilization are then reacted with a solution of ethanolamine. Finally, the surface is subjected to an acid wash to remove noncovalently adsorbed protein. The entire

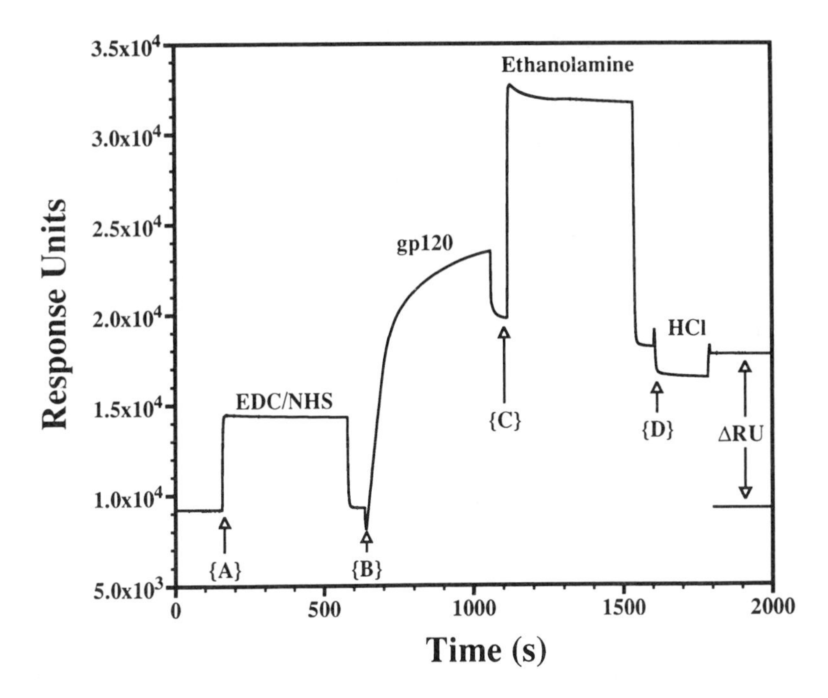

Figure 5 Sensorgram obtained on immobilization of HIV-1 gp120 onto sensor chip CM5. During the immobilization protocol, a constant flow of HBS of 5 μL/min was maintained. At {A}, 30 μL of a 1:1 mixture of 0.1 *M* EDC and 0.1 *M* NHS (both prepared in water) was injected. The signal represents a change in refractive index from that of HBS. At {B}, 30 μL of a 20-μg gp120/mL acetate buffer (10 m*M* sodium acetate, pH 4.7) was injected. The curve observed represents the covalent immobilization of gp120 onto the surface via reaction of the primary amines with the NHS-ester activated surface, detected in real time. At {C}, 30 μL of a 1 *M* ethanolamine (pH 8.5) solution was injected to react with residual NHS esters on the sensor chip surface, and at {D}, 12 μL of 0.1 *M* HCl was injected to wash noncovalently bound gp120 from the surface. The degree of immobilization, in response units, is taken as the difference in response between the underivatized, naked layer and the layer after HCl washing, as indicated by the arrows (ΔRU).

immobilization procedure, which typically takes less than 30 min, can be controlled by parameters such as protein concentration, protein solution ionic strength and pH, reagent (EDC–NHS) concentration, and reaction times. Reproducibility studies for the immobilization of three different proteins onto 150 sensor chips shows a precision of better than 4% relative standard deviation. This value includes errors in the preparation of reagents, variability in the sensor chip per se, as well as instrument performance. Depending on the stability of the immobilized ligand, the sensor surface can be regenerated and used for a number of analytical determinations (see below).

Recently, several alternative chemistries for immobilization of ligands onto the sensor chip surface were described (14). In each case the carboxymethylated dextran surface was activated with a mixture of EDC and NHS as described above. Subsequent derivatization of the surface chemistry was performed to generate amine-derivatized, hydrazino-derivatized, sulfhydryl-derivatized, or maleimide-derivatized surfaces, allowing for the immobilization of a variety of functionalized ligands. In addition, the introduction of these chemistries should allow for the site-directed immobilization of ligands, in particular antibodies, onto the sensor chip surface, which has been demonstrated to be of value in chromatographic systems with respect to molar binding capacities. Indeed, the carboxymethylated dextran surface of the sensor chip can be considered as a microchromatographic matrix, and numerous chemical approaches to the immobilization of ligands should be possible, assuming that the reagents used are compatible with the hardware of the instrument.

E. Assay Configurations

1. *Direct Binding*

The simplest assay format on BIAcore is to observe directly the binding of soluble ligate to immobilized ligand, where the ligand may be an antibody, antigen, receptor, binding protein, peptide, or the like. An example of this type of analysis is shown in Figure 6, which demonstrates the binding of soluble T-cell receptor CD4 (sCD4) to an immobilized monoclonal antibody, L-71. In this type of assay it is most important to have purified ligand for immobilization, although it is possible to immobilize partially (>75%) purified ligands. The immobilized ligand layer acts as a baseline from which ligate binding can be determined and quantitated. Since the change in *resonance units* is directly proportional to the change in mass at the sensor chip surface (13), there are limitations to the direct binding assay in that the smaller the mass of the ligate, the lower the response on binding. The lower limit of detection of this instrument configuration is approximately 1000 mass units. Note in Figure 4 that after binding, the ligate may be desorbed to allow regeneration of the binding (ligand) layer. Depending on the stability of the immobilized ligand layer, the surface may be used for numerous analyses. Table 1 summarizes some stability data obtained for sCD4 binding to immobilized MoAb L-71 through 100 cycles.

1. *Indirect Binding*

The indirect binding configuration adds versatility to the BIAcore instrument. An example of how an indirect assay may be configured is shown in Figure 7. In this example, a capture antibody (rabbit antimouse Fc; RaMFc) is firstly immobilized.

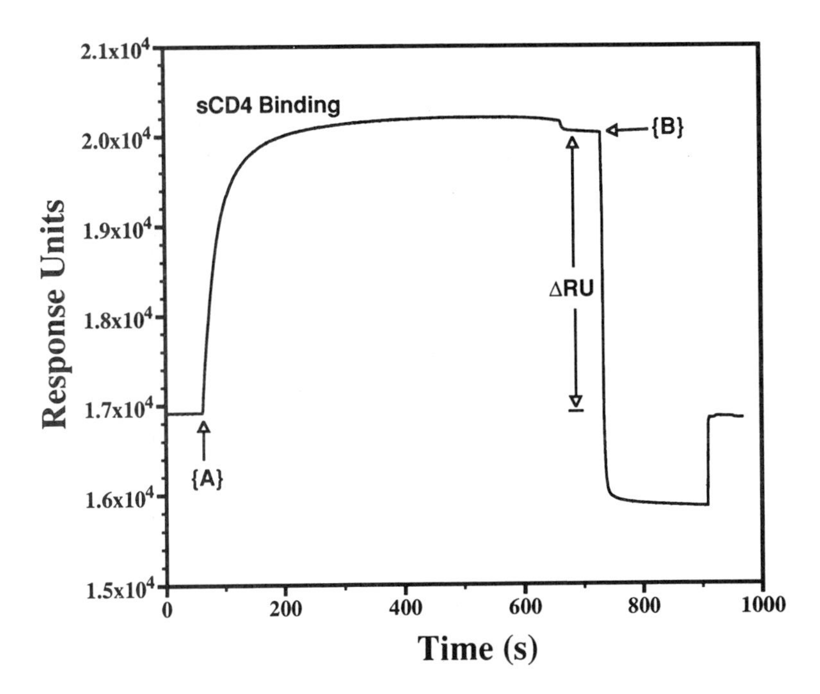

Figure 6 Direct binding assay configuration. MoAb L-71 was immobilized as described in the legend to Figure 5. At {A}, purified sCD4 was injected across the layer and allowed to interact with the immobilized L-71. The surface was then regenerated by injection of 0.1 *M* phosphoric acid {B}. The total bound sCD4 is indicated by the arrows (ΔRU).

Table 1 Effect of Repeated Binding/Regeneration Cycles on the Binding of sCD4 to Immobilized MoAb L-71

Ligate concentration (μg/mL sCD4)	Percent maximal binding activity	
	After 50 cycles	After 100 cycles
Baseline	99.27	98.86
5	98.07	94.05
10	96.34	92.06
25	94.13	89.84
100	93.74	88.47

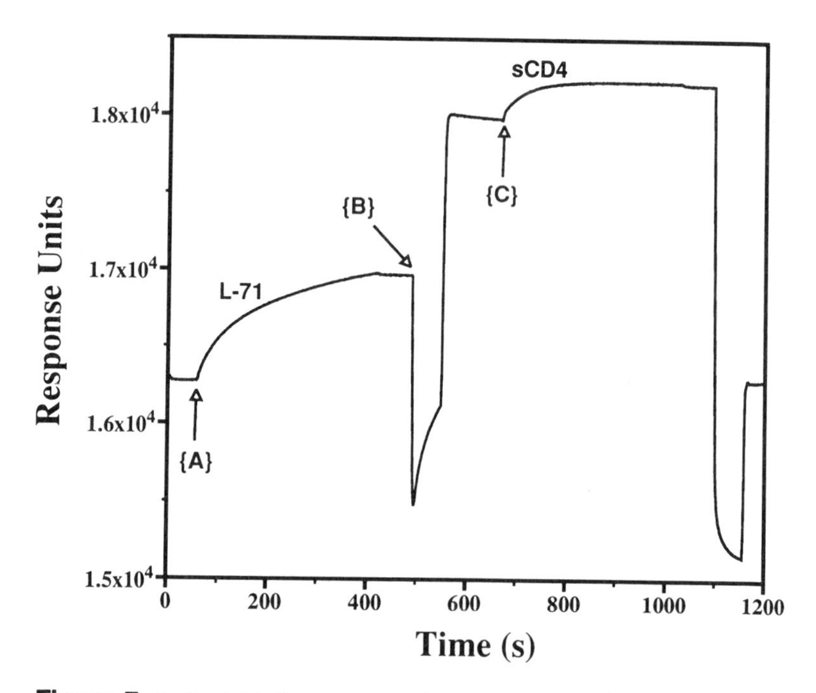

Figure 7 Indirect binding assay configuration. In this example, a capture antibody (rabbit antimouse Fc; RaMFc) is first immobilized as described in the legend to Figure 5. At {A}, a solution of MoAb L-71 is injected and the MoAb is captured by the RaMFc. At {B} a nonspecific MoAb was injected to block remaining RaMFc sites, and at {C}, purified sCD4 was injected and allowed to interact with the immobilized MoAb L-71. Regeneration results in a RaMFc layer ready for subsequent analyses.

Second, a solution of a specific mouse monoclonal antibody is injected and captured by the RaMFc. In this case the mouse monoclonal antibody need not be purified and can in fact be unfractionated hybridoma culture supernatant (see below). This clearly adds a degree of flexibility to the design of indirect assay configurations. In a third step in this example, the antigen of interest is injected and binding to the captured monoclonal antibody is observed. As described above, the layer may be regenerated with an appropriate solvent. In this case, however, the regenerated surface is a RaMFc layer, which can be used for capturing the same or a different monoclonal antibody. Therefore, the RaMFc layer per se adds versatility to the system.

3. *Signal Enhancement: The Use of Second Antibodies*

Signal enhancement may be used in either the direct or indirect assay configurations described above. An example of signal enhancement is shown in Figure 8. In

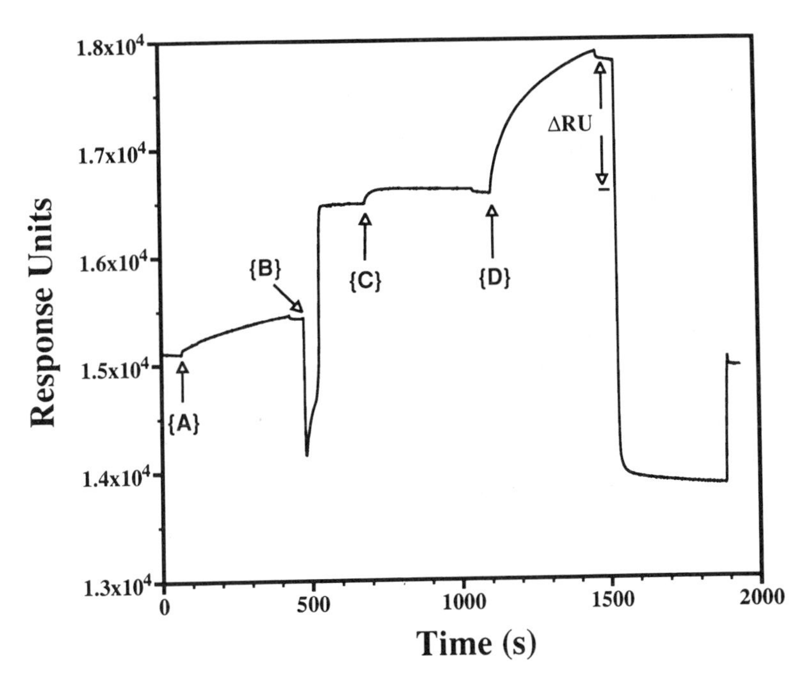

Figure 8 Signal enhancement: use of second antibodies. In the sensorgram shown, RaMFc has been immobilized. At {A}, an anti-p24 MoAb is injected. At {B} the remaining RaMFc sites are blocked using a nonspecific MoAb. At {C}, a solution of p24 is injected across the layer, and at {D} the signal is *enhanced* by injecting a polyclonal antibody against p24. Finally, the layer is regenerated to produce RaMFc.

this situation, after binding of the ligate, a second antibody, usually a polyclonal, is injected to bind to the already bound ligate. This secondary binding causes an increase in the response of the detector and therefore an increase in the sensitivity of detection of ligate. Polyclonal antibodies are more suited for use as enhancer antibodies since multiple antibody binding per ligate molecule is possible, allowing for maximal signal enhancement. In addition, the enhancer antibody preparation need not be purified.

Various permutations and combinations of these three basic assay configurations are possible and are discussed in more detail below. Note that none of the interactants is labeled and in many cases impure sample preparations may be used. Also, very small quantities of samples are required. Since the immobilized ligand surface may be used for at least 100 binding–regeneration cycles, BIAcore is also very reagent efficient.

III. APPLICATIONS OF SPR

A. Analyte Detection–Concentration Determinations

Let us consider the sensorgram shown in Figure 9. In this example a RaMFc antibody has been immobilized onto the surface of the sensor chip. At the beginning of this analysis cycle with a continuous flow of buffer over the surface, the resonance signal is 20,350 RU. At {A} crude hybridoma cell culture supernatant is injected by switching one of the sample loops into the buffer flow. The instantaneous rise in the resonance signal is due to the high refractive index of the culture media compared to the continuous buffer flow. At {B} the sample plug has passed and the system returns to constant buffer flow. At this point the total RU is 21,550, indicating that 1200 RU of monoclonal antibody (MoAb) from the hybridoma medium bound to the immobilized RaMFc. At {C} a solution containing the antigen is injected across the surface and one sees a gradual increase in the resonance signal, indicating binding of the antigen to the immobilized MoAb. When buffer flow replaces the antigen solution at {D}, one observes the dissociation of the antigen from the MoAb. At {E} a regeneration solvent is injected across the surface and results in a resonance signal of 20,350 RU equivalent to the original RaMFc modified layer. The surface is at this stage ready for another analysis cycle. This example serves to demonstrate several features of the technology worth stressing: (a) The selectivity of the sensor surface can be controlled by the operator through the choice of the immobilized ligand; (b) concentration determination from crude samples can be made if a standard curve is constructed;

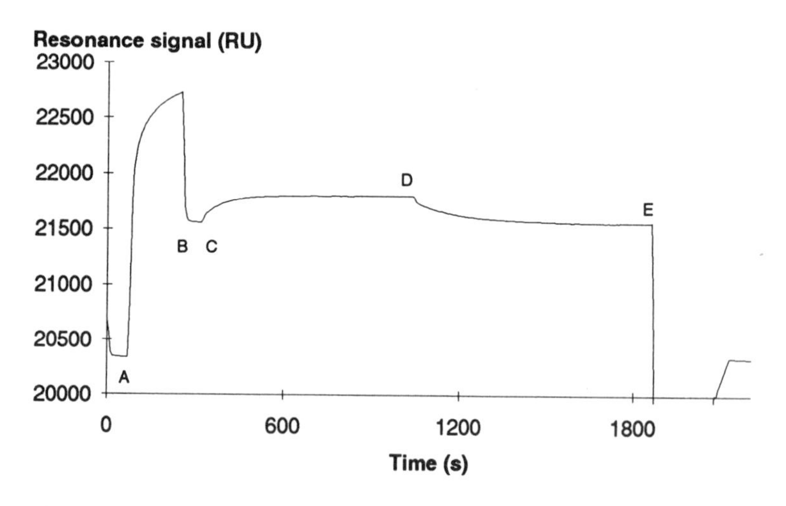

Figure 9 Analyte detection using crude samples.

(c) various unlabeled molecules can be injected in series across a single layer and each interaction is quantifiable, allowing one to calculate the stoichiometry of each interaction; (d) both association and dissociation events can be observed and quantitated, forming the basis for kinetic analysis of any given interaction; and (e) once regeneration conditions have been optimized for a particular system, the surface is generally stable for at least 50 to 100 cycles, in some cases more (see Table 1). As with affinity chromatographic matrices, the choice of elution or regeneration solvents is largely empirical. However, most of the commonly employed eluants are compatible with the instrument. We have had particular success with 0.1 *M* phosphoric acid as an eluant, both in biospecific affinity chromatography and on BIAcore (15).

The instrumentation is especially suited for analyte concentration determinations in small numbers of samples since each analysis cycle (i.e., binding and regeneration of the surface takes only 5 to 10 min. In addition, since the biospecific interaction of ligand–ligate is observed directly, analyses may be performed on crude samples, including serum, urine, or cell culture supernatants.

Figure 10 presents an example of the quantitative aspects of analyte concentration determinations. The example shows the generation of a standard curve for the quantitation of recombinant, soluble CD4 (sCD4) interacting with an immobilized

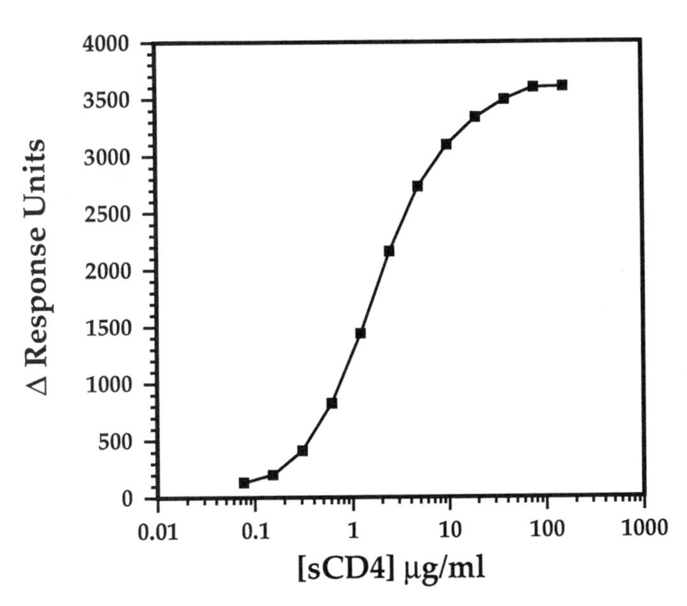

Figure 10 Analyte concentration determinations. The data presented depict a standard curve generated from binding of sCD4 to immobilized MoAb L-71.

MoAb, L-71. In this experiment, MoAb L-71 was immobilized directly onto the sensor chip surface, not via a capture antibody as detailed for the experiment described in Figure 9. Using the direct binding assay configuration described above, the sensitivity of detection of this assay was approximately 80 ng sCD4/mL buffer (approximately 1.5 p*M* sCD4), comparable to other assay systems. If, however, a secondary polyclonal antibody is employed for signal enhancement, the sensitivity of detection is approximately 5 to 10 ng sCD4/mL buffer (approximately 100 to 200 f*M* sCD4). The *absolute* sensitivity of detection of analyte will, clearly, depend on the ligand–ligate pair under investigation. We have also used the BIAcore for the detection and quantitation of sCD4 directly from unfractionated cell culture supernatants (15).

Aside from the quantitative aspects of BIAcore the ability to *detect* very low concentrations of specific proteins present in a complex mixture, in a matter of minutes, is of particular value. Areas of application include selection of cell cultures expressing recombinant proteins or hybridomas expressing MoAbs of interest.

B. Epitope Mapping Studies

Epitope mapping using MoAbs is a powerful tool for examining surface topology of macromolecules. The binding of each MoAb defines one specific site, or epitope, on the antigen. Epitope mapping commonly involves testing the ability of pairs of MoAbs to bind simultaneously to the antigen and is usually performed with such techniques as RIA or ELISA. A requirement for both of these methodologies is to label at least one of the MoAbs, so that it can be specifically detected. This in turn requires purification of the MoAbs and substantial time in labeling procedures. The SPR detector is particularly suited for such studies as epitope mapping (16), but in a label-free mode. In addition, purification of the MoAbs is not necessary, and indeed, crude hybridoma culture supernatants may be employed. An example of such a study is shown in Figure 11. In this example, a MoAb immobilized onto the surface of the sensor chip was used to capture its antigen, the HIV-1 nucleocapsid protein p24. Hybridoma culture supernatants were then injected across the surface and binding of the MoAbs assessed. Numerous supernatants may be injected in succession, without regeneration of the layer. By varying the order of injection of the various hybridoma supernatants, a thorough epitope map may be constructed. An alternative approach is to assess the ability of synthetic peptides or fragments of the antigen to compete for binding of the MoAb to the intact antigen. Again, such studies can be performed on BIAcore.

C. Kinetic Analyses: Theoretical Aspects

Since the SPR detector is a continuous, real-time detector, the possibility for assessing the kinetics of interaction exits. As noted earlier, the interaction

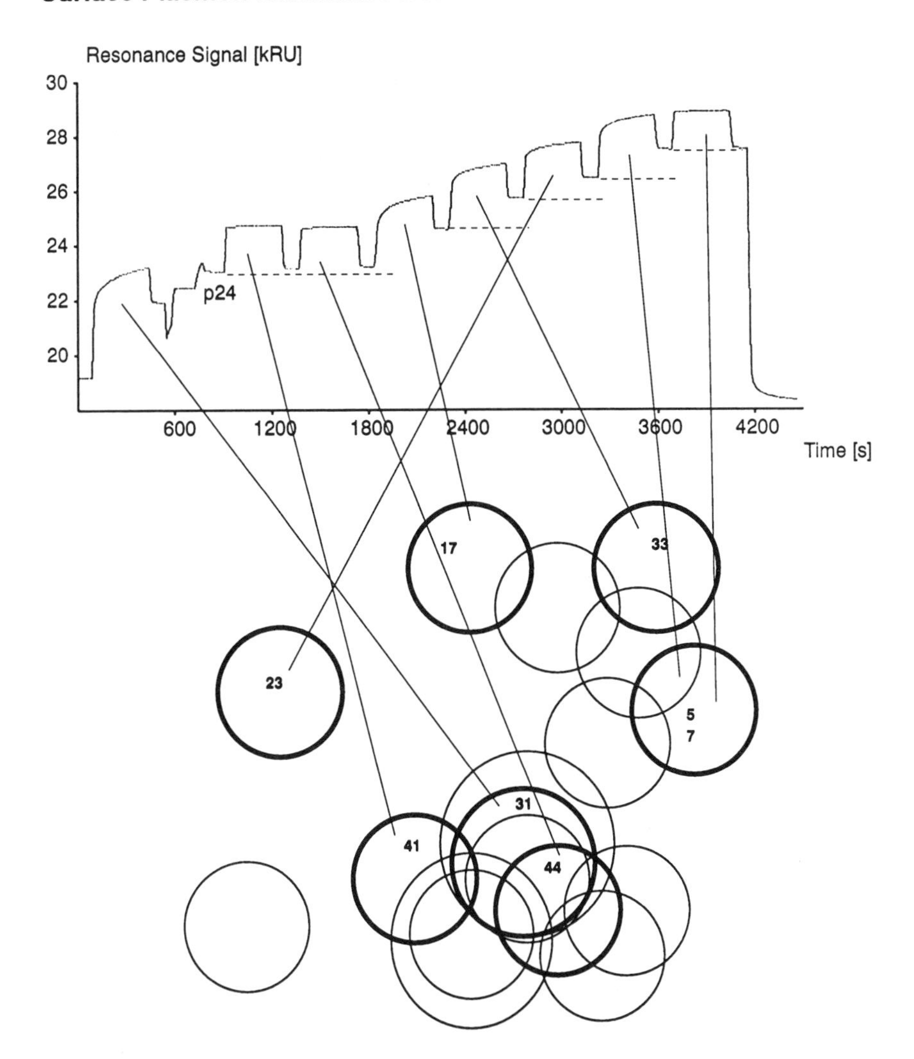

Figure 11 Epitope mapping. This sensorgram depicts the ability to observe sequential binding of a number of MoAbs to p24. A monoclonal anti-p24 was immobilized onto the sensor chip and used to capture purified p24. Subsequently, a number of hybridoma culture supernatants were assessed for their ability to bind to the captured p24.

observed is between an immobilized ligand and a soluble ligate and involves transport of the ligate to the surface immobilized ligand and the binding event itself. Initially, the rate of interaction (binding) of ligate to ligand may be mass-transport limited. As binding proceeds and the surface concentration of free ligand decreases, the binding will be limited by the association per se. To minimize the contribution of mass transfer effects on the observed binding rate, low surface concentrations of immobilized ligand are used for kinetic studies. This will also limit the effect of steric hindrance due to crowding of the surface.

For a one-to-one interaction in solution between reactants A and B to form the complex AB, the rate of formation of AB at time t may be written as

$$\frac{d[\mathrm{AB}]}{dt} = k_{\mathrm{ass}}[\mathrm{A}]_t[\mathrm{B}]_t - k_{\mathrm{diss}}[\mathrm{AB}]_t \tag{1}$$

where k_{ass} is the *association rate constant* in M^{-1} s^{-1} and k_{diss} is the *dissociation rate constant* in s^{-1}. The concentration of reactant B at time t $[\mathrm{B}]_t$, equals the initial concentration of B, $[\mathrm{B}]_0$, minus the concentration of the formed complex AB at time t (i.e., $[\mathrm{B}]_t = [\mathrm{B}]_0 - [\mathrm{AB}]_t$). Substituting into Eq. (1) gives

$$\frac{d[\mathrm{AB}]}{dr} = k_{\mathrm{ass}}[\mathrm{A}]_t([\mathrm{B}]_0 - [\mathrm{AB}]_t) - k_{\mathrm{diss}}[\mathrm{AB}]_t \tag{2}$$

In the BIAcore, one of the interactants is immobilized onto the surface and the other is continuously replenished from the injection volume flowing over the surface. The resonance signal observed, R, corresponds to the formation of AB complexes, and the *maximum* response, R_{max}, will be proportional to the surface concentration of *active* ligand. In the case of BIAcore, therefore Eq. (2) becomes

$$\frac{dR}{dt} = k_{\mathrm{ass}}C(R_{\mathrm{max}} - R_t) - k_{\mathrm{diss}}R_t \tag{3}$$

where dR/dt is the rate of formation of surface-associated complexes (i.e., the derivative of the observed response curve); C, which is essentially constant, is the concentration of ligate in solution; R_{max} is the *capacity* of the immobilized ligand surface expressed in *resonance units*; and $(R_{\mathrm{max}} - R_t)$ is equivalent to the number of unoccupied surface binding sites at time t. Note that since terms in R appear on both sides of the equation, the response value R_t can be used directly without conversion to *absolute* concentrations of formed complexes at the sensor chip surface. Rearranging Eq. (3) gives

$$\frac{dR}{dt} = k_{\mathrm{ass}}CR_{\mathrm{max}} - (k_{\mathrm{ass}}C + k_{\mathrm{diss}})R_t \tag{4}$$

From a plot of dR/dt versus R_t, therefore, k_{ass} and k_{diss} can be calculated. However, this requires experimental determination of R_{max}, which can only be assessed from injections of ligate at very high concentrations. Clearly, this has practical limitations.

From Eq. (4), the slope, k_s, from a dR/dt versus R_t plot can be written as

$$k_s = k_{\mathrm{ass}}C + k_{\mathrm{diss}} \tag{5}$$

Therefore, determination of k_s at a number of ligate concentrations (C) allows one to plot a straight line of k_s versus C where the slope corresponds to k_{ass} and the intercept corresponds to k_{diss}.

The process of dissociation may also be observed directly, after the sample plug has passed the surface. At this point, the ligate concentration, C, falls to zero and dissociation of ligate from the immobilized ligand is observed according to

$$\frac{dR}{dt} = -k_{diss}R_t \tag{6}$$

or

$$\ln \frac{R_0}{R_n} = k_{diss}(t_n - t_0) \tag{7}$$

where R_n is the response at time t_n and R_0 is the response at an arbitrary starting time, t_0. The dissociation rate constant, k_{diss}, can therefore be obtained as the slope from a plot of $\ln(R_0/R_n)$ versus $(t_n - t_0)$.

D. Kinetic Analyses: Practical Considerations

Kinetic analyses of MoAb–antigen interactions using the methodology described above have been described (17) and affinity constants determined on BIAcore agree well with those determined by ELISA or solution methods. As an example of kinetic analysis of macromolecular interactions, we describe the interaction of insulin-like growth factor I (IGF-I) and insulin-like growth factor binding protein I (IGFBP-I) (18). IGFBP-I was immobilized onto the sensor chip surface using NHS-ester chemistry as described above. The binding of IGF-I to the immobilized IGFBP-I was then investigated in the concentration range 10 to 320 n*M* IGF-I. Figure 12 shows an overlay plot of the binding curves obtained. Linearized plots [dR/dt versus R; Eq. (4)] are shown in Figure 13.

A comparative study was also performed between native IGF-I (nIGF-I), a truncated form of IGF-I (tIGF-I) lacking three amino acid residues at the N terminus and a disulfide mismatched IGF-I (mmIGF-I). Plots of k_s versus C [Eq. (5)] are shown in Figure 14. From the parallelism of the plots for nIGF-I and tIGF-I, the association rate constants, k_{ass}, are obviously similar and were calculated to be 5.8×10^5 and 5.7×10^5 M^{-1} s^{-1}, respectively. For the disulfide mismatched form, mmIGF-I, k_{ass} was calculated to be 4.6×10^4 M^{-1} s^{-1}, clearly decreased from the other two forms of IGF-I. On the other hand, the dissociation rate constants, determined as the intercept on the *y*-axis in Figure 14, indicate low but similar values for nIGF-I and mmIGF-I, whereas tIGF-I appears to show significantly faster dissociation. However, since the data used for this analysis were obtained from the *association phase* of the sensorgrams, the dominance of

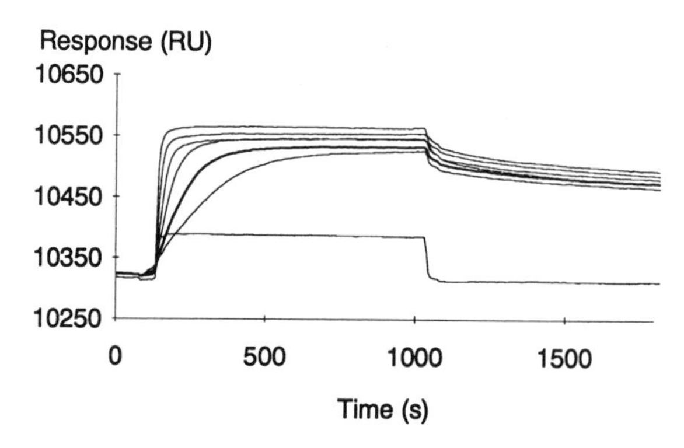

Figure 12 Interaction of soluble IGF-I with immobilized IGFBP-I. Overlay plot of sensorgrams showing the concentration-dependent interaction of soluble IGF-I with immobilized IGFBP-I. Injections were performed at 2 μL/min using IGF-I concentrations of 10, 20, 40, 80, 160, and 320 n*M* (from bottom to top). Dissociation of the formed complexes is also observed.

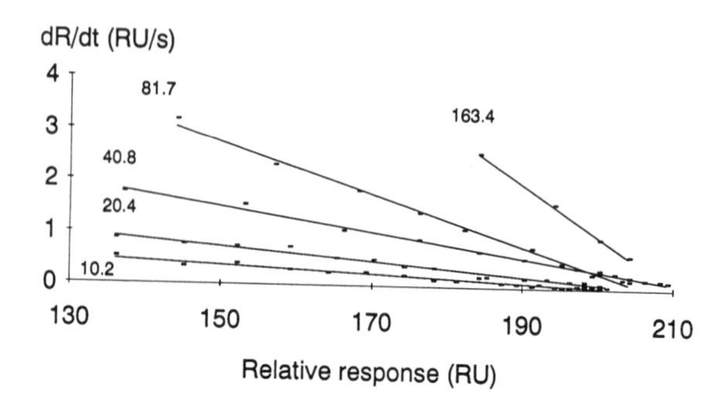

Figure 13 *dR/dt* versus *R* plots for the interaction of soluble IGF-I with immobilized IGFBP-I; linearized plots according to Eq. (4), for the sensorgrams depicted in Figure 12.

the association rate constant on the progress of binding introduces substantial uncertainty in calculating *dissociation rate constants* from a plot of k_s versus C (Figure 14).

A direct comparison of sensorgrams obtained using identical concentrations of the three variants of IGF-I, as shown in Figure 15, provides a simple means of *qualitative* comparison of their *relative* kinetic and affinity properties. During the injection phase, it can be seen that both nIGF-I and tIGF-I reach a steady state, and

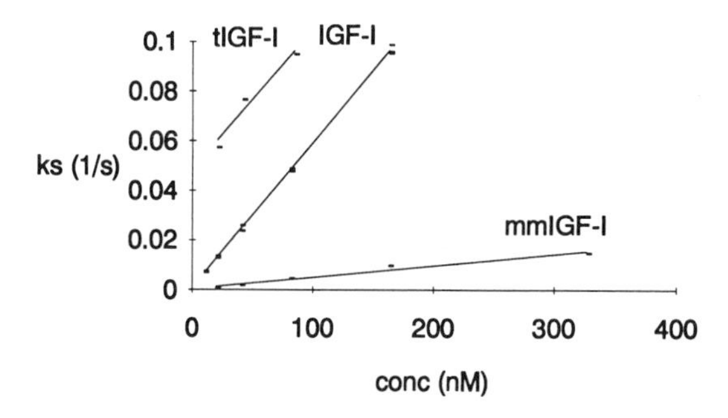

Figure 14 Kinetic analysis of the interaction of various forms of IGF-I with immobilized IGFBP-I. Plots of the concentration dependence of slopes [k_s versus C; Eq. (5)] of the binding of native IGF-I, truncated IGF-I (tIGF-I), and disulfide mismatched IGF-I (mmIGF-I) to immobilized IGFBP-I. Slopes (k_s) were derived from plots of dR/dt versus R [Eq. (4)]. The slopes of the plots shown correspond to the association rate constants, k_{ass}, and the intercepts on the y-axis correspond to the dissociation rate constants, k_{diss}. The parallellism of the plots for IGF-I and tIGF-I indicate similar k_{ass} values, whereas the intercepts indicate a similar but low k_{diss} value for IGF-I and mmIGF-I.

since the *absolute* response at steady state differs for the two variants, it can be concluded that nIGF-I has a higher affinity. This qualitative assessment of affinities directly from sensorgrams obtained under identical conditions can be extremely informative. It can also be seen in Figure 15 that the apparent dissociation rate of tIGF-I from the IGFBP-I derivatized surface is significantly greater than that observed for nIGF-I. Plots of $\ln(R_n - R_0)$ versus $(t_n - t_0)$ for the three IGF-I variants are shown in Figure 16. The reasons for the lack of linearity, and therefore varying k_{diss}, in these plots is unknown but may reflect heterogeneity in one or both of the interactants, or, indeed, reassociation of dissociated ligate, which could be expected to change with time as the number of unoccupied IGFBP-I sites differs. In any event, a thorough kinetic analysis of this system is not possible. However, a qualitative assessment of this interaction is possible, and from Figures 15 and 16 it can be concluded that tIGF-I shows a significantly greater dissociation from the immobilized IGFBP-I than does nIGF-I or mmIGF-I, which appear similar.

E. Caveats

The example above serves to demonstrate several aspects of analysis of BIAcore binding data which may be operative in other solid–solution-based systems, such

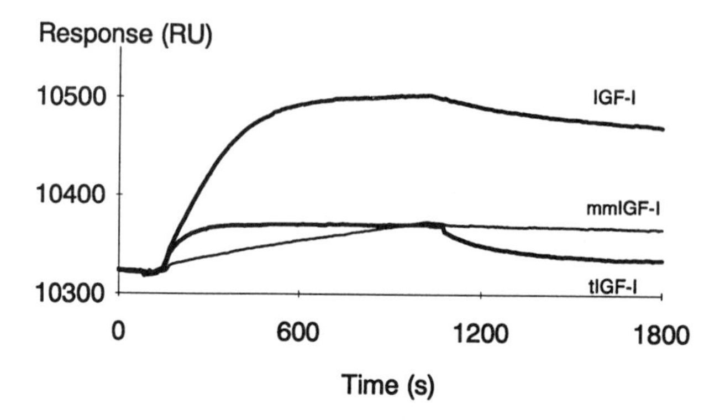

Figure 15 Sensorgrams for the binding of IGF-I, tIGF-I, and mmIGF-I to immobilized IGFBP-I. Sensorgrams for the three forms of IGF-I, recorded at equivalent concentrations of 10 n*M*. From the *levels* of the steady state responses, which during the injection are only reached for the native IGF-I and for tIGF-I, it can be concluded that native IGF-I has a higher affinity. It is also clear from these sensorgrams that tIGF-I has a comparatively high dissociation rate.

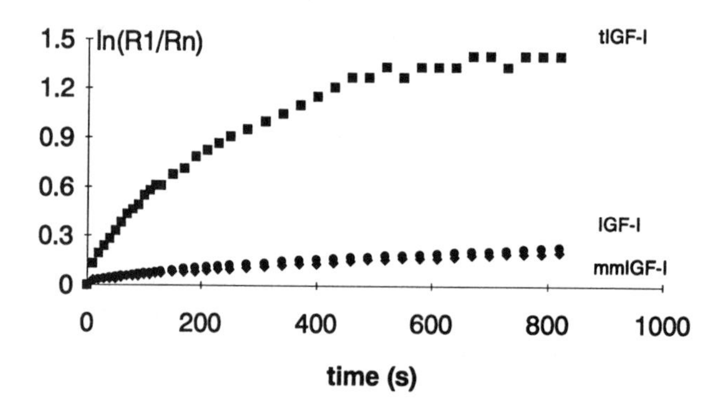

Figure 16 Dissociation plots for IGF-I, tIGF-I, and mmIGF-I. Dissociation plots, according to Eq. (7), obtained from the sensorgrams shown in Figure 15. During the 15-min interval shown, only the tIGF-I form dissociates completely from the immobilized IGFBP-I. It is also obvious from these data that the dissociation rate constants for IGF-I and mmIGF-I are similar and significantly lower than that for tIGF-I.

as analytical affinity chromatography. First, the rate constants and affinity constants derived from such data are *apparent* constants only and are highly dependent on the conditions of the assay and the methodology used to obtain such data. This is not unique to BIAcore data, however, as all methods for determining kinetic and/or affinity constants are similarly artificial. A second, important consideration is that *relative* kinetic and affinity data can be obtained on BIAcore in a very short time frame. In some situations, such as selection of MoAbs, *relative* data may be sufficient. In that light, such *relative* data can be obtained from crude samples without the need to purify the ligate. Third, it is clear from the example above that quantitative kinetic analyses are not possible in all cases, which may reflect an oversimplification of the kinetic model. For example, the model assumes a one-to-one interaction and would clearly be invalid for multiple stoichiometric interactions. In addition, the major assumption of this model is that mass transfer contributions to the observed binding rates are minimal. This may be too general an assumption. With these reservations noted, however, it is again worth stressing that at the very least, relative kinetic data can be obtained using this instrument.

F. Determination of Conditions for Affinity Chromatography

Since the sensor chip surface may be considered as a microchromatographic matrix, a valuable application of this instrumentation in the analytical mode is for defining optimal conditions for affinity chromatography. The chemistry of immobilization of the ligand, elution of bound ligate, and overall stability of the immobilized ligand can thus be assessed in microscale. An example of screening elution solvents for their ability to disrupt the complex formed between immobilized MoAb and soluble complement receptor 1 (sCR1) is depicted in Figure 17. As can be seen from these data, many and varied elution solvents can be rapidly screened under identical conditions for their ability to desorb ligate from immobilized ligand. It can be seen that 0.1 *M* glycine–HCl (pH 1.8), a common immunoaffinity eluant, was totally ineffective in this system, whereas 0.1 *M* phosphoric acid was shown to be the most effective solvent for elution of sCR1. Further studies on the effect of elution flow rate allowed us to define conditions for immunoaffinity chromatography. In the chromatographic mode, 0.1 *M* glycine–HCl was again shown to be ineffective at eluting bound sCR1. However, this proved to be a beneficial wash step for the purification of sCR1 (D. J. O'Shannessy et al., unpublished results).

The stability, or reusability, of biospecific affinity columns is an important consideration in the development of a purification scheme. Such studies are usually time consuming and reagent rich. However, taking advantage of the robotics unit designed into BIAcore allows these analyses to be performed in an

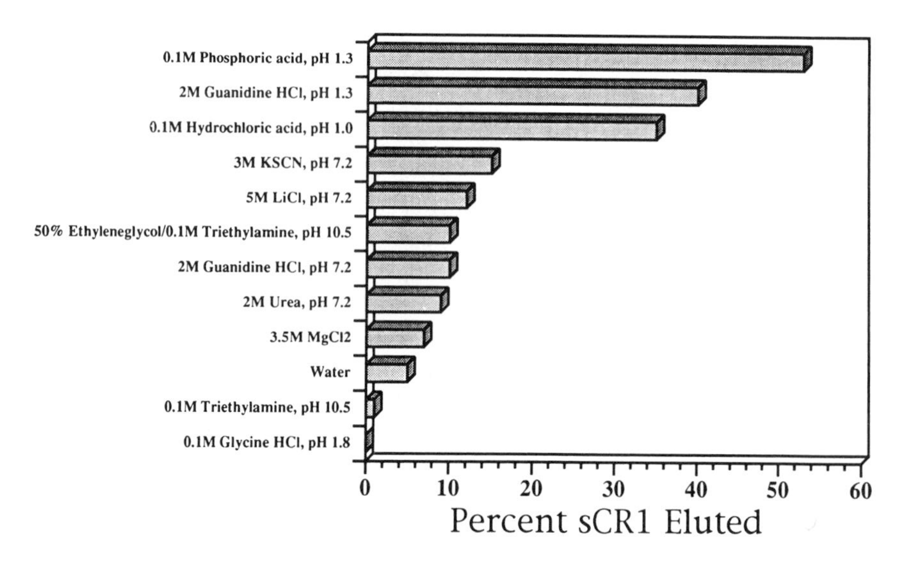

Figure 17 Elution solvent screening. In this example, a MoAb specific for soluble complement receptor I (sCR-I) was immobilized onto the sensor chip surface as described. sCR-I was allowed to bind to the MoAb and various solvents assessed for their ability to disrupt the complex. The percent sCR-I eluted was determined from the ΔRU before and after injection of elution solvent.

automated fashion and with very small sample requirements. The result is a significant savings in samples and, importantly, time, over the use of conventional chromatography to obtain the same amount of information. As an example, a stability study of binding of sCD4 to immobilized HIV-1 envelope glycoprotein, gp120, is shown in Figure 18. These data suggested that immobilized gp120 was extremely stable and could be used for at least 100 binding/regeneration cycles. Subsequent experiments in the chromatographic mode showed this to be the case (not shown).

IV. CONCLUDING REMARKS

The foregoing has been a brief overview of the principles and applications of SPR in affinity technologies. SPR per se is not a new concept but it was only recently that commercial instruments became available for evaluation. The instrumentation includes a number of features that may result in more widespread use of the technology. As described above, SPR detection is very sensitive and, as presented in the BIAcore format, is also versatile with respect to assay configuration, sample

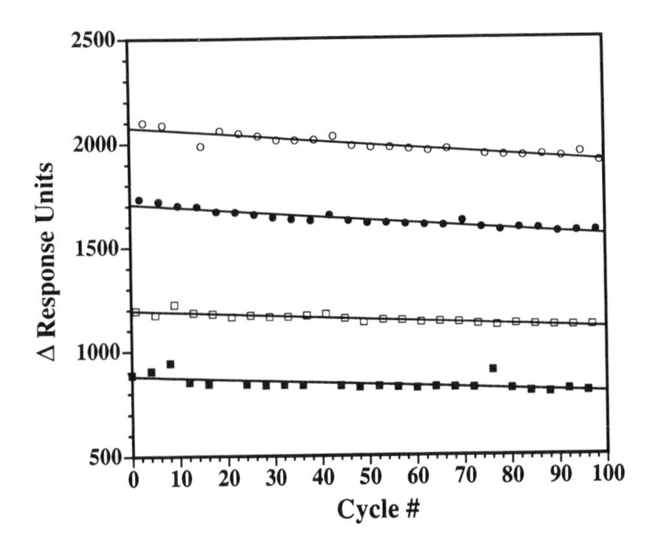

Figure 18 Analysis of stability of immobilized ligands. In this example, the HIV-1 envelope glycoprotein gp120 was immobilized onto the sensor chip surface. The binding of sCD4 (100 μg/mL, ○; 25 μg/mL, ●; 10 μg/mL, □; and 5 μg/mL, ■) was analyzed over 100 binding/elution cycles. Elution was performed with 0.1 *M* phosphoric acid. The data represent the ΔRU obtained on binding of sCD4 as a function of cycle number.

acceptance (serum, urine, cell culture supernatants, etc.), and user interfacing (e.g., the ability to program methods). Two clear advantages of SPR detection for investigations of macromolecular interactions compared to methods described previously are that it is a label-free detection principle and that detection is continuous and in real time. From a qualitative point of view, therefore, this translates into significant time savings over conventional analyses such as ELISA, since the interaction is observed directly, in a matter of minutes, without the need for secondary or tertiary labeled interactants. In addition, the sensitivity of SPR has been shown to rival that of enzyme-linked assays and radioimmunoassays. Importantly, SPR detection is amenable to the use of crude, unpurified samples such as serum and cell culture supernatants. While only a single sample may be analyzed at any given time, the robotics unit built into the instrument coupled with the ability to program injection and elution events as well as sample mixing, and so on, allows total automation of sample handling and analysis of multiple samples with minimal hands-on time.

The applications of SPR presented above to exemplify particular areas in which SPR detectors could have significant impact. Most important of these is kinetic analysis of macromolecular interactions since most other techniques will only allow evaluation of *apparent* dissociation constants, K_d. A thorough evaluation of a macromolecular interaction requires determination of the kinetic rate constants since K_d is merely a ratio of these constants (i.e., $K_d = k_{ass}/k_{diss}$). Therefore, interactants could have similar K_d values but significantly different kinetic rate constants. Knowledge of the kinetic rate constants allows for more accurate selection of, for example, MoAbs for immunoaffinity chromatographic purposes. We are now just beginning to explore other applications of SPR detectors, and it is the belief of the authors that this technology will become commonplace in research laboratories.

REFERENCES

1. M. N. Kronick and W. A. Little, *J. Immunol. Methods, 8*:235 (1975).
2. B. Liedberg, C. Nylander, and I. Lundström, *Sensors Actuators, 4*:299 (1983).
3. D. C. Cullen, R. G. V. Brown, and C. R. Lowe, *Biosensors, 3*:211 (1987).
4. H. Raether, *Surface Plasmons on Smooth and Rough Surfaces and on Gratings*, Springer-Verlag, Berlin, 1988.
5. R. P. H. Kooyman, H. Kolkman, J. Van Gent, and J. Greve, *Anal. Chim. Acta, 213*:35 (1988).
6. C. S. Mayo and R. B. Hallock, *J. Immunol. Methods, 120*:105 (1989).
7. D. C. Cullen and C. R. Lowe, *Sensors Actuators, 31*:576 (1990).
8. S. Löfås and B. Johnsson, *J. Chem. Soc. Chem. Commun.*, 1526 (1990).
9. J. W. Attridge, P. B. Daniels, J. K. Deacon, G. A. Robinson, and G. P. Davidson, *Biosensors Bioelectron., 6*:201 (1991).
10. W. Lukosz, *Biosensors Bioelectron., 6*:215 (1991).
11. B. Johnsson, S. Löfas, and G. Lindquist, *Anal. Biochem., 198*: (1991).
12. S. Sjölander and C. Urbaniczky, *Anal. Chem.* (1991).
13. E. Stenberg, B. Persson, H. Roos, and C. Urbaniczky, *J. Colloid Interface Sci., 143*:513 (1991).
14. D. J. O'Shannessy, M. Brigham-Burke, and K. Peck, submitted for publication.
15. M. Brigham-Burke, J. R. Edwards, and D. J. O'Shannessy, submitted for publication.
16. L. G. Fägerstam, Å. Frostell, R. Karlsson, M. Kullman, A. Larsson, M. Malmqvist, and H. Butt, *J. Mol. Recog., 3*:208 (1990).
17. R. Karlsson, A. Michaelsson, and L. Mattsson, (1991) *J. Immunol. Methods, 145*:229 (1991).
18. L. G. Fägerstam, Å. Frostell-Karlsson, R. Karlsson, B. Persson, and I. Rönnberg, *J. Chromatogr., 597*:397 (1992).

10

Determination of Binding Constants by Quantitative Affinity Chromatography

Current and Future Applications

Donald J. Winzor

University of Queensland, Brisbane, Queensland, Australia

Craig M. Jackson

American Red Cross Blood Services, Detroit, Michigan

I. INTRODUCTION

The unique advantages of affinity chromatography as a preparative procedure (1) were quickly incorporated into techniques for evaluating quantitatively the stoichiometries and binding constants of biospecific interactions (2–4). In this review we attempt to present quantitative affinity chromatography (QAC) as a technique with the unique advantages that it possesses (5–10) and to indicate its versatility and applicability to situations where other techniques fail or are not readily applicable. First, we indicate the scope and power of QAC to characterize biospecific interactions. Although the focus is not exclusively on interactions between macromolecules and macromolecular systems, several such examples are used in this review to illustrate the power and capability of QAC for quantifying these systems. Second, we describe the general approaches for the performance of QAC measurements to determine binding constants and indicate the relationship of the experimental parameters that are measured in QAC to those of the more familiar binding techniques. In that section we also describe and develop the algebraic expressions that are used to obtain binding affinities from QAC experiments for a selected set of situations that are likely to be encountered in the investigation of more complex systems.

II. SCOPE AND POWER OF QUANTITATIVE AFFINITY CHROMATOGRAPHY

The scope and power of QAC are indicated by both the breadth of the types of biological systems that can be investigated and the range of affinities that can be measured. Some examples of biospecific phenomena that have been investigated are the interactions of enzymes with substrates, inhibitors, and effectors; of hormones with receptors; of antigens with antibodies; of proteins with lipids, drugs, and metabolites; and of glycoproteins with lectins. Although the quantitative expressions for analysis of QAC experiments were initially developed for description of ligand-facilitated and ligand-retarded elution of solute from an affinity matrix in a conventional chromatography experiment, these expressions have also been applied directly to studies in complex biological mixtures. For example, quantitative affinity chromatography has been used to measure the binding constants for the interaction of NADH with the various lactate dehydrogenase isoenzymes in a crude tissue extract (11). Other QAC studies have measured the metabolite-dependent binding of enzymes to naturally occurring affinity matrices such as muscle myofibrils (12–14) and erythrocyte membranes (15,16). Quantitative affinity chromatography methods can also be used to investigate the multitude of nucleic acid and protein interactions responsible for gene expression as well as for the transfer of genetic information during cell division.

The range of affinities that can be characterized by QAC methods is broad. Quantitative affinity chromatography was envisaged initially (2–4,17) as being of particular value for characterizing relatively weak interactions (binding constants of $10^3\ M^{-1}$ and lower), situations that prove troublesome in conventional methods such as equilibrium dialysis because of difficulties encountered in measuring the difference between total and free ligand concentrations. Subsequent studies (18–20) have established the potential of quantitative affinity chromatography for characterizing interactions at the other end of the energy spectrum—those for which the binding constant is so large that the range of ligand concentration required for their study is too low for experimental detection, let alone quantitative measurement, in a reaction mixture. QAC has been used to investigate the binding of long-chain fatty acids to albumin that is characterized by association constants in the range 10^3 to $10^8\ M^{-1}$ (21), and association constants for enzyme–substrate complex formation, which are typically in the range 10^3 to $10^6\ M^{-1}$.

A. General Description of a QAC Procedure

Quantitative affinity chromatography measurements can be performed in a variety of ways, each with advantages and disadvantages that are determined by the biological components, the analytical methods available, and the affinities of the interacting species.

Any QAC system comprises a matrix, which either by virtue of its intrinsic chemical properties, or as the result of covalent attachment of a specific reactant group, interacts selectively with particular molecules in solution. The matrix-linked reactant (designated X) is thus chosen to obtain the selective interaction with the desired molecules of interest (A).

After a decision has been made about the most appropriate affinity matrix for a particular biospecific interaction, the next question to be addressed concerns the experimental protocol to be adopted for the quantitative affinity chromatography study. Basically, there are three choices: column chromatography, partition equilibrium studies, or a recycling partition technique.

The relationship between matrix-linked reactant (X) and the solute (A) is illustrated in Figure 1. Upon mixing a solution of solute with the affinity matrix an equilibrium distribution of the solute between adsorbed and solution states is established. This distribution, or partition, of solute between matrix and solution is the first fundamental property to be quantified for determining the stoichiometry of binding, which is manifested in the analysis as an effective concentration of matrix sites, and the equilibrium constant (k_{AX}) governing the interaction. The partition of solute between matrix and solution states can be influenced by all physical and chemical variables which change the properties of the matrix, the matrix-bound reactant, and/or the solute. Such influences include nonspecific electrostatic effects from the presence of charge on the matrix, or adsorption of solute to the matrix that is independent of the presence of the specific immobilized reactant. Precautions are therefore required to ensure that those factors are maintained constant throughout the measurements. These effects may also be responsible for the differences in affinity observed between some types of affinity chromatography techniques and techniques that measure binding of components in solution.

When the goal is to use affinity chromatography for purification of the solute only, this measurement with a relatively small amount of matrix may by itself be

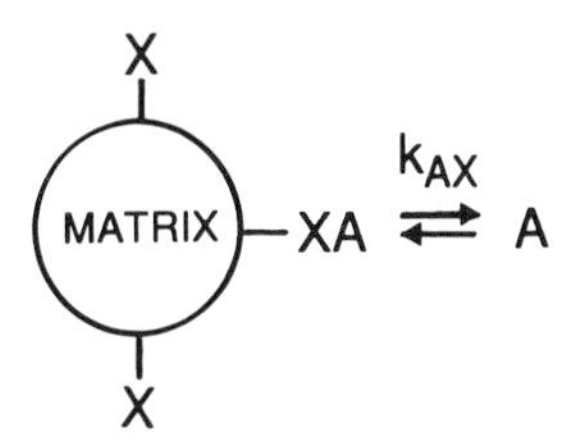

Figure 1 Schematic representation of the interaction of solute (acceptor), A, with affinity matrix sites, X, governed by intrinsic binding constant k_{AX}.

adequate for designing a preparative affinity chromatography isolation procedure. However, restriction of interest to the interaction between matrix-bound reactant and solute greatly underutilizes the power of affinity chromatography for characterizing binding phenomena. The potential of quantitative affinity chromatography extends far beyond its use as a means to optimize the purification of macromolecules. For example, by examining the distribution of solute between matrix-bound and solution states in the presence of ligand (S) which competes with matrix sites (X) for the solute (A), it is possible to characterize the interaction of solute with ligand in the solution phase. It is this combined information that makes QAC so powerful and useful—not only in macromolecule isolation but also in the characterization of macromolecule–ligand and macromolecule–macromolecule interactions.

B. Summary of Terms and Symbols

Constituent *Constituent* is the term applied to a chemical species in the system independent of its state [i.e., irrespective of whether it is present as free species or as complex(es) with other species]. Specific terms given to constituents in QAC are partitioning solute (acceptor), ligand, and matrix. The *constituent concentration* is the total concentration of that constituent in the system.

Partitioning solute The *partitioning solute* is the constituent of an affinity chromatographic system whose partition between the liquid and solid phases is being monitored. Commonly, but not obligatorily, the partitioning solute is a macromolecular species: by analogy with equilibrium dialysis, it is designated as A and is sometimes referred to as the *acceptor.*

Ligand The *ligand* is a soluble chemical species with which the partitioning solute interacts and is designated by S. In many instances the same species may also be covalently attached to the insoluble matrix, but the sites thus created will be termed *affinity matrix sites* and assigned the symbol X.

Affinity matrix The *affinity matrix* is the chromatographic matrix bearing sites, X, that exhibit specific interaction with the partitioning solute, A.

Affinity system *Affinity system* describes the equilibrium coexistence of affinity matrix, partitioning solute, and the ligand that modulates the interaction between affinity matrix and partitioning solute.

Partition The distribution of the solute (acceptor) between the matrix (stationary phase) and the surrounding solution (mobile phase) is termed *partition.* It is the partition between the two phases of the system that enables the binding affinity to be measured or the differences in affinity to be exploited for separating solutes.

Concentration terms In order to describe these affinity systems comprising stationary and liquid phases, several concentration symbols are clearly required. A summary of those used in this chapter follows.

$\overline{C}_A$ Constituent (or total) molar concentration of partitioning solute (A) in the mobile phase. When a ligand, S, is also present, $\overline{C}_A$ includes the concentration(s) of any AS_i complex(es) as well as that of free A.

$\overline{\overline{C}}_A$ Total concentration of A constituent in the system, including that bound to matrix. $\overline{\overline{C}}_A = Q_A/V_A^*$, where Q_A is the total mass of A constituent present and V_A^* is the volume accessible to A (i.e., the volume in which A is distributed).

$\overline{C}_A^\alpha$ Constituent concentration of A in the plateau (α phase) in a frontal chromatography experiment; it is also the constituent concentration of solute in the large volume of sample that is applied in this procedure.

$\overline{C}_S$ Constituent concentration of ligand in the mobile phase (the counterpart of $\overline{C}_A$). An α superscript is used to denote this concentration in the plateau phase of a frontal chromatographic experiment.

$\overline{\overline{C}}_S$ Total concentration of S in the system (as for $\overline{\overline{C}}_A$).

C_S Concentration of unbound (free) ligand in the mobile (liquid) phase.

$\overline{\overline{C}}_X$ Effective total concentration of matrix sites, X, in the system, which may differ substantially from the corresponding concentration determined analytically because of inaccessibility of some sites to the partitioning solute.

Volume terms Several volume symbols are also used.

$\overline{V}_A$ Elution volume of constituent A in conventional column chromatographic procedures; also, the apparent distribution volume in batch or recycling partition experiments. In frontal chromatography it is obtained (see Figure 2a) from the expressions $\overline{V}_A = [\Sigma(V\ \Delta\overline{C}_A)]/\overline{C}_A$, or $\overline{V}_A = [V' - \Sigma(\overline{C}_A\ \Delta V)]/\overline{C}_A^\alpha$. In these expressions $\overline{C}_A$ is the constituent concentration at volume V in the elution profile for an experiment with an applied concentration $\overline{C}_A^\alpha$ of solute. For a symmetrical boundary $\overline{V}_A$ is the effluent volume at which $\overline{C}_A = \overline{C}_A^\alpha/2$. In zonal chromatography this elution volume is obtained rigorously (see Figure 2b) as

$$V_A = V_p + \frac{\sum_{0}^{V_p}(C_A\,\Delta V) - \sum_{V_p}^{\infty}(\overline{C}_A\,\Delta V)}{\overline{C}_p}$$

where $\overline{C}_p$ is the constituent concentration at volume V_p, the effluent volume corresponding to the peak in the elution profile. For a symmetrical zone the two areas defined by the integrals self-cancel, whereupon $V_A = V_p$; for a skewed peak, however, this equality does not apply.

V_A^* Volume of the system accessible to A, which is the elution volume of A under conditions where no interaction of solute with affinity matrix sites occurs. Under

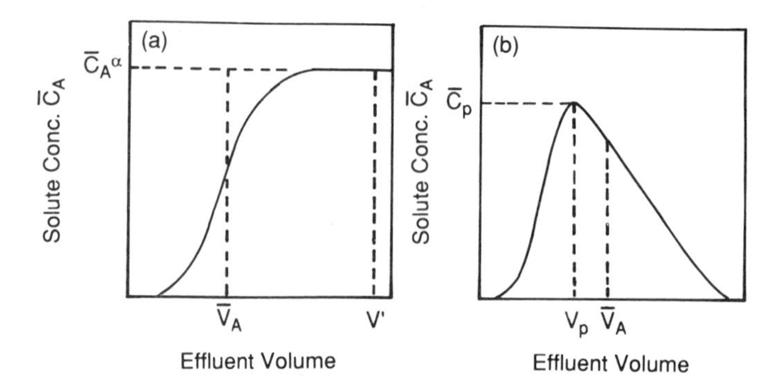

Figure 2 Identification of the constituent elution volume, $\overline{V}_A$, of the partitioning solute in elution profiles obtained by (a) frontal chromatography and (b) zonal chromatography. $\overline{V}_A$ is calculated from the advancing frontal profile on the basis that the rectangle with area $(V' - \overline{V}_A)\overline{C}_A^{\alpha}$ defines the area under the curve up to the effluent volume V'. In zonal chromatography $\overline{V}_A$ is calculated as the effluent volume that divides the eluted zone into two regions with equal area. (See the text for evaluation procedures.)

those conditions this volume in partition equilibrium studies may be determined from the amount of A added (Q_A) and the relationship $V_A^* = Q_A/\overline{C}_A$.

Binding parameters The interactions giving rise to a particular affinity chromatographic system are described in terms of the following binding parameters.

r Binding function or ratio (22,23), (i.e., moles of ligand bound per mole of acceptor): in the affinity chromatographic context the affinity matrix is the acceptor and the partitioning solute is the ligand.

k_{AX} Intrinsic association equilibrium constant (22) for the interaction of solute (A) with matrix sites (X).

$\overline{k}_{AX}$ Corresponding constitutive (apparent) equilibrium constant that is determined experimentally when ligand (S) is also present.

k_{AS} Intrinsic association constant for the binding of soluble ligand (S) to the partitioning solute (A).

k_{XS} Intrinsic association constant for the binding of soluble ligand (S) to affinity matrix sites (X).

k_{ASX} Intrinsic association constant for the interaction of AS complex with affinity matrix sites (X).

k_{XSA} Intrinsic association constant for the binding of partitioning solute (A) to XS, a complex between ligand and a matrix site.

III. CHARACTERIZATION OF THE SOLUTE–MATRIX INTERACTION

To use QAC for the characterization of interactions it is first necessary to identify the two primary equilibria governing the biospecific partition of solute that is measured. In instances where one of these interactions is a chemical equilibrium between partitioning solute and the immobilized reactant sites on the affinity matrix, the second usually reflects competition from a ligand that interacts reversibly with either the solute or the matrix-bound reactant. When affinity chromatography is being used to purify the solute, advantage may be taken of this second equilibrium reaction to elute the solute. Although characterization of both equilibria is the ultimate aim of quantitative affinity chromatography, the presentation and development of the fundamental quantitative expressions are rendered more straightforward by restricting consideration initially to the solute–matrix interaction.

Understanding the quantitative relationships that describe the experimental variable involved in QAC analysis is aided by a mental picture of the apparatus to be used and of the components and interactions that are present in the system. To create this picture, imagine the following. Beads of affinity matrix (the stationary phase) are contained within a vessel (a column in conventional affinity chromatography or simply a beaker in partition equilibrium studies) and are surrounded by liquid (the mobile or solution phase). Each of the components of the system occupies a portion of the total volume. In that regard the matrix occupies a volume within which the matrix-bound reactant sites are located—the situation that leads to the commonly used expression of matrix capacity in terms of moles of immobilized reactant per milliliter of hydrated gel. However, for thermodynamic characterization of the solute–matrix interaction the most convenient definition of total matrix-site concentration ($\overline{\overline{C}}_X$) is based on consideration of these sites to be distributed in the same volume as that occupied by solute. $\overline{\overline{C}}_X$ is thus an effective parameter to be evaluated from the analysis, and is not one that can be deduced necessarily from the analytical composition of the matrix.

The volume terms of most importance for analysis of data from QAC experiments are the total volume, V_t, and V_A^*, the volume accessible to the partitioning solute. Whereas the former is usually obtained by conducting an experiment with a very small solute that has access to effectively the entire volume of the system, V_A^* must be determined from experiments in which interaction of the solute with matrix is completely suppressed. Experimental measurement of the total solute concentration in the mobile (solution) phase, $\overline{C}_A$, is also required, this being a parameter that is readily obtained by analysis of the liquid phase. Interpretation of results from partition equilibrium experiments requires knowledge of the total solute concentration in the system, $\overline{\overline{C}}_A$, a parameter obtained by division of the amount of solute present by the accessible volume (V_A^*).

A. Association Constant for the Solute–Matrix Interaction

The first step in a QAC study is, as noted above, determination of the binding constant for the solute–matrix interaction, k_{AX}, and the associated parameter $\overline{\overline{C}}_A$, the effective total concentration of sites on the affinity matrix. In a basic formulation of the algebraic expressions that describe affinity chromatographic studies, the partitioning solute was considered to be univalent in its interaction with matrix (2–4,24–30). Any multivalency of the matrix-bound reactant is incorporated by defining the strength of the interaction in terms of an intrinsic (22) or site-binding (17) constant, k_{AX} (Figure 1), in conformity with the concept for conventional binding studies (22,23).

The binding function, r, for a univalent partitioning solute is also defined in the conventional way (22,23) as

$$r = \frac{\overline{\overline{C}}_A - \overline{C}_A}{\overline{\overline{C}}_X} \tag{1}$$

where $\overline{C}_A$ denotes the constituent concentration of partitioning solute in the liquid phase (also the free concentration C_A in this situation with no ligand present). $\overline{\overline{C}}_A$ and $\overline{\overline{C}}_X$ are the respective total concentrations of solute and matrix sites in the system. Since $\overline{\overline{C}}_X$ is a parameter to be evaluated from the analysis, Eq. (1) is used to evaluate the product $r\overline{\overline{C}}_X$ as the difference between $\overline{\overline{C}}_A$ and $\overline{C}_A$.

In column chromatography the quantity $\overline{\overline{C}}_A$ is not determined. Instead, the experimental parameter measured is the constituent elution volume of solute, $\overline{V}_A$ (see Section III.C), the dependence of which upon solute concentration in the liquid phase ($\overline{C}_A$) provides the information required for elucidation of the binding curve. The relationship between elution volumes and concentrations is recognized by noting (31) the mass conservation requirement, $\overline{V}_A\overline{C}_A = V_A^*\overline{\overline{C}}_A$. Stated in an alternative way, the ratio of the accessible volume (V_A^*) to the constituent elution volume ($\overline{V}_A$) defines the proportion of the solute in the liquid phase. That is,

$$\frac{V_A^*}{\overline{V}_A} = \frac{\overline{C}_A}{\overline{\overline{C}}_A} \tag{2}$$

where again $\overline{C}_A$ and $\overline{\overline{C}}_A$ denote the liquid-phase and total solute concentrations, respectively. On the grounds that Eq. (1) may be arranged as

$$r\overline{\overline{C}}_X = \frac{\overline{\overline{C}}_A}{\overline{C}_A} - 1 \tag{3}$$

its combination with Eq. (2) allows binding to be defined in terms of elution volumes by means of the expression

$$r\overline{\overline{C}}_X = \left(\frac{\overline{V}_A}{V_A^*} - 1\right)\overline{C}_A \tag{4}$$

This equation may now be combined with the Scatchard (23) linear transform of the rectangular hyperbolic relationship for solute binding to a single class of matrix sites with intrinsic affinity constant k_{AX}, namely,

$$\frac{r\overline{\overline{C}}_X}{\overline{C}_A} = k_{AX}\overline{\overline{C}}_X - k_{AX}r\overline{\overline{C}}_X \tag{5a}$$

to give

$$\frac{\overline{V}_A}{V_A^*} - 1 = k_{AX}\overline{\overline{C}}_X - k_{AX}\left(\frac{\overline{V}_A}{V_A^*} - 1\right)\overline{C}_A \tag{5b}$$

as its chromatographic counterpart. A plot of $(\overline{V}_A/V_A^*) - 1$ versus $[(\overline{V}_A/V_A^*) - 1]\overline{C}_A$ is thus seen to be a Scatchard plot of binding data recorded in column-chromatographic format. Since the intrinsic association constant for solute–matrix interaction may be evaluated from the slope, it follows that the effective total concentration of matrix sites ($\overline{\overline{C}}_X$) may be inferred from the ordinate intercept ($k_{AX}\overline{\overline{C}}_X$) or, indeed, directly from the abscissa intercept. An alternative linear transform of the binding equation is (4)

$$\frac{1}{\overline{V}_A - V_A^*} = \frac{1}{k_{AX}V_A^*\overline{\overline{C}}_X} + \frac{\overline{C}_A}{V_A^*\overline{\overline{C}}_X} \tag{5c}$$

which allows k_{AX} and $\overline{\overline{C}}_X$ to be evaluated from the linear dependence of $1/(\overline{V}_A - V_A^*)$ upon $\overline{C}_A$. In these expressions it should be noted that $\overline{\overline{C}}_X$ is the effective (notional) total concentration of matrix sites, X, uniformly distributed in the volume accessible to partitioning solute, V_A^* (4).

B. Extensions to Systems with Multivalent Solutes

In many affinity chromatographic systems the partitioning solute is not univalent in its interaction with affinity matrix. Consequently, in evaluating the intrinsic affinity constant, k_{AX}, it is necessary to take this multivalency into account. For the interaction of an f-valent partitioning solute (A) with matrix sites (X), the binding function, r_f, should be defined (32) as

$$r_f = \frac{\overline{\overline{C}}_A^{1/f} - \overline{C}_A^{1/f}}{\overline{\overline{C}}_X} \tag{6a}$$

or, in chromatographic terms, as

$$r_f\bar{\bar{C}}_X = \bar{C}_A^{1/f}\left[\left(\frac{\bar{V}_A}{V_A^*}\right)^{1/f} - 1\right] \tag{6b}$$

Provided that a single intrinsic association constant, k_{AX}, governs all solute–matrix interactions, the general counterpart of the Scatchard analysis then becomes (32)

$$\frac{r_f\bar{\bar{C}}_X}{\bar{C}_A^{1/f}} = \frac{\bar{\bar{C}}_A^{1/f} - \bar{C}_A^{1/f}}{\bar{C}_A^{1/f}} = k_{AX}\bar{\bar{C}}_X - fk_{AX}\bar{\bar{C}}_X r_f\bar{\bar{C}}_A^{(f-1)/f} \tag{7}$$

Combination of Eqs. (6b) and (7) then yields

$$\left(\frac{\bar{V}_A}{V_A^*}\right)^{1/f} - 1 = k_{AX}\bar{\bar{C}}_X - fk_{AX}\left(\frac{\bar{V}_A}{V_A^*}\right)^{(f-1)/f}\bar{C}_A\left[\left(\frac{\bar{V}_A}{V_A^*}\right)^{1/f} - 1\right] \tag{8}$$

Equation (8) is readily rearranged to the form

$$\frac{1 - (V_A^*/\bar{V}_A)^{1/f}}{(V_A^*/\bar{V}_A)^{1/f}} = k_{AX}\bar{\bar{C}}_X - fk_{AX}\frac{\bar{V}_A}{V_A^*}\bar{C}_A\left[1 - \left(\frac{V_A^*}{\bar{V}_A}\right)^{1/f}\right] \tag{9}$$

which is an expression more in keeping with the manner in which the original equations for quantitative affinity chromatography of multivalent solutes were couched (17,31). A linear plot of $(\bar{\bar{C}}_A^{1/f} - \bar{C}_A^{1/f})/\bar{C}_A^{1/f}$ versus $(\bar{\bar{C}}_A^{1/f} - \bar{C}_A^{(1/f)})\bar{\bar{C}}_A^{(f-1)/f}$ or one of its corresponding chromatographic counterparts, Eq. (8) or (9), is thus a requirement for equivalence and independence of matrix sites for a partitioning solute that is multivalent.

An obvious prerequisite for application of Eqs. (6) to (9) to affinity chromatographic data is the assignment of a magnitude to the solute valence, *f*. On that score it must be clearly understood that any attempt to avoid such assignment by resort to a conventional Scatchard analysis, Eq. (5a) or (5b), merely means that unity has been chosen as the most appropriate valence. Pronounced curvilinearity of such plots, when *f* is known, signifies the existence of matrix–site heterogeneity and/or cooperativity (31,33,34). The important point to emanate from the considerations above is that Eqs. (6) to (9) provide the means of characterizing the interaction of partitioning solute with matrix in terms of the intrinsic binding constant (k_{AX}) and also the effective concentration of matrix sites ($\bar{\bar{C}}_X$). However, it is stressed that the affinity constant so determined refers to the interaction of solute with an immobilized ligand, and that the chemical modification associated with such immobilization may give rise to a situation in which the magnitude of k_{AX} differs considerably from the value for reaction between solute and ligand in solution.

C. Experimental Procedures for Evaluating k_{AX}

A quantitative affinity chromatography experiment to determine k_{AX} can be performed in any one of three ways. The choices are: by column chromatography, by a partition equilibrium technique, or by a recycling partition technique.

1. *Column Chromatography*

Of the three procedures mentioned above, column chromatography has usually been the technique used for characterizing ligand binding by virtue of biospecific interaction of one reactant, here designated the partitioning solute, with an affinity matrix. Apart from the fact that the experiments may need to be conducted on a relatively small scale (frequently, columns have a total volume considerably less than 1 mL), the procedures for setting up an affinity chromatography column do not differ substantially from those for any other column technique.

A decision needs to be made whether to use frontal affinity chromatography (4,24–26), the procedure of choice from the theoretical viewpoint; or whether to opt for the simplifying combination of circumstances that allow characterization of the interaction by the conventional zonal chromatographic technique (2,3,27).

2. *Zonal Affinity Chromatography*

Most quantitative affinity chromatographic studies have employed zonal analysis, perhaps because of its correspondence to conventional chromatography used in purification procedures. In this method a small zone of solute is applied to a column equilibrated with buffer or buffer–ligand mixture (2,3,27–30). As the zone migrates through the chromatographic bed it undergoes continual dilution because of axial dispersion, with the result that the concentration of the eluted zone is considerably smaller than that of applied solution. For example, Figure 3 illustrates the effect of NADH on elution profiles obtained (35) by zonal affinity chromatography of rat liver lactate dehydrogenase (0.1 mL, 180 nM) on a 1-mL column (0.5 × 1.28 cm) of affinity matrix, 10-carboxydeclyamino-Sepharose.

In principle, the elution volume, $\overline{V}_A$, should be determined as the median bisector of the zone (volume bisecting the area under the curve). However, this position becomes progressively removed from the peak position, V_p, with increasing asymmetry of the zone. Furthermore, inspection of Figure 3 reveals that the location of $\overline{V}_A$ in a zonal elution profile can become difficult because of poor definition of the trailing edge of skewed zones. Consequently, the effluent volume corresponding to the eluted peak, V_p, has invariably been substituted for $\overline{V}_A$, even though it is a relatively poor estimate in some instances. From Figure 4, which shows the effect of cytidine monophosphate on elution profiles in zonal affinity chromatography of ribonuclease on a column of uridine-5′-(Sepharose-4-aminophenylphosphoryl)-2′(3′)-phosphate (28), the problem encountered in Figure 3 is not unique. The availability of microcomputer-based data acquisition

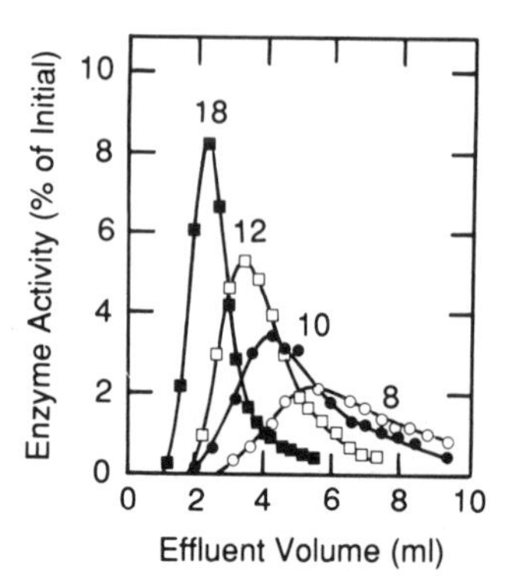

Figure 3 Elution profiles obtained in zonal chromatography of rat liver lactate dehydrogenase (0.1 mL, 0.18 μM) on a 1.0-mL column of 10-carboxydecylamino-Sepharose in the presence of the indicated concentrations (μM) of NADH. (Adapted from Ref. 35.)

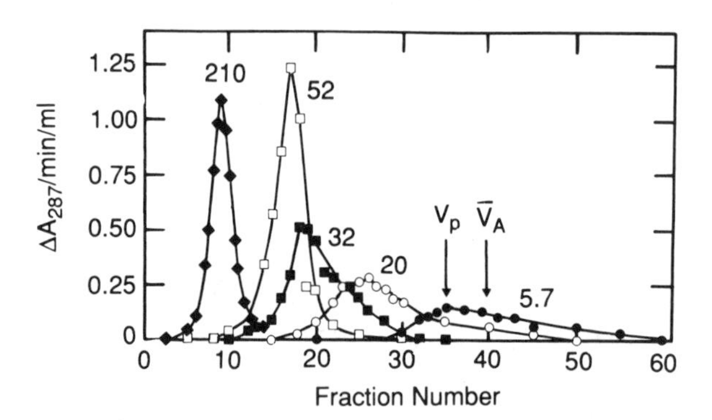

Figure 4 Zonal elution profiles for ribonuclease A on a column of uridine-5′-(Sepharose-4-aminophenylphosphoryl)-2′(3′)-phosphate preequilibrated with the indicated concentrations (μM) of cytidine-2′-monophosphate, the ordinate being expressed in terms of enzymic activity against cytidine-2′-3′-monophosphate. V_p denotes the elution volume taken as the peak of the eluted zone, whereas $\overline{V}_A$ is the median bisector of the zone for enzyme in the presence of 5.7 m*M* ligand. (Adapted from Ref. 28.)

systems that can be attached to devices for continuous monitoring of solute concentration now allows more accurate determination of $\overline{V}_A$ with little additional effort—a development that will hopefully herald the less frequent use of V_p as a substitute for $\overline{V}_A$.

3. *Frontal Affinity Chromatography*

The technique for frontal affinity chromatography differs from its conventional zonal counterpart only in regard to the applied volume of solution, which must be sufficiently large to ensure that the elution profile contains a plateau region (α phase) in which the concentrations of all soluble reactants equal those in the solution applied to the column. This type of QAC procedure is illustrated in Figure 5, with the results of an experiment on the effect of NADH concentration (C_S) on elution profiles obtained in the frontal affinity chromatography of rat liver lactate dehydrogenase (9 n*M*) on a 0.1-mL column (0.08 × 5.0 cm) of 10-carboxydecylamino-Sepharose (35). On the grounds that enzyme activity has been employed as the assay procedure specific for the total concentration of partitioning solute, $\overline{C}_A$, the elution volume of this constituent, $\overline{V}_A$, is given by the median bisector of the profile, which may be obtained from the relationship

$$\overline{V}_A = \frac{\Sigma (V \Delta \overline{C}_A^{\alpha})}{\overline{C}_A^{\alpha}} \tag{10}$$

where V is the mean elution volume corresponding to an increment in solute concentration, $\Delta\overline{C}_A$; and where the limits of the summation are the solvent plateau

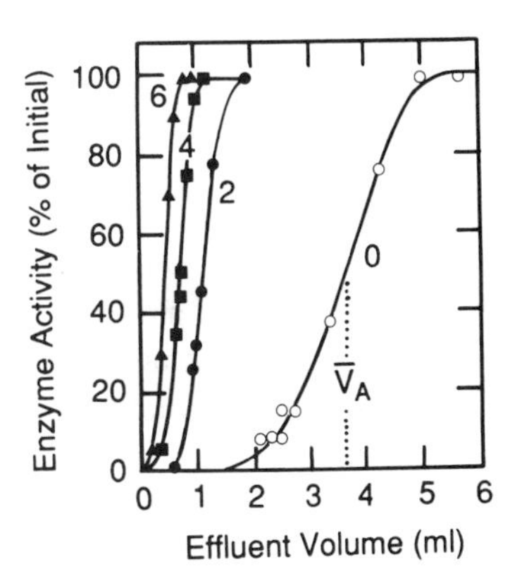

Figure 5 Advancing elution profiles obtained in frontal chromatography of rat liver lactate dehydrogenase (9 n*M*) on a 0.1-mL column of 10-carboxydecylamino-Sepharose in the presence of indicated concentrations (μ*M*) of NADH. (Adapted from Ref. 35.)

($\overline{C}_A = 0$) preceding the boundary and the plateau of original composition ($\overline{C}_A = \overline{C}_A^\alpha$). For the illustrated experiments (Figure 5), in which the column effluent was divided into fractions with volume ΔV, a more convenient expression for evaluating this position of the equivalent sharp boundary is

$$\overline{V}_A = V' - \frac{\Sigma(\overline{C}_A \Delta V)}{\overline{C}_A^\alpha} \tag{11}$$

where $\overline{C}_A$ is the constituent concentration of solute (enzyme) in a fraction, and V' corresponds to a selected effluent volume within the plateau region of original composition (α phase): V' is also the upper limit of the summation. In instances where the boundary exhibits essentially no asymmetry (skewness), $\overline{V}_A$ may be approximated by the effluent volume at which $\overline{C}_A = \overline{C}_A^\alpha/2$.

From the theoretical viewpoint the important feature of $\overline{V}_A$ from a frontal experiment is that it describes the affinity chromatographic behavior of the partitioning constituent in a solute–ligand mixture with the specified composition of the applied solution. $\overline{V}_A$ therefore reflects unequivocally the equilibrium state for a biphasic reaction mixture comprising constituent concentrations $\overline{C}_A^\alpha$ and $\overline{C}_S^\alpha$ of solute and ligand, respectively, in the liquid phase, and an effective total concentration $\overline{\overline{C}}_X$ of affinity matrix sites. Although the last-named parameter may be of unknown magnitude at the beginning of the experiment, the fact remains that it must be a constant for a series of experiments with the same column. The composition of the reaction mixture to which $\overline{V}_A$ refers is thus defined in a frontal chromatographic experiment.

4. *Partition Equilibrium Experiments*

The elution volume, $\overline{V}_A$, derived from a chromatographic experiment is merely an indirect measure of the distribution of solute concentration between liquid and matrix phases. An alternative method of defining this equilibrium position is to conduct a series of partition equilibrium experiments in which the concentrations of partitioning solute in the liquid phase are determined for mixtures with known total concentrations of solute. Separate mixtures containing known amounts of affinity matrix and partitioning solute, A (with or without soluble ligand, S), are allowed to equilibrate at the temperature of interest to establish chemical equilibrium. At that stage a sample of the supernatant is obtained by filtration (4) or by centrifugation (12–16) of each mixture at the same temperature. The concentration of partitioning solute in the liquid phase, $\overline{C}_A$, is then determined by any appropriate spectrophotometric, radiochemical, or enzymatic means. When ligand is present, this constituent concentration of partitioning solute in the liquid phase includes the contributions of any soluble solute–ligand complexes as well as free solute. The total solute concentration, $\overline{\overline{C}}_A$, can be determined by dividing the amount of partitioning solute added by V_A^*, the volume accessible to partitioning solute in the absence of any interaction with affinity matrix (4).

A disadvantage of such experiments is the need for *precise control of the amount of affinity matrix added* to each reaction mixture (4,6). Meeting this requirement can be difficult to fulfill in instances where the affinity matrix is being dispensed as a concentrated slurry. This difficulty has been obviated (17,31) by resort to a recycling partition technique (36) in which the liquid phase of a stirred slurry of affinity matrix and solute is monitored spectrophotometrically to obtain $\overline{C}_A$ (Figure 6). Since further aliquots of solute (or ligand) may be added after establishment of $\overline{C}_A$, all mixtures may be prepared by successive additions of more solute or ligand to the one sample of affinity matrix, thereby avoiding the need for any assumption about the identity of matrix amounts in a series of equilibrating mixtures. This recycling partition technique has another advantage in that attainment of partition equilibrium after each addition is recognized by time independence of $\overline{C}_A$ in the continuously monitored liquid phase. Furthermore, incorporation of an on-line data acquisition system to monitor the concentration of partitioning solute in the liquid phase has enabled this equilibrium position to be evaluated from the form of the progress curve toward equilibrium, a development that decreases significantly the time required to obtain the necessary data for characterizing an interaction by quantitative affinity chromatography (20).

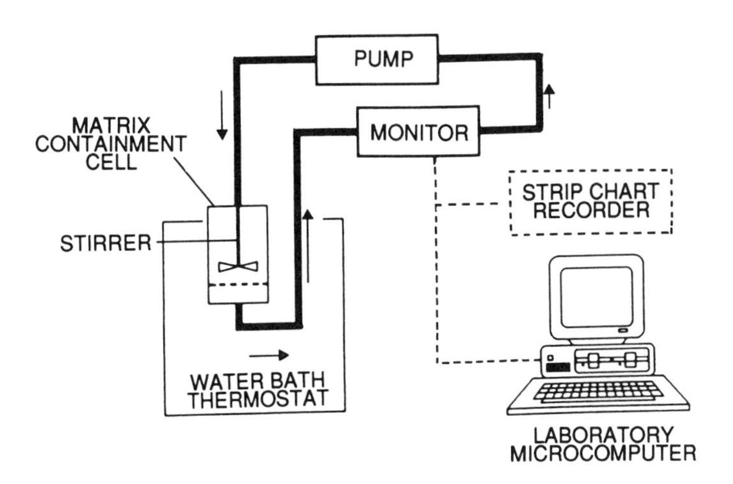

Figure 6 Schematic representation of the current recycling partition equilibrium system. (Adapted from Ref. 20.)

Even in instances where quantitative affinity chromatography is being used to characterize a solute–ligand interaction in solution, quantification of the solute–matrix interaction forms an integral part of the analysis, which, strictly speaking, requires quantitative description of all equilibria involved in the affinity chromatographic process. We now consider the possible consequences of soluble ligand on the affinity chromatographic behavior of the partitioning solute.

IV. CHARACTERIZATION OF SOLUTE–LIGAND INTERACTIONS

In affinity chromatography the effect of including a ligand, S, is either to facilitate desorption of the solute because of interactions that are competitive with solute partitioning, or to retard desorption of the solute through formation of complexes that enhance the interaction with matrix. For systems with only two types of interaction there are two competitive situations: one in which the interactions of ligand and matrix with solute are mutually exclusive, and the other in which the competition is between solute and ligand for matrix sites. On the other hand, solute-retarded desorption occurs when the formation of a solute–ligand complex is a prerequisite for interaction with matrix sites; and also when the partition of solute reflects interaction with a liganded matrix site. These four situations, illustrated in Figure 7, will be considered in turn.

A. Analyses in Terms of Free Ligand Concentration

Because binding constants are defined as appropriate ratios of equilibrium concentrations of product and reactant species, the expressions for analyzing the effects of a soluble ligand on the affinity chromatographic behavior of a partitioning solute were developed initially in terms of the free concentration of ligand in the liquid phase, C_S. That approach is also adopted in this initial consideration of the four potential situations.

1. *Case I: Competition Between Matrix and Ligand for Solute Sites*

Consider, first, the commonly encountered affinity chromatographic situation in which the matrix site, X, is essentially an immobilized form of the soluble ligand, S, that is to be added. It is envisaged that like the X site, S is univalent in its interaction with A. Clearly, the partition behavior of the solute requires description in terms of two phenomena: the solute–matrix interaction governed by intrinsic association constant k_{AX} discussed above, and a solute–ligand interaction, governed by a corresponding binding constant, k_{AS}. However, despite the fact that there are now two interactions contributing to the overall extent of partitioning, Eq. (6a) or (6b) may still be used for the determination of an *effective* binding function, r_{eff}, for any mixture with defined concentrations of solute constituent

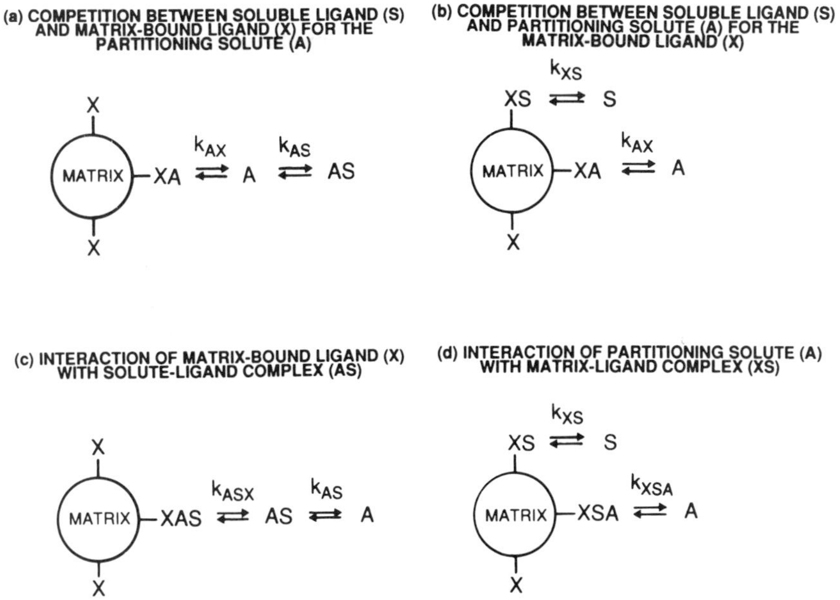

(a) COMPETITION BETWEEN SOLUBLE LIGAND (S) AND MATRIX-BOUND LIGAND (X) FOR THE PARTITIONING SOLUTE (A)

X
MATRIX —XA $\overset{k_{AX}}{\rightleftharpoons}$ A $\overset{k_{AS}}{\rightleftharpoons}$ AS
X

(b) COMPETITION BETWEEN SOLUBLE LIGAND (S) AND PARTITIONING SOLUTE (A) FOR THE MATRIX-BOUND LIGAND (X)

XS $\overset{k_{XS}}{\rightleftharpoons}$ S
MATRIX —XA $\overset{k_{AX}}{\rightleftharpoons}$ A
X

(c) INTERACTION OF MATRIX-BOUND LIGAND (X) WITH SOLUTE-LIGAND COMPLEX (AS)

X
MATRIX —XAS $\overset{k_{ASX}}{\rightleftharpoons}$ AS $\overset{k_{AS}}{\rightleftharpoons}$ A
X

(d) INTERACTION OF PARTITIONING SOLUTE (A) WITH MATRIX-LIGAND COMPLEX (XS)

XS $\overset{k_{XS}}{\rightleftharpoons}$ S
MATRIX —XSA $\overset{k_{XSA}}{\rightleftharpoons}$ A
X

Figure 7 Schematic representation of the four possible affinity chromatographic situations involving an interplay of two chemical equilibria.

($\overline{C}_A$) and free ligand (C_S) in the liquid phase. By conducting a series of such experiments with a fixed concentration of free ligand and a range of solute concentrations ($\overline{C}_A$), the expressions used above for ligand-free solutions, Eqs. (7) to (9), may be employed to describe the partitioning of solute in terms of an effective association constant, $\overline{k}_{AX}$. The task then is to relate this effective equilibrium constant to k_{AX} and k_{AS}.

Inasmuch as Eqs. (8) and (9) are merely chromatographic manifestations of the multivalent Scatchard expression, Eq. (7), it is evident that the effective equilibrium constant, $\overline{k}_{AX}$, is being determined as if $\overline{C}_A$ were actually the equilibrium concentration of free partitioning solute in the liquid phase. The concentration of free solute, C_A, is less than $\overline{C}_A$ as the result of solute–ligand complex formation, the quantitative description of their interrelationship being given (22) by

$$C_A = \frac{\overline{C}_A}{(1 + k_{AS}C_S)^f} \tag{12}$$

Since $\bar{C}_A^{1/f}$ and $C_A^{1/f}$ are required for the respective determinations of $\bar{k}_{AX}$ and k_{AX}, it follows that the two solute–matrix affinity constants are related by the expression (12,17)

$$k_{AX} = \bar{k}_{AX}(1 + k_{AS}C_S) \tag{13a}$$

Provided that k_{AX} is first determined from experiments conducted in the absence of ligand, as described above, the measurement of $\bar{k}_{AX}$ for a known concentration of free ligand (C_S) clearly allows estimation of the solute–ligand affinity constant, k_{AS}, from Eq. (13a). Alternatively, both equilibrium constants may be obtained, albeit with less precision, from a transform of Eq. (13a) such as a plot of $\bar{k}_{AX}$ versus $\bar{k}_{AX}C_S$ [see Eq. (13b)].

$$\bar{k}_{AX} = k_{AX} - \bar{k}_{AX}k_{AS}C_S \tag{13b}$$

2. *Case II: Competition Between Solute and Ligand for Matrix Sites*

Facilitated desorption also results from competition by the ligand (S) for matrix sites. An example of this situation is a comparative study of the binding of various saccharides to immobilized concanavalin A (37,38). In this instance the two equilibria to be characterized are the solute–matrix interaction governed by intrinsic affinity constant k_{AX}, and a matrix–ligand interaction with binding constant k_{XS}.

From the viewpoint of interpreting the magnitude of $\bar{k}_{AX}$ deduced from r_{eff} [Eq. (7)], the absence of a solute–ligand interaction means that the source of the discrepancy between $\bar{k}_{AX}$ and k_{AX} is not the result of a difference between C_A and $\bar{C}_A$, which is, indeed, the free concentration of partitioning solute in the liquid phase. However, $\bar{k}_{AX}$ is calculated on the basis that all matrix sites are available for reaction with A, whereas some are occupied by ligand. This leads to the relationship $\bar{k}_{AX}\bar{\bar{C}}_X = k_{AX}C_X$ with $C_X = \bar{\bar{C}}_X/(1 + k_{XS}C_S)$ as the expression for the concentration of free matrix sites (C_X) in terms of its total counterpart ($\bar{\bar{C}}_X$). The relationship between the two solute–matrix affinity constants is therefore

$$k_{AX} = \bar{k}_{AX}(1 + k_{XS}C_S) \tag{14}$$

Comparison of Eqs. (13a) and (14) confirms the earlier observation (39) that these two competitive cases are formally identical. Consequently, although the magnitude of the second equilibrium constant is unequivocal, its identification as k_{AS} or k_{XS} must rely on chemical considerations that dictated the selection of the immobilized reactant as affinity matrix.

3. *Case III: Requirement of a Solute–Ligand Complex for Interaction with Matrix*

We now turn to systems exhibiting ligand-retarded desorption of the partitioning solute as the result of ternary complex formation between ligand, solute, and matrix site. Here again there are two situations to consider. One involves the reaction of soluble solute–ligand complexes with matrix sites, with k_{AS} and k_{ASX} as the relevant intrinsic binding constants. The other involves reaction of matrix–ligand complex with partitioning solute, and thus the two equilibrium constants are designated k_{XS} and k_{XSA}. Attention is restricted initially to the former ordered reaction mechanism wherein solute–ligand complex formation precedes the interaction with matrix. An example of this situation is encountered in the two affinity chromatographic systems for which ligand-retarded elution has been quantified: the *N*-acetylglucosamine-retarded desorption of galactosyltransferase from α-lactalbumin–Sepharose (2), and the NADH-dependent retardation of lactate dehydrogenase on oxamate-Sepharose (11).

The absence of a direct solute–matrix interaction means that there is no information, apart from V_A^*, to be gained from partition studies of solute alone. Both intrinsic affinity constants (k_{AS} and k_{ASX}) must therefore be evaluated from affinity chromatographic experiments conducted in the presence of ligand. In that regard, evaluation of r_{eff} [Eq. (7)] for a series of solute–ligand mixtures comprising varying concentrations of solute ($\overline{C}_A$) in the presence of a constant free concentration of ligand (C_S) again allows the evaluation of $\overline{k}_{AX}$ via Eq. (7), (8), or (9). Furthermore, by repeating such measurements at a series of C_S values, the magnitude of k_{ASX} could be evaluated as the value of $\overline{k}_{AX}$ in the limit of infinite free ligand concentration, where there is effectively no free partitioning solute, and where $\overline{C}_A$ may be identified with C_{AS_f}, the concentration of fully saturated solute–ligand complex. The corresponding intrinsic affinity constant is therefore k_{ASX}.

At lower free ligand concentrations the smaller magnitude of $\overline{k}_{AX}$ reflects the existence of unliganded A sites with no affinity for matrix. On the grounds that the fractional saturation of A sites with ligand is $k_{AS}C_S/(1 + k_{AS}C_S)$ (22), the relationship between $\overline{k}_{AX}$ and k_{ASX} becomes (39)

$$\overline{k}_{AX} = \frac{k_{ASX}k_{AS}C_S}{1 + k_{AS}C_S} \tag{15a}$$

or in Scatchard linearized format, as

$$\frac{\overline{k}_{AX}}{C_S} = k_{ASX}k_{AS} - k_{AS}\overline{k}_{AX} \tag{15b}$$

The slope and ordinate intercept of the linear Scatchard plot (or its double-reciprocal equivalent) yield both k_{AS} and k_{ASX}. Extrapolation of $\bar{k}_{AX}$ to infinite free ligand concentration is therefore unnecessary, but does provide an additional consistency check on the interpretation of the experiment.

4. *Case IV: Requirement of a Matrix–Ligand Complex for Interaction with Solute*

Application of the corresponding arguments to the situation in which the partitioning solute reacts with matrix–ligand complex leads to a formally identical expression (39):

$$\bar{k}_{AX} = \frac{k_{XSA} k_{XS} C_S}{1 + k_{XS} C_S} \tag{16}$$

Consequently, identical magnitudes of the binding constants for binary and ternary complex formation emanate from analyses of results via Eq. (15) or (16). In the event that ternary complex formation does proceed by an ordered mechanism, that knowledge allows identification of the appropriate pair of parameters (k_{AS} and k_{ASX}, or k_{XS} and k_{XSA}).

For a random mechanism it is immaterial which of the two descriptions is used, because a thermodynamic study, by definition, cannot be used to distinguish between alternative pathways of ternary complex formation. A system that would, in principle, present this dilemma is the analysis of protein–nucleic acid interactions by filtration through nitrocellulose, which binds protein (S) as well as nucleoprotein complex (AS). On thermodynamic grounds it is not possible to decide which of the two pathways is responsible for immobilization of the nucleic acid as nucleoprotein–matrix complex; but the slowness of complex dissociation virtually guarantees that ternary complex is formed solely between existing protein–nucleic acid complexes and matrix (i.e., that k_{AS} and k_{ASX} are the two pertinent equilibrium constants—the interpretation inherent in nitrocellulose filter assays) (40–42).

5. *Binary and Ternary Complex Formation*

Thus far the involvement of partitioning solute in binary and ternary complex formation with matrix sites has been considered to be mutually exclusive; but clearly there are situations in which all three equilibria (k_{AS}, k_{AX}, k_{ASX}, or k_{XS}, k_{AX}, k_{XSA}) may be operative. Although the required general theoretical expressions have been formulated (39), their application was limited to the theoretical demonstration that the binding of solute via binary and ternary complex formation introduces curvilinearity into plots of results according to the above linear transforms of binding data for systems in which only one form of solute–matrix interaction operates. This inference has been exploited experimentally to establish the noncompetitive nature of the inhibitory effect of Ca(II) binding on the aldolase–myofibril interaction, a system for which the binding of enzyme to the various liganded states of the myofibrillar matrix proved amenable to quantitative characterization (14).

B. Analysis in Terms of Total Ligand Concentration

A limitation of the analyses above is their reliance on expressions for $\bar{k}_{AX}$ in terms of free ligand concentration, a factor that restricts their application to studies of relatively weak interactions for which the total ligand concentration approximates the required free concentration (17), or to studies of high-affinity interactions for which the concentration of free ligand may be established by prior equilibrium dialysis or its gel chromatographic counterpart (18,19,31). Such an approach is clearly inapplicable to situations in which ligand and partitioning soluble are both macromolecular; and furthermore, there are certainly technical problems associated with ensuring that dialysis equilibrium has been attained (19). Indeed, as the strength of a solute–ligand interaction increases, the volume of extremely dilute ligand solution required for attainment of dialysis equilibrium with a detectable concentration of solute becomes prohibitively excessive. This limitation has now been removed by the recent realization (20,43,44) that the theoretical expressions for quantitative affinity chromatography are also amenable to analytical solution in terms of total ligand concentration.

1. *Competition for Solute Sites (Case I)*

In instances where the competition is between ligand and matrix sites for partitioning solute, the total ligand concentration in the liquid phase ($\bar{C}_S$) may be introduced into the quantitative analysis by taking advantage of the expression (22)

$$\bar{C}_S = C_S + fk_{AS}C_AC_S\,(1 + k_{AS}C_S)^{f-1} \tag{17a}$$

or on substitution of Eq. (12) for the free solute concentration,

$$\bar{C}_S = C_S + \frac{fk_{AS}\bar{C}_AC_S}{1 + k_{AS}C_S} \tag{17b}$$

A combination of Eq. (17b) with Eq. (13a) then leads to the expression (20)

$$R - 1 = k_{AS}\left[\bar{C}_S - \frac{(R-1)\,f\bar{C}_A}{R}\right] \tag{18}$$

where $R = k_{AX}/\bar{k}_{AX}$ has been substituted for the experimentally measurable ratio of intrinsic affinity constants for the solute–matrix interaction in the absence and presence of ligand. By measuring R at a series of total ligand concentrations, the value of the solute–ligand intrinsic binding constant (k_{AS}) may be determined from the slope of a plot of $(R-1)$ versus $[\overline{C}_S - (R-1)f\overline{C}_A/R]$. The application of this approach is illustrated in Section V.E.

2. *Competition for Matrix Sites (Case II)*

In the competitive situation with k_{AS} and k_{AX} the operative equilibrium constants, the total concentration of ligand in the liquid phase is C_S, and hence Eq. (14) continues to provide the relationship for analyzing frontal chromatographic data. However, in partition equilibrium experiments only the total ligand concentration, $\overline{\overline{C}}_S$, is of known magnitude. $\overline{\overline{C}}_S$ is introduced by noting that

$$\overline{\overline{C}}_S = C_S(1 + k_{XS}C_X) \tag{19}$$

To avoid the problem of measuring the effective concentration of free matrix sites, C_X, this parameter is eliminated by its replacement with the expression

$$C_X = \frac{\overline{\overline{C}}_A^{1/f} - \overline{C}_A^{1/f}}{k_{AX}\overline{C}_A^{1/f}} \tag{20}$$

which is a rearranged form of the definition of the intrinsic binding constant for the interaction of f-valent A with a univalent matrix site, X. Equation (20) also follows from Eq. (7) on noting that the right-hand side thereof describes the product $k_{AX}C_X$ (18). Combination of Eqs. (14), (19), and (20) then yields the relationship

$$(R-1)\,k_{AX}\,\overline{C}_A^{1/f} = k_{XS}[k_{AX}\overline{C}_A^{1/f}\overline{\overline{C}}_S - (R-1)(\overline{\overline{C}}_A^{1/f} - \overline{C}_A^{1/f})] \tag{21}$$

For this type of ligand-facilitated desorption the magnitude of k_{XS} may therefore be obtained from the slope of a plot of $(R-1)k_{AX}\overline{C}_A^{1/f}$ versus $[k_{AX}\overline{C}_A^{1/f}\overline{\overline{C}}_S - (R-1)(\overline{\overline{C}}_A^{1/f} - \overline{C}_A^{1/f})]$.

3. *Creation of a Solute Site (Case III)*

For ligand-retarded desorption that reflects the interaction of solute–ligand complexes with matrix sites, the theoretical expression for $\overline{k}_{AX}$, Eq. (15), is amenable to analytical solution in terms of total ligand concentration in the liquid phase, $\overline{C}_S$, provided that k_{ASX} is first obtained as the extrapolated value of $\overline{k}_{AX}$ in the limit of infinite ligand concentration ($1/\overline{C}_S \rightarrow 0$). With that proviso the ratio of solute–matrix affinity constants, $Q = k_{ASX}/\overline{k}_{AX}$, becomes an experimentally determinable parameter that is related to the free ligand concentration, C_S, by Eq. (15). Specifically,

$$Q = \frac{1 + k_{AS}C_S}{k_{AS}C_C} \tag{22a}$$

or

$$\frac{1}{Q-1} = k_{AS}C_S \tag{22b}$$

The total ligand concentration in the liquid phase, $\bar{C}_S$, is now introduced by combining Eq. (17b) with Eq. (22a) to give the relationship $C_S = \bar{C}_S - f\bar{C}_A/Q$, whereupon Eq. (22b) becomes

$$\frac{1}{Q-1} = k_{AS}\left(C_S - \frac{f\bar{C}_A}{Q}\right) \tag{23}$$

The remaining equilibrium constant (k_{AS}) may be obtained as the slope of a plot of $1/(Q-1)$ versus $\bar{C}_S - f\bar{C}_A/Q$.

Equation (23) is suitable for use with frontal affinity chromatographic data, since $\bar{C}_S$ and $\bar{C}_A$ refer to the composition of the applied solution. In partition equilibrium studies, however, the only ligand concentration available from the amount added is again $\bar{\bar{C}}_S$, the total ligand concentration in the biphasic system. As in the previous case, we need a relationship in terms of $\bar{\bar{C}}_S$ if this type of affinity system is to be characterized from partition equilibrium experiments. The differences between the total concentration of ligand in the system, $\bar{\bar{C}}_S$, and that in the liquid phase, $\bar{C}_S$, reflects the concentration of ligand associated with matrix in the form of ASX complexes; and clearly, this concentration difference must equal the concentration of A sites that are involved in such complexes.

Evaluation of the effective association constant for the solute–matrix interaction via Eqs. (7) to (9) is based on the definition of this intrinsic constant as (17,18)

$$\bar{k}_{AX} = \frac{[1-(\bar{C}_A/\bar{\bar{C}}_A)^{1/f}]f\bar{\bar{C}}_A}{C_X\,[(\bar{C}_A/C_A)^{1/f}]f\bar{\bar{C}}_A} \tag{24}$$

On the grounds that the numerator of Eq. (24) corresponds to the concentration of A sites complexed with matrix (17), we may therefore write

$$\left[1-\left(\frac{\bar{C}_A}{\bar{\bar{C}}_A}\right)^{1/f}\right]f\bar{\bar{C}}_A = \bar{\bar{C}}_S - \bar{C}_S \tag{25}$$

whereupon $\bar{C}_S$, the parameter required for evaluation of k_{AS} via Eq. (23), is given by

$$\bar{C}_S = \bar{\bar{C}}_S - (\bar{\bar{C}}_A^{1/f} - \bar{C}_A^{1/f})f\bar{\bar{C}}_A^{(f-1)/f} \tag{26}$$

4. *Creation of a Matrix Site (Case IV)*

For situations in which solute adsorption reflects the interaction with matrix–ligand complex, absence of a solute–ligand interaction in the liquid phase ensures that Eq. (16) may be applied directly to frontal affinity chromatography results because $C_S = \overline{C}_S$. It therefore remains to obtain expressions in terms of $\overline{\overline{C}}_S$ to render partition studies of such systems amenable to quantitative analysis.

The first point to note is that the intrinsic constant for ternary complex formation, k_{XSA}, may again be obtained as the limiting value of $\overline{k}_{AX}$ as $\overline{C}_S \rightarrow \infty$, thereby allowing the dependence of the effective equilibrium constant for the solute–matrix interaction ($\overline{k}_{AX}$) upon ligand concentration to be expressed in terms of the ratio $Q = k_{XSA}/\overline{k}_{XA}$. Specifically,

$$\frac{1}{Q-1} = k_{XS}C_S \tag{27}$$

which follows from Eq. (22) and the analogous forms of Eqs. (15) and (16) for the two affinity systems involving ternary complex formation. Second, the concentration of ligand present as ternary complex continues to equal the concentration of bound solute and is therefore given by the left-hand side of Eq. (25). Third, by analogy with Eq. (24), the intrinsic association constant for ternary complex formation, k_{XSA}, may be written

$$k_{XSA} = \frac{[1 - (\overline{C}_A/\overline{\overline{C}}_A)^{1/f}]f\overline{\overline{C}}_A}{C_{XS}[(\overline{C}_A/\overline{\overline{C}}_A)^{1/f}]f\overline{\overline{C}}_A} \tag{28}$$

which allows the concentration of matrix–ligand complex, C_{XS}, to be expressed [cf. Eq. (20)] as

$$C_{XS} = \frac{\overline{\overline{C}}_A^{1/f} - \overline{C}_A^{1/f}}{k_{XSA}\overline{C}_A^{1/f}} \tag{29}$$

The ligand concentration required for substitution into Eq. (26) is therefore

$$C_S = \overline{C}_S = \overline{\overline{C}}_S - (\overline{\overline{C}}_A^{1/f} - \overline{C}_A^{1/f})\left(fC_A^{(f-1)/f} + \frac{1}{k_{XSA}\overline{C}_A^{1/f}}\right) \tag{30}$$

whereupon k_{XS} may also be evaluated from partition equilibrium studies of systems within this category.

V. ILLUSTRATION OF VARIOUS EXPERIMENTAL PROCEDURES

Having presented the basic quantitative expressions for characterization of the interactions responsible for biospecific elution–adsorption, we now illustrate their application to experimental results obtained by a variety of procedures and offer comparisons as a basis for selecting or rejecting particular procedures for the systems being characterized. Consideration is first given to the analysis of results obtained by frontal chromatography.

A. Characterization by Frontal Chromatography

There have been only two reports of completely rigorous characterization of solute–ligand interactions by frontal affinity chromatography: the NADH-enhanced desorption of lactate dehydrogenase from trinitrophenyl-Sepharose (18), and the methylglucoside-facilitated elution of concanavalin A from Sephadex G-50—an example of an affinity matrix in which the glucosyl residues of the Sephadex act as the immobilized ligand (X) (45). For each of these systems a series of frontal chromatographic experiments was first performed with a range of concentrations of partitioning solute (protein) in the absence of ligand, thereby allowing the intrinsic affinity constant for the solute–matrix interaction (k_{AX}) to be evaluated via one of Eqs. (7) to (9). Corresponding analyses of the concentration dependence of elution volume obtained by frontal chromatography of mixtures containing a fixed concentration, C_S, of free ligand then yielded the effective solute–matrix affinity constant ($\bar{k}_{AX}$), which is related to k_{AX} and the solute–ligand binding constant (k_{AS}) by Eq. (13a). This experimental protocol is amplified in the following consideration of results for the NADH-facilitated elution of lactate dehydrogenase from trinitrophenyl-Sepharose (18).

1. *Rigorous Analysis of Frontal Affinity Chromatographic Results*

In the study of the lactate dehydrogenase–NADH interaction by quantitative affinity chromatography on trinitrophenyl-Sepharose (18), mixtures of rabbit muscle lactate dehydrogenase (0.4 to 5.4 μM) and predetermined concentrations of free NADH (0 to 20 μM) were prepared by zonal gel chromatography of concentrated enzyme solution (5 mL, 40 μM) on a column of Sephadex G-25 (2.0 × 21 cm) preequilibrated with 0.067 M phosphate buffer (pH 7.2) containing the required free concentration of coenzyme (C_S). Protein eluting at the void volume was suitably diluted with more of the buffer–NADH solution to give the required concentration of partitioning solute (enzyme), $\bar{C}_A$. The same buffer–NADH solution, which corresponds to the dialysate in equilibrium dialysis (46), was used to pre-equilibrate the column of trinitrophenyl-Sepharose (0.9 × 9.5 cm) prior to application of sufficient mixture to generate an elution profile with a plateau (α phase) in which the total concentration of solute equaled that of the mixture ($\bar{C}_A$). The elution volume, $\bar{V}_A$, in each experiment was then obtained as the median bisector of the advancing elution profile via Eq. (11), and the results were analyzed in terms of Eq. (8). V_A^*, the column volume accessible to lactate dehydrogenase, was taken as the elution volume of enzyme in buffer supplemented with 4 M NaCl to suppress the solute–matrix interaction.

Results obtained in coenzyme-free solutions as well as in the presence of NADH are presented in Figure 8, where the value of 4 used for the valence of the enzyme is in keeping with the tetrameric nature of lactate dehydrogenase. The series of experiments conducted in the absence of NADH (●) suffices to establish

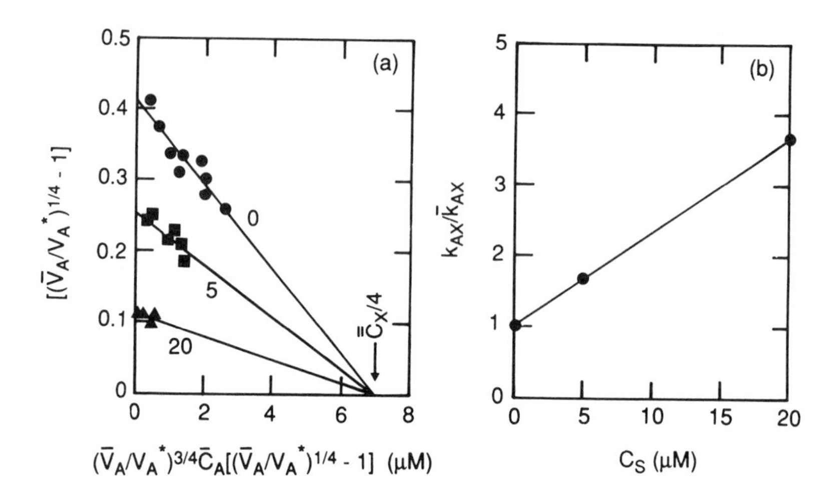

Figure 8 Characterization of the interaction between NADH and rabbit muscle lactate dehydrogenase by frontal chromatography of the enzyme on trinitrophenyl-Sepharose. (a) Multivalent Scatchard plot for determination, via Eq. (8), of the constitutive association constant ($\bar{k}_{AX}$) for the enzyme–matrix interaction in the absence of coenzyme and in the presence of free concentrations, C_S, of 5 μM and 20 μM NADH. (Adapted from Ref. 18.) (b) Plot of the resultant $\bar{k}_{AX}$ values in accordance with Eq. (13a) to obtain k_{AS} for the enzyme–NADH interaction.

values of $1.5 \times 10^4\ M^{-1}$ for k_{AX} and 28 μM for the effective total concentration of matrix sites from the slope ($4k_{AX}$) and abscissa intercept ($\overline{\overline{C}}_X/4$), respectively, of this plot of results in accordance with Eq. (8). Advantage is then taken of the fact that the same concentration of matrix sites ($\overline{\overline{C}}_X$) must also apply to the other series of experiments with 5 μM (■) and 20 μM (▲) NADH. As predicted by Eq. (13a), these values exhibit a linear dependence upon free coenzyme concentration (Figure 8b), the slope of which signifies a magnitude of $1.3 \times 10^5\ M^{-1}$ for k_{AS}, the intrinsic binding constant for the interaction of NADH with rabbit muscle lactate dehydrogenase under these conditions (18).

The foregoing procedure for quantifying the solute–matrix interaction in the absence of ligand has also been used to characterize the interaction of fatty acids with immobilized bovine serum albumin (47), the binding of 1,8-anilinonaphthalenesulfonic acid to α-lactalbumin (48), and the interaction of Ca(II) ions with immobilized casein submicelles (49). In each of these cases the partitioning solute was univalent. The curvilinear Scatchard plot obtained in the first study (47) provides evidence of an affinity system in which the solute–matrix interaction is not describable by a single intrinsic constant, k_{AX}—the assumption inherent in the suggested procedures for evaluating solute–ligand binding constants. However, even in situations where there is nonconformity with this assumption, frontal affinity chromatography may, in principle, still be used to characterize the solute–ligand interaction. Instead of using the results obtained in the absence of ligand to evaluate k_{AX} and $\bar{\bar{C}}_X$, they would merely be used as a calibration plot ($\overline{V}_A$ versus $\overline{C}_A$) that related the concentration (or amount) of adsorbed solute to a corresponding free concentration (C_A) in the liquid phase. The latter concentration in an applied solute–ligand mixture with total solute concentration $\overline{C}_A$ could then be identified on the basis of the calibration plot and the measured $\overline{V}_A$ for the mixture. Although no such frontal affinity chromatographic studies have been reported, the feasibility of the approach has been amply demonstrated in partition equilibrium studies designed to characterize the interactions of methyl orange with bovine serum albumin (36), of lysomyristoylphosphatidylcholine with myelin basic protein (50), and of heparin with antithrombin (43) on the basis of the effect of ligand addition on the amount of partitioning solute adsorbed to affinity matrix; methyl orange to Sephadex G-25 (36), lipid to Bio-Gel P-2 (50), and antithrombin to heparin–Sepharose (43).

2. *Approximate Analyses of Frontal Chromatographic Experiments*

There have been several quantitative investigations in which the affinity constant for a solute–ligand interaction has been deduced from the effect of ligand concentration on the magnitude of $\overline{V}_A$ in frontal affinity chromatographic experiments with a single concentration, $\overline{C}_A$, of partitioning solute (11,24–26). The method does not exhibit the rigor of the approach above because of its reliance (51) upon reexpression of Eq. (8) as

$$\left(\frac{\overline{V}_A}{V_A^*}\right)^{1/f} - 1 = \bar{k}_{AX}\bar{\bar{C}}_X\left[1 - B\left(1 - \left(\frac{V_A^*}{\overline{V}_A}\right)^{1/f}\right)\right] \tag{31}$$

and its consequent truncation (18) to

$$\left(\frac{\overline{V}_A}{V_A^*}\right)^{1/f} - 1 = \bar{k}_{AX}\bar{\bar{C}}_X \tag{32}$$

on the grounds that $B = f(\overline{V}_A/V_A^*)\overline{C}_A/\bar{\bar{C}}_X = f\bar{\bar{C}}_A/\bar{\bar{C}}_X$ may well be sufficiently small for the term omitted on the right-hand side to be of negligible magnitude in relation to unity.

This approximate procedure is first illustrated by the consideration of results from a frontal study of the effect of benzamidine (10 to 500 μ*M*) on the elution volume of β-trypsin (0.2 μ*M*) on a column of GlyGlyArg–Sepharose (25). For this situation involving competition between ligand and matrix sites for the single active site on trypsin ($f = 1$) the relationship between $\bar{k}_{AX}$ and the solute–matrix affinity constant, k_{AX}, is given by Eq. (13). Furthermore, the small concentration of enzyme used ($\overline{C}_A = 0.2\ \mu M$) ensures the essential identity of C_S and $\overline{C}_S$. On this basis, Eq. (32) for the benzamidine-facilitated desorption of trypsin from GlyGlyArg–Sepharose becomes

$$\frac{\overline{V}_A - V_A^*}{V_A^*} = \frac{k_{AX}\overline{\overline{C}}_X}{1 + k_{AS}C_S} \tag{33}$$

Figure 9 presents results inferred from Fig. 2 of Ref. 25 in terms of two linear transforms of Eq. (33). The advantage of the linear transform presented in Figure 9a (18) is that the magnitude of k_{AS} follows directly from the slope, the product of the effective total concentration of matrix sites ($\overline{\overline{C}}_X$) and the solute–matrix affinity constant (k_{AX}) being given by the ordinate intercept. On the other hand, adoption of the second linear transform (4) in Figure 9b requires k_{AS} to be evaluated as the ratio of the slope $(1/V_A^* k_{AS} k_{AX}\overline{\overline{C}}_X)$ and the ordinate intercept $(1/V_A^* k_{AX}\overline{\overline{C}}_X)$. In instances where $\overline{\overline{C}}_X$ is deliberately made large to ensure conformity with the requirement that $f\overline{C}_A << \overline{\overline{C}}_X$, the ordinate intercept may well be indistinguishable from zero (18). For the present system, however, that difficulty was not encountered, and similar estimates of $6.4 \times 10^4\ M^{-1}$ and $7.0 \times 10^4\ M^{-1}$ for k_{AS} therefore emanate from the respective analyses presented in Figure 9a and b. The slight disparity between estimates merely reflects the different weighing accorded each experimental point in the two linear transforms of Eq. (33).

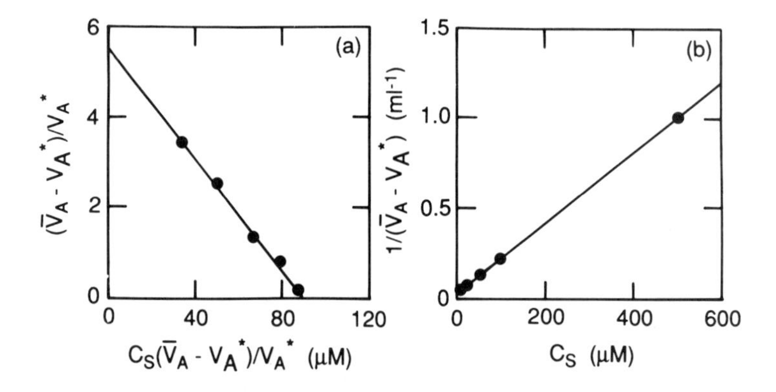

Figure 9 Approximate evaluation of the binding constant for the interaction of benzamidine with β-trypsin by frontal affinity chromatography of enzyme–benzamidine mixtures on a column of GlyGlyArg–Sepharose, the results, inferred from Fig. 2 of Ref. 25, being plotted in accordance with two linear transforms of Eq. (33).

Although Eq. (33) seemingly justifies adoption of the foregoing approach, the validity of the method relies on the acceptability of Eq. (32) as a truncated form of Eq. (31). Rigorous consideration of the procedure (4) shows that the slope of Figure 9a is $k_{AS}/(1 + k_{AX}\overline{C}_A)$ and that the ordinate intercept of Figure 9b is $(1 + k_{AX}\overline{C}_A)/V_A^* k_{AX}\overline{\overline{C}}_X$. For the present system there is reason to believe that $k_{AS} \approx k_{AX}$ (25,26), whereupon the error inherent in the approximate estimate of k_{AS}, 1.3 to 1.4%, is of no experimental significance.

A second such example of approximate analysis that merits consideration is the NADH-retarded desorption of lactate dehydrogenase from oxamate–Sepharose (11). For this system entailing the interaction of solute–ligand complex with affinity matrix sites, the expression relating $\overline{k}_{AX}$ and the solute–matrix affinity constant, k_{ASX}, is Eq. (15), whereupon the equivalent of Eq. (33) becomes

$$\left(\frac{\overline{V}_A}{V_A^*}\right)^{1/f} - 1 = \frac{k_{ASX}\overline{\overline{C}}_X k_{AS} C_S}{1 + k_{AS} C_S} \tag{34}$$

Although a value of 4 for f is indicated by the subunit structure of lactate dehydrogenase, the results of the frontal affinity chromatographic study (11) with $\overline{C}_A = 0.9\ \mu M$ signify conformity to a linear transform of Eq. (34) with $f = 1$ (Figure 10a). This type of situation occurs when steric factors preclude multiple attachment of solute to matrix despite the fact that f is not unity (4,11,52). Under those conditions Eq. (34) becomes (4)

$$\frac{\overline{V}_A}{V_A^*} - 1 = \frac{f k_{ASX}\overline{\overline{C}}_X k_{AS} C_S}{1 + k_{AS} C_S} \tag{35}$$

in which case the ordinate intercept of the double-reciprocal linear transform (Figure 10a) defines $1/4k_{ASX}\overline{\overline{C}}_X$ for the present system.

A second feature of these results for the affinity chromatographic behavior of lactic dehydrogenase–NADH mixtures on oxamate-Sepharose is the fact that the total enzyme concentration (0.9 μM) is not sufficiently small to allow adoption of the approximation that $\overline{C}_S \approx C_S$. Consequently, although Figure 10a suffices to establish the magnitude of $4k_{ASX}\overline{\overline{C}}_X$ (i.e., 24.8), the ratio of the slope to the ordinate intercept does not define k_{AS}. Resort is therefore made to Eq. (23), which expresses the ratio of solute–matrix affinity constants.

$$Q = \frac{k_{ASX}\overline{\overline{C}}_X}{k_{AX}\overline{\overline{C}}_X} = \frac{24.8}{(\overline{V}_A/V_A^*) - 1}$$

in terms of $\overline{C}_S$. The slope of the resultant linear plot of results in accordance with Eq. (23), Figure 10b, yields an intrinsic association constant of $9 \times 10^5\ M^{-1}$ for the enzyme–NADH interaction. This value of k_{AS} was also obtained previously (11) by the more tedious procedure of evaluating C_S by iterative solution of the

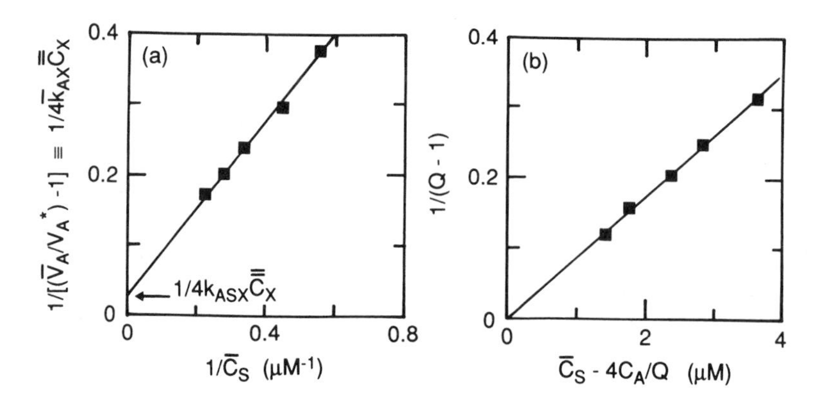

Figure 10 Approximate evaluation of the binding constant for the interaction of NADH with lactate dehydrogenase by frontal affinity chromatography of enzyme–NADH mixtures on a column of oxamate–Sepharose. (a) Analysis of the results, inferred from Fig. 1 of Ref. 11, on the basis of the double-reciprocal linear transform of Eq. (35) with C_S taken as the total ligand concentration, $\overline{C}_S$. (b) Consequent evaluation of the enzyme–NADH binding constant via Eq. (23).

quadratic equation that relates $\overline{C}_S$, $\overline{C}_A$, k_{AS}, and C_S—using the apparent value of k_{AS} from Figure 10a as an initial estimate of the solute–ligand affinity constant.

B. Zonal Studies: Their Advantages and Limitations

From the considerations above it is evident that the approximate analyses of frontal affinity chromatographic data place no demands on the magnitude of $\overline{C}_A$ apart from a requirement that $f\overline{\overline{C}}_A$ be much smaller than $\overline{\overline{C}}_X$—a point emphasized by the lack of $\overline{C}_A$ terms in the quantitative expressions [Eqs. (33) to (35)] for characterization of systems in which C_S, the concentration of free ligand, is of known magnitude. Consequently, the inability to ascribe a unique value of $\overline{C}_A$ to the ever-changing solute concentration in a migrating zone poses no impediment to the application of Eqs. (33) and (34) to elution volumes ($\overline{V}_A$) obtained by the conventional zonal chromatographic technique. Indeed, if an approximate analysis is all that is contemplated, there is no great advantage in opting for frontal chromatography, which is far more demanding in terms of the amount of partitioning solute required for quantitative affinity chromatographic studies of ligand binding. This consideration has undoubtedly been the deciding factor that has led to a situation where zonal affinity chromatographic studies have greatly outnumbered their frontal counterparts for the characterization of ligand binding. Many of those studies have entailed the use of zonal affinity chromatography to quantify the binding of modifiers, inhibitors or substrates to enzymes (2,3,18,27–30,53,54); others have employed the zonal technique to comment on drug binding by proteins (55), antigen–antibody systems (52), and neurophysin–neuropeptide interactions (56–62). A most interesting application has entailed the use of immobilized ribonuclease S-peptide to examine the ability of peptides coded by the antisense strand of DNA to interact with those coded by the sense strand (63).

1. *Routine Characterization of Solute–Ligand Interactions*

Three illustrative applications of the zonal technique for the characterization of ligand binding by quantitative affinity chromatography are presented in Figure 11 and are considered below.

1. The first panel (Figure 11a) summarizes results of the initial quantitative affinity chromatographic study (2), in which the binding constant for the interaction of *N*-acetylglucosamine with galactyosyltransferase was evaluated from the retarding effect of this modifier of the lactose synthetase system on elution of the enzyme from α-lactalbumin–Sepharose. On the basis that the retarded desorption reflects the univalent interaction of enzyme–modifier complex with the matrix-bound α-lactalbumin, the results are plotted in accordance with the double-reciprocal linear transform of Eq. (35). A binding constant of 200 M^{-1} for the

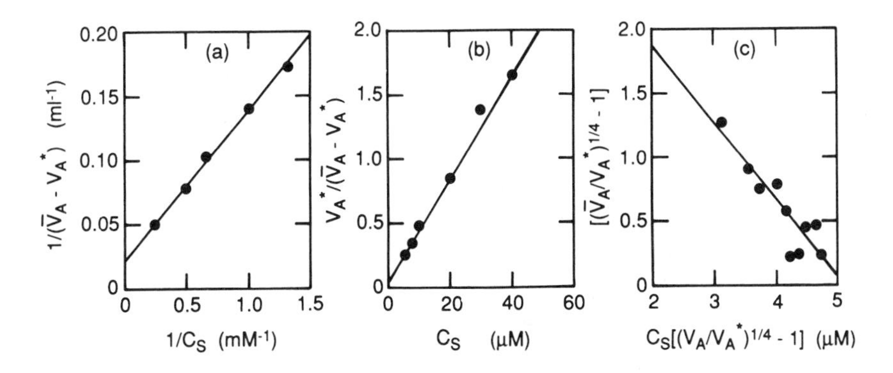

Figure 11 Evaluation of binding constants for solute–ligand interactions by zonal affinity chromatography. (a) Analysis of results for the *N*-acetylglucosamine-retarded desorption of lactate dehydrogenase from α-lactalbumin–Sepharose. (Adapted from Ref. 2.) (b) Analysis of data for 3′-(*p*-aminophenylphosphate)-5′-phosphate-facilitated elution of *Staphylococcus* nuclease from Sepharose to which the same ligand had been immobilized. (Adapted from Ref. 27.) (c) Allowance for solute tetravalency in the analysis of the NADH-facilitated desorption of lactate dehydrogenase from trinitrophenyl-Sepharose. (Adapted from Ref. 18.)

interaction of *N*-acetylglucosamine with galactosyltransferase is obtained from the ratio of the slope to ordinate intercept (2).

2. Figure 11b is taken from the first reported characterization of solute–ligand binding that is competitive with the solute–matrix interaction (3,27). These results for the thymidine 3′-(*p*-aminophenyl phosphate) 5′ phosphate-facilitated elution of staphylococcal nuclease from a column of thymidine-(*p*-Sepharose-aminophenyl phosphate) 5′-phosphate are plotted in accordance with the semi-reciprocal linear transform of Eq. (33). A value of $4.0 \times 10^4\ M^{-1}$ for k_{AS} is obtained from the ratio of slope to ordinate intercept.

3. Finally, Figure 11c illustrates another such example where competitive elution, namely, the NADH-facilitated elution of lactate dehydrogenase from trinitrophenyl-Sepharose, is being quantified by zonal chromatography (18). For this system, however, tetravalency of the partitioning enzyme must be taken into account. A value of $6.0 \times 10^5\ M^{-1}$ for k_{AS} is obtained from this linear transform (18) of Eq. (32) with $\bar{k}_{AX} = k_{AX}/(1 + k_{AS}C_S)$.

Despite its greater popularity, the quantitative characterization of ligand binding by zonal affinity chromatography does have its limitations. First, there is the undeniable criticism that the analysis is only approximate (see above). Second, the inability of the analysis to provide a value for the effective total concentration of matrix sites $\bar{\bar{C}}_X$ means that there is no means of assessing the validity of the inherent assumption that $f\bar{\bar{C}}_A << \bar{\bar{C}}_X$, particularly in instances where no information is available on the likely magnitude of the appropriate solute–matrix affinity constant (k_{AX} or k_{AXS}) incorporated into the product with $\bar{\bar{C}}_X$ that is defined by the ordinate intercept. Third, the lack of a $\bar{C}_A$ value in zonal studies precludes the use of analyses in terms of quantitative expressions based on $\bar{C}_S$; the free ligand concentration, C_S, to which $\bar{V}_A$ refers must therefore be of known magnitude. This is achieved by applying the small zone of partitioning solute to an affinity column preequilibrated with ligand-supplemented buffer. In such experiments, which are essentially of Hummel and Dreyer (64) design, C_S is taken as the concentration of ligand used for preequilibrating the column. However, such identification of the preequilibrating ligand concentration with C_S is only valid for systems in which the partitioning solute and any soluble complex(es) with ligand exhibit the same exclusion chromatographic behavior on the affinity gel (65). In zonal studies of systems in which partitioning solute and ligand are both macromolecular, the preequilibrating concentration of ligand is likely to underestimate C_S if the affinity matrix is based on a porous gel such as Sepharose (65).

Of these limitations the enforced assumption that $C_X \cong \bar{\bar{C}}_X$ is probably the most critical. An obvious way to guarantee the validity of this approximation is to employ an affinity matrix with a very large concentration of immobilized reactant residues. Such affinity matrices are commonly used for solute purification, and the only factor mitigating against their use for characterizing solute–ligand interactions seems to be the assertion (e.g., Ref. 54) that preparative affinity columns are unsuitable for quantitative assessment of solute–ligand binding constants. This assertion, which clearly contradicts the present inference that a preparative affinity matrix should be the one of choice for quantitative zonal studies, stems from the fact that the semireciprocal linear transform of Eq. (33) was being used for the evaluation of k_{AS} as the ratio of the slope to the ordinate intercept obtained in plots of $1/(\overline{V}_A - V_A^*)$ versus C_S. Since a large value of $\overline{\overline{C}}_X$ results in the ordinate intercept being indistinguishable from zero, it was recommended (54) that a lower concentration of immobilized reactant be employed to allow better delineation of the ordinate intercept. Although such action certainly increases the precision with which the magnitude of the ordinate intercept may be determined, it also increases the likelihood that Eq. (33) is not a valid approximation of the complete expression, Eq. (31) with $f = 1$. As demonstrated previously (10,18), the solution to this dilemma is not to repeat the experiment with a lower value of $\overline{\overline{C}}_X$ but rather to employ the linear transform of Eq. (33) that yields k_{AS} as the slope (i.e., the transform used in Figure 11c).

Despite its theoretical limitations, the zonal technique of quantitative affinity chromatography does enjoy the undeniable attraction of being more economical in terms of solute requirements. Furthermore, it also has some unique advantages for the characterization of solute–ligand interactions, as the following two applications serve to illustrate.

2. *Characterization of Solute–Ligand Interactions in a Tissue Extract*

If the degree of biospecificity exhibited by the matrix for a solute is sufficiently great, quantitative affinity chromatography has the potential for characterization of solute–ligand interactions in an unfractionated tissue extract. As an example we cite the determination of equilibrium constants for the interaction of NADH with the various isoenzymes of lactate dehydrogenase by zonal affinity chromatography of a crude mouse tissue extract on oxamate–Sepharose (11), a matrix selected because of the unique specificity of this immobilized pyruvate analog for lactate dehydrogenase–NADH complexes. A tissue extract obtained from the homogenate of the heart, liver, and kidneys of a mouse was first brought into dialysis equilibrium with the appropriate NADH concentration by zonal gel chromatography on a Sephadex G-25 column equilibrated with the same buffer–NADH solution that was used for preequilibration of the oxamate-Sepharose column. The resulting extract (2 mL, 0.5 to 1.0 μM enzyme) was applied to the column and then eluted with more of the same buffer–NADH solution. In the absence of NADH the elution profile exhibited a single, symmetrical zone of enzyme activity (Figure 12a, upper profile) that coincided with the elution volume (V_A^*) of all other protein in the extract. However, the presence of 2.5 μM NADH resulted not only in pronounced retardation of lactate dehydrogenase activity but also in partial resolution of the five isoenzyme zones (Figure 12a, lower elution profile). Plots of such results for four NADH concentrations (C_S) in accordance with Eq. (35) are shown in Figure 12b, from which binding constants of 2×10^5 M^{-1}, 3×10^5 M^{-1}, 4×10^5 M^{-1}, 7×10^5 M^{-1}, and 2×10^6 M^{-1} are obtained for the interactions of NADH with the M_4, M_3H, M_2H_2, MH_3 and H_4 isoenzymes, respectively, of lactate dehydrogenase.

Frontal experiments on the tissue extract would almost certainly have yielded an essentially unresolved advancing boundary, and hence a single value of k_{AS} representing a weighted mean of the five separate constants. In contrast, the zonal study provided estimates of all five binding constants that are certainly of the right order of magnitude. From the very nature of the experiment, namely zonal affinity chromatography of a crude tissue extract, it would be overly optimistic to expect to characterize these enzyme–coenzyme reactions more quantitatively than has been achieved.

3. *Characterization of a Protein–Protein Interaction*

Thus far the emphasis in this review of quantitative affinity chromatography has centered on its use for characterizing interactions between a partitioning macromolecular solute and a small ligand. It transpires that zonal affinity chromatography can play a useful role in the characterization of interactions in which the competing ligand is also macromolecular (51). Since the quantitative expressions presented in Section IV are independent of ligand size, no new theory is required for the analysis of affinity chromatographic data reflecting the effect of a

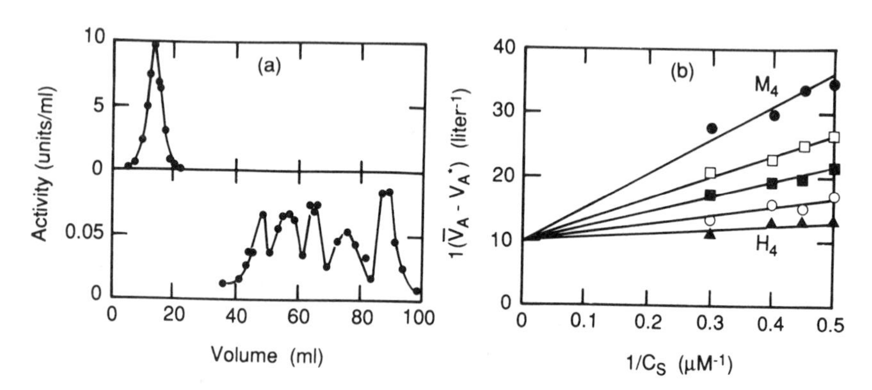

Figure 12 Zonal affinity chromatography of a mouse tissue extract on oxamate–Sepharose. (a) Elution profiles obtained in the absence of NADH (upper pattern) and in the presence of 2.5 μM coenzyme. (b) Dependence of elution volume of the various isoenzymes on NADH concentration, plotted in accordance with the double-reciprocal linear transform of Eq. (35). (Adapted from Ref. 11.)

macromolecular ligand on the desorption of solute. However, some adaptation of methodology is required to take into account the larger size of the ligand.

The first point to note is that use of the rigorous frontal chromatographic procedure based on quantitative expressions in terms of free ligand concentration is precluded by inability to establish the magnitude of C_S by prior dialysis, but that this difficulty may now be overcome by use of the expressions couched in terms of total ligand concentration (Section IV.B). Second, acceptance of an approximate characterization of the solute–ligand interaction introduces the possibility of employing expressions such as Eqs. (31) to (35) in conjunction with elution volumes ($\overline{V}_A$) obtained by zonal chromatography on an affinity column preequilibrated with defined concentrations, C_S, of ligand. Third, it is important to realize that all of the quantitative expressions in Section IV entail an inherent assumption that V_A^* describes not only the accessible volume for partitioning solute but also that for solute–ligand complexes—a requirement that can obviously be met by choosing an affinity matrix that excludes partitioning solute and hence all solute–ligand complexes from the matrix phase. The feasibility of this approach has been illustrated (51) by quantitative characterization of the interaction between concanavalin A and a glycoprotein, ovalbumin, by zonal affinity chromatography on Sephadex G-50.

Results of those zonal affinity chromatography experiments, which involved the application of a small zone (50 μL) of the lectin solution (94 μ*M*) to a Sephadex G-50 column (0.9 × 12.5 cm) preequilibrated with buffer supplemented with ovalbumin (1.2 to 14.0 μ*M*), are plotted in Figure 13a. The format is in accord with the expression

$$\left(\frac{\overline{V}_A}{V_A^*}\right)^{1/2} - 1 = k_{AX}\overline{\overline{C}}_X - k_{AS}C_S\left[\left(\frac{\overline{V}_A}{V_A^*}\right)^{1/2} - 1\right] \tag{36}$$

which is obtained by combining Eq. (32) with Eq. (13a) for affinity chromatography of bivalent concanavalin A (66) in the presence of a concentration C_S of univalent ligand that competes with matrix for solute sites. Consideration of ovalbumin to be univalent in its interaction with the lectin is in keeping with the fact that this glycoprotein possesses a single, relatively short, carbohydrate chain attached to an asparagine side chain (67). Although the results clearly conform with the linear dependence predicted by Eq. (36), the fact that interaction of concanavalin A with the carbohydrate matrix of the Sephadex G-50 is confined to the bead surface makes it necessary to justify experimentally the use of an analytical expression that applies only to systems for which $f\overline{\overline{C}}_A \ll \overline{\overline{C}}_X$. An independent estimate of the ordinate intercept of Figure 13a, $k_{AX}\overline{\overline{C}}_X$, was therefore obtained by frontal chromatography of concanavalin A alone on the same Sephadex G-50 column. By plotting those results in accordance with Eq. (8), a

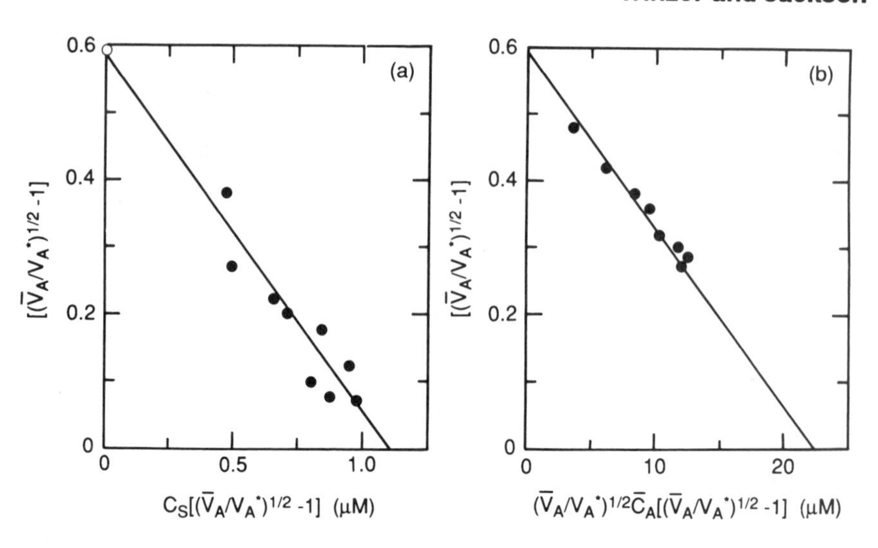

Figure 13 Characterization of the interaction between concanavalin A and ovalbumin by affinity chromatography on Sephadex G-50. (a) Zonal chromatography of the lectin on a column preequilibrated with different concentrations of ovalbumin, the results being plotted in accordance with Eq. (36). (b) Evaluation of the parameter $k_{AX}\bar{\bar{C}}_X$ as the ordinate intercept of results from frontal runs on concanavalin A alone plotted essentially in accordance with Eq. (8): this value is also indicated in Fig. 13a. (Adapted from Ref. 51.)

value of 0.59 for $k_{AX}\bar{\bar{C}}_X$ is obtained from the ordinate intercept (Figure 13b). Use of the truncated quantitative expression [Eq. (36)] for analysis of the zonal data is clearly justified by the concordance of the ordinate intercept inferred from those results (●) and the unequivocal estimate of $k_{AX}\bar{\bar{C}}_X$ (○) obtained from Figure 13b. From the slope of the line in Figure 13a it may therefore be concluded that an intrinsic association constant of 5.3 (± 0.8) × 10^5 M^{-1} describes the interaction of ovalbumin with concanavalin A under the conditions of the investigation (pH 5.5, $I = 0.5$).

C. Subunit-Exchange Chromatography

The specificity of monomer–monomer interactions in self-associating proteins has led to the development of subunit-exchange chromatography as a means of isolating and characterizing such proteins (68–70). Briefly, the technique entails covalent linkage of protein monomer to a matrix so that the protein is retarded during chromatography under conditions where monomeric and polymeric states coexist in self-association equilibrium. From the quantitative viewpoint this is an unusual example of affinity chromatography in that the partitioning solute (protein) is also the competing ligand. Since the concentration of competing ligand clearly cannot be varied independently of that of the partitioning solute, completely new theoretical expressions are required.

In subunit-exchange chromatography the two competing equilibria for monomeric solute, M, undergoing reversible self-association to dimeric state, D, are

$$2M \rightleftharpoons D \qquad K = \frac{C_D}{C_M^2} \tag{37a}$$

$$M + X \rightleftharpoons MX \qquad L = \frac{C_{MX}}{C_M C_X} \tag{37b}$$

where X continues to denote the matrix site (immobilized monomer), and K and L are the respective association constants for the formation of soluble and immobilized dimers. Provided that the same accessible volume, V_A^*, applies to monomeric and dimeric states of the solute, the quantitative expression for solute retardation may be written (71)

$$\frac{\overline{C}_A(\overline{\overline{C}}_X - \overline{\overline{C}}_A + \overline{C}_A)}{\overline{\overline{C}}_A - \overline{C}_A} = L^{-1} + \frac{2KL^{-2}(\overline{\overline{C}}_A - \overline{C}_A)}{\overline{\overline{C}}_X - \overline{\overline{C}}_A + \overline{C}_A} \tag{38}$$

In Eq. (38) $\overline{C}_A = C_M + 2C_D$ refers to the base-molar concentration of partitioning solute (weight concentration divided by monomeric molecular weight) in the liquid phase, whereas $\overline{\overline{C}}_A$ also includes solute that has formed immobilized dimer with X. Provided that $\overline{\overline{C}}_X$, the effective total concentration of covalently bound monomer, is first determined as the limiting value of $(\overline{\overline{C}}_A - \overline{C}_A)$ as $\overline{C}_A \rightarrow \infty$ (or $1/\overline{C}_A \rightarrow 0$), the magnitudes of both dimerization constants are obtainable from a plot of $\overline{C}_A(\overline{\overline{C}}_X - \overline{\overline{C}}_A + \overline{C}_A)/(\overline{\overline{C}}_A - \overline{C}_A)$ versus $(\overline{\overline{C}}_A - \overline{C}_A)/(\overline{\overline{C}}_X - \overline{\overline{C}}_A + \overline{C}_A)$.

Frontal chromatographic data may be analyzed via Eq. (38) by introducing Eq. (2) to express concentrations in terms of elution volumes. Such consideration of this mass conservation leads to the expression $(\overline{\overline{C}}_A - \overline{C}_A) = (\overline{V}_A - V_A^*)\overline{C}_A/V_A^*$, where $\overline{V}_A$ is the weight-average elution volume of solute present at a concentration $\overline{C}_A$ in the mobile phase. In that regard, zonal data such as those reported by Swaisgood and Chaiken (59,61) for neurophysin dimerization are clearly not amenable to rigorous quantitative interpretation, because there is no defined value of $\overline{C}_A$ to which the measured elution volume refers.

Results from a frontal chromatographic investigation of α-chymotrypsin dimerization by subunit-exchange chromatography (70) are summarized in Figure 14, where the left-hand panel (Figure 14a) establishes the magnitude of $\overline{\overline{C}}_X$. Although the consequent replot of data (Figure 14b) conforms with the prediction [Eq. (38)] of a linear relationship, the ordinate intercept ($1/L$) is not defined with sufficient

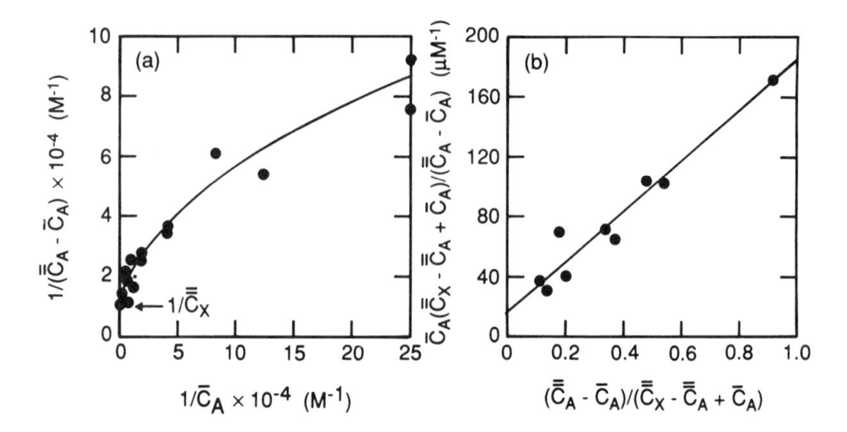

Figure 14 Characterization of the dimerization of α-chymotrypsin by subunit-exchange chromatography on α-chymotrypsin–Sepharose. (a) Double-reciprocal plot for evaluation of $\bar{\bar{C}}_X$ as the maximal capacity of the column for enzyme (i.e., as $\bar{\bar{C}}_A - \bar{C}_A$ as $1/\bar{C}_A \rightarrow 0$). (b) Consequent plot of the results in accordance with Eq. (38). (Adapted from Ref. 71.)

accuracy for unequivocal determination of the association constant for dimer formation with immobilized monomer. However, combination of the slope, $2K/L^2$, with the dimerization constant (K) of 15,000 M^{-1} deduced by sedimentation equilibrium under comparable conditions (72,73) gives a value of 13,000 ($\pm$ 2000) M^{-1} for L (71). Immobilization of α-chymotrypsin has therefore led to no significant change in its ability to form dimers, this being a basic assumption in the original analysis of the results (70) in terms of a dimerization constant for α-chymotrypsin in solution.

By demonstrating that covalent immobilization of monomeric enzyme has not affected its dimerization characteristics, Figure 14 has certainly vindicated the earlier use of these subunit-exchange chromatography results to characterize the dimerization behavior of α-chymotrypsin in solution. However, it does not establish the viability of subunit-exchange chromatography as a method of evaluating self-association constants, because there is no guarantee that the inherent basic assumption, verified for this system, applies universally. Clearly, subunit exchange chromatography is not a serious challenger to such techniques as sedimentation equilibrium or frontal gel chromatography, which provide far less equivocal characterization of protein self-association.

D. Partition Studies: The Aldolase–Myofibril System

From the viewpoint of considering the interaction of partitioning solute with matrix sites as a binding process, the study of such adsorption by partition equilibrium experiments provides the most direct mode of attack on the problem of quantifying the equilibria responsible for the biospecific phenomenon. First used in this context for characterizing the glucose-facilitated desorption of lysozyme from Sephadex G-100 in order to illustrate the newly developed theory of quantitative affinity chromatography (4), the simple partition equilibrium technique has been used most frequently to study biospecific interactions of glycolytic enzymes with two naturally occurring affinity matrices: the muscle myofibril (12–14,74–77) and the erythrocyte membrane (15,16). Of the systems that have been investigated, the interaction of aldolase with myofibrils has received the greatest attention, and is therefore the one used here to exemplify the type of information that can ensue by regarding the characterization of such interactions as a problem encompassed by quantitative affinity chromatography.

The biological advantages of metabolite-dependent myofibrillar adsorption of enzymes as a means of regulating glycolytic flux were being extolled long before such enzyme ambiquity (78) was shown to be a physiologically significant phenomenon. Indeed, five years elapsed before a combination of partition equilibrium studies and quantitative affinity chromatographic theory provided the physicochemical evidence that established the coexistence of aldolase in free and myofibril-bound states under conditions physiological with respect to pH and ionic strength (12). Those results, reported in Table 1 of Ref. 12, are summarized graphically in Figure 15. In keeping with the tetrameric subunit structure of aldolase, a valence (f) of 4 has been used in this plot of the experimental data according to Eq. (7) with $\overline{\overline{C}}_X$ replaced by the corresponding weight concentration of myofibrils [$r'_4 = (\overline{\overline{C}}_A^{1/4} - \overline{C}_A^{1/4})]/\overline{\overline{c}}_X$, where $\overline{\overline{c}}_X$ is the myofibril concentration in mg/mL. An intrinsic binding constant of $3.8 \times 10^5\ M^{-1}$ for the aldolase–myofibril interaction is obtained from the slope ($-4k_{AX}$) of Figure 15, while a myofibrillar capacity of 71 nmol/g is signified by the abscissa intercept. The concept of metabolite dependence of the equilibrium between soluble and adsorbed enzyme states is supported by the demonstrations that the interactions of phosphate (12) and substrate (79) with aldolase are competitive with myfibrillar adsorption of the enzyme. In the context of quantitative affinity chromatography, phosphate has been shown (12) to effect a decrease in the effective association constant for the aldolase–myofibril interaction ($\bar{k}_{AX}$) in a manner consistent with Eq. (13a). Moreover, involvement of the active site in myofibrillar adsorption is implicated by the fact that the slope of the linear dependence of $k_{AX}/\bar{k}_{AX}$ upon phosphate concentration, C_S, yielded a value of k_{AS} that equals the magnitude of the

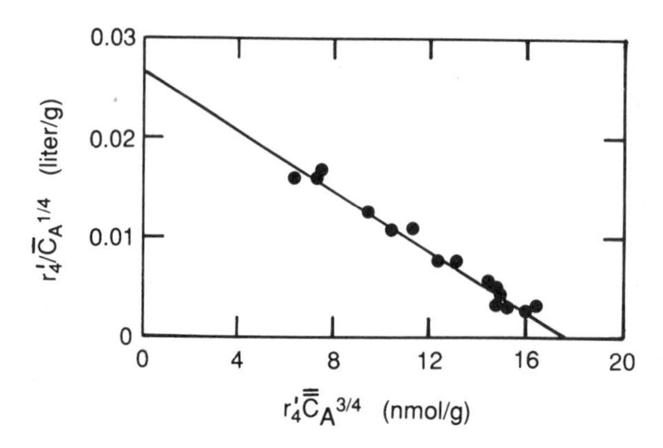

Figure 15 Multivalent Scatchard analysis of partition equilibrium data (Table 1 of Ref. 12) for the interaction of aldolase with rabbit muscle myofibrils (pH 6.8, $I = 0.16$).

equilibrium constant obtained enzyme kinetically for the competitive inhibition of aldolase by phosphate (17).

An interesting application of partition studies to quantify a physiologically significant biphasic equilibrium is found in a study of the binding of thrombin to fibrin (80). To overcome the problem of defining the concentration ($\bar{\bar{C}}_X$) of clotted fibrin in each reaction mixture, the fibrin was formed in situ by thrombin action on fibrinogen, a protein that could be pipetted readily into the individual tubes. Separation of fibrin-bound thrombin from free thrombin in the liquid phase was then achieved by brief centrifugation of the clotted mixture to provide a small volume of supernatant thrombin solution for determination of its concentration ($\bar{C}_A$) by enzymatic activity.

The examples above serve to demonstrate that the power of quantitative affinity chromatography as an analytical method is not restricted to the characterization of interactions in a chromatographic system contrived by the experimenter through design/choice of an immobilized reactant for an affinity matrix. Clearly, quantitative affinity chromatography has much to offer as a means of studying the effector-mediated binding of solutes to cellular receptors and subcellular assemblies. For example, the control of hormonal responses involving cellular receptors is clearly a phenomenon that should prove amenable to quantitative characterization by the application of the QAC theory and methodology.

E. Recycling Partition Studies of Tight Binding

The recycling partition equilibrium technique (36) was first used in the context of quantitative affinity chromatography to characterize the phosphate-facilitated desorption of aldolase from phosphocellulose (17) and has found subsequent application in studies of the NADH-dependent desorption of dehydrogenases from blue-Sepharose (31,33) and of galactose-enhanced desorption of a lectin from Sepharose (51). In the first (17) and last (51) of those studies the ligand–solute interactions were sufficiently weak for the approximation $C_S \approx \overline{C}_S$ to be made, while in the other two investigations (31,33) the free concentration of NADH was established by prior equilibrium dialysis. Instead of concentrating upon those studies, in which the recycling partition technique merely afforded a convenient way of gathering data for analysis by procedures already demonstrated, attention is turned to a recent investigation (20) of the inhibition by high-affinity heparin of the binding of antithrombin III to heparin–Sepharose. For this system the macromolecular nature of the ligand and the tightness of solute–ligand binding required the results to be interpreted in terms of a quantitative affinity chromatography expression based on total ligand concentration, Eq. (18).

Measurement of the partition of antithrombin III between heparin–Sepharose and solution phases was aided by using an Amicon concentration cell with its suspended magnetic stirrer and polyethylene frit (but without a membrane) as the stirred cell shown in Figure 6. Circulation of the mobile phase by means of a peristaltic pump through an ultraviolet UV monitor afforded not only an easy means of determining the concentration of antithrombin in the liquid phase ($\overline{C}_A$) but also of ascertaining that the interaction was at equilibrium. The initial total accessible volume of the system, $(V_A^*)_0$, was determined by weighing the tared Amicon cell and heparin–Sepharose slurry prior to determining V_A^* for several additions of aliquots of antithrombin III under conditions (3 M NaCl) that suppressed completely its binding to affinity matrix. Since mass conservation dictates that the amount of antithrombin present is given by the product $V_A^*\overline{C}_A$, each successive addition of antithrombin provided a value of V_A^* that differed from $(V_A^*)_0$ by the summed volume of added antithrombin aliquots. After draining away the liquid phase, the same slurry of heparin-Sepharose was washed and reequilibrated with buffer in readiness for the binding study. By comparing the weight of the assembly at this stage with its counterpart in the determination of $(V_A^*)_0$ above, the corresponding accessible volume for the binding study could be determined unequivocally. Because the affinity matrix is not removed from the cell, the amount that is present in the calibration experiment for determining the accessible volume is also present in the binding study, whereupon the binding data refer to that fixed amount of matrix, which can be expressed as the product of an initial accessible volume, $(V_A^*)_0$, and the corresponding total matrix site concentration, $(\overline{\overline{C}}_X)_0$.

The only procedural change required in graphical analysis of data is the incorporation of the ratio $V_A^*/(V_A^*)_0$ into the ordinate and abscissa parameters of the usual plots to accommodate the progressive increase in volume and hence decrease in matrix-site concentration ($\overline{\overline{C}}_X$). This aspect of the procedure is shown in Figure 16a, which summarizes the determination of k_{AX} for the antithrombin–matrix interaction at three temperatures. Figure 16b presents the analyses, via Eq. (18), of the $\bar{k}_{AX}$ values obtained in the second half of the experiment, where aliquots of high-affinity heparin were added to the slurry in order to evaluate k_{AS}. The values of k_{AX} ranged between 13.0 (15°C) and 3.0 (35°C) μM^{-1}, whereas the corresponding binding constants for the heparin–antithrombin III interaction (80 to 10 μM^{-1}) were significantly higher (20).

The fact that measurement of $\overline{C}_A$ can be made readily by the recycling partition technique, even in instances where the mobile phase must be sampled for measurement of $\overline{C}_A$ by external assay (50), makes the recycling QAC technique the method of choice for many systems. Application of the technique to systems that exhibit widely differing binding affinities merely requires a change in the amount of affinity matrix in relation to the volume of mobile phase and

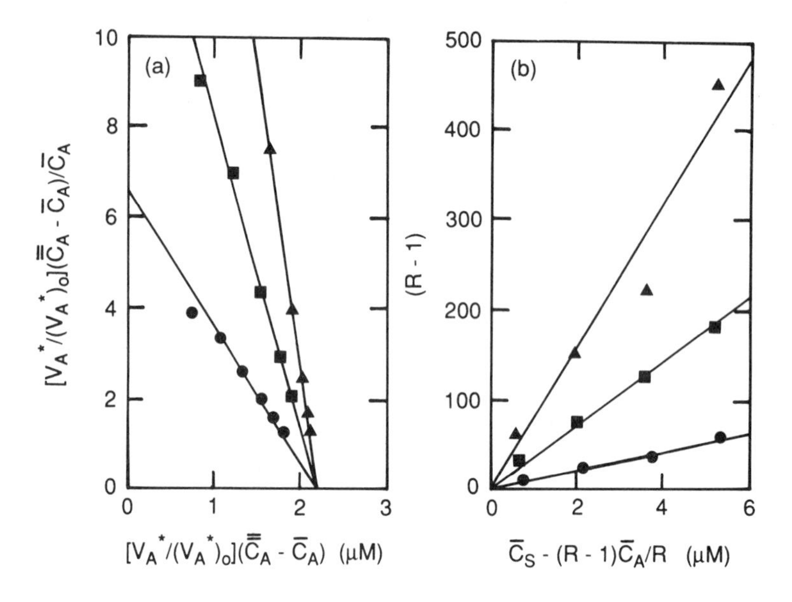

Figure 16 Evaluation of the binding constant for the interaction of high-affinity heparin with antithrombin III by the recycling partition QAC technique with heparin–Sepharose as affinity matrix. (a) Characterization of the temperature dependence of the binding constant (k_{AX}) for the interaction of antithrombin with affinity matrix. (b) Characterization of the corresponding dependence of the binding constant (k_{AS}) for the reaction between antithrombin III and high-affinity heparin in solution, the results being plotted in accordance with Eq. (18). (Adapted from Ref. 20.)

appropriate adjustment of the concentration range for competing ligand: modifications that are undeniably simple. Furthermore, the convenience of performing the required calculations by means of a microcomputer spreadsheet program (20) further increases the attractiveness of this method of determining binding parameters, particularly for interactions in which both reactants are macromolecular.

VI. QUANTITATIVE AFFINITY CHROMATOGRAPHY: ITS POTENTIAL

The versatility of QAC may well be its greatest potential. As noted in the examples reported here, the technique has been used to characterize the binding of small molecules to macromolecules, of macromolecules to macromolecules, and of macromolecules to cells and subcellular assemblies. The ability of QAC methods to encompass a wide range of affinities, particularly now that theoretical developments have provided expressions in terms of total ligand concentration (Section IV.B), has expanded dramatically the range of experimental systems that may readily be investigated.

When research problems in cellular and molecular biology reach the state of requiring distinction between molecules on the quantitative basis of the extent to which they bind to cells and subcellular organelles, QAC will have much to offer. In that regard it is not unrealistic to envisage that QAC studies have the potential to provide quantitative description of the initial stages of lymphocyte activation (30), or, indeed, to characterize the interactions responsible for biospecific separation of cells by the FACS procedure. The ability to use simple partition techniques, particularly the recycling adaptation, on a microscale makes available to the cell biologist the type of binding information that was previously regarded as being obtainable only by the physical biochemist.

Whether it be for basic structure–function information or for use in industrial processes, engineering of enzymes by site-directed mutagenesis requires quantitative evaluation of the alterations in properties of the mutant protein. In that regard QAC methods are ideally suited to assessment of quantitative differences in binding affinity and specificity. Clearly, the binding of a chosen partitioning solute to a single affinity matrix can be used to screen a panel of competing solutes; and hence efficiently and definitively provide quantitative information about the altered functional characteristics that result from mutagenesis.

The development of clinical assay procedures based on immunological reactivity has traditionally followed an empirical path that restricts the amount of information that may be obtained from solid-phase radioimmunoassay and ELISA measurements. Recent considerations of these techniques in the context of quantitative affinity chromatography (81–84) have opened up the possible clinical use of such immunochemical assays to comment on pathological abnormalities that reflect deficient functioning of the solute being assayed as well as those reflecting its presence in diminished concentration.

Finally, the availability of a range of QAC methods provides the experimenter with the opportunity to carry out the binding experiment in the way most suited to the particular system under consideration. There is a choice between elution volume measurements in either zonal or frontal affinity chromatography, and concentration measurements in partition studies. In instances where the amounts of materials are limited, the technique that is most sparing of the reactant in shortest supply can be selected without compromising the quality of the binding data obtained. Furthermore, the combination of QAC methods with continuous monitoring of concentrations and microcomputer-based on-line data acquisition (e.g., Ref. 20) provides practical simplification and efficiency. To that end, the recent emergence of a variety of very powerful curve-fitting programs can do nothing but increase further the ease with which QAC can be used to evaluate binding constants for solute–ligand interactions.

REFERENCES

1. P. Cuatrecasas and C. B. Anfinsen, *Annu. Rev. Biochem., 40*:259 (1971).
2. P. Andrews, B. J. Kitchen, and D. J. Winzor, *Biochem. J., 135*:897 (1973).
3. B. M. Dunn and I. M. Chaiken, *Proc. Natl. Acad. Sci. USA, 71*:2382 (1974).
4. L. W. Nichol, A. G. Ogston, D. J. Winzor, and W. H. Sawyer, *Biochem. J., 143*:435 (1974).
5. I. M. Chaiken, *Anal. Biochem., 97*:1 (1979).
6. D. J. Winzor, in *Affinity Chromatography: A Practical Approach* (P. D. G. Dean, W. S. Johnson, and F. A. Middle, eds.), IRL Press, Oxford, 1985, p. 149.
7. I. M. Chaiken, *J. Chromatogr., 376*:11 (1986).
8. H. E. Swaisgood and I. M. Chaiken, in *Analytical Affinity Chromatography* (I. M. Chaiken, ed.), CRC Press, Boca Raton, Fla., 1987, p. 65.
9. G. Fassina and I. M. Chaiken, *Adv. Chromatogr., 27*: 247 (1987).
10. D. J. Winzor and J. de Jersey, *J. Chromatogr., 492*:377 (1989).
11. R. I. Brinkworth, C. J. Masters, and D. J. Winzor, *Biochem. J., 151*:631 (1975).
12. M. R. Kuter, C. J. Masters, and D. J. Winzor, *Arch. Biochem. Biophys., 225*:384 (1983).
13. S. J. Harris and D. J. Winzor, *Arch. Biochem. Biophys., 275*:185 (1989).
14. S. J. Harris and D. J. Winzor, *Biochim. Biophys. Acta, 999*:95 (1989).
15. G. E. Kelley and D. J. Winzor, *Biochim. Biophys. Acta, 778*:67 (1984).
16. S. J. Harris and D. J. Winzor, *Biochim. Biophys. Acta, 1038*:306 (1990).
17. L. W. Nichol, L. D. Ward, and D. J. Winzor, *Biochemistry, 20*:4856 (1981).
18. D. A. Bergman and D. J. Winzor, *Anal. Biochem., 153*:380 (1986).
19. M. C. Waltham, J. W. Holland, P. F. Nixon, and D. J. Winzor, *Biochem. Pharmacol., 37*:541 (1988).
20. P. J. Hogg, C. M. Jackson, and D. J. Winzor, *Anal. Biochem., 192*:303 (1991).

21. A. A. Spector, *J. Lipid Res., 16*:165 (1975).
22. I. M. Klotz, *Arch. Biochem., 9*:109 (1946).
23. G. Scatchard, *Ann. N.Y. Acad. Sci., 51*:660 (1949).
24. K. Kasai and S. Ishii, *J. Biochem. (Tokyo), 77*:261 (1977).
25. K. Kasai and S. Ishii, *J. Biochem. (Tokyo), 84*:1051 (1978).
26. K. Kasai and S. Ishii, *J. Biochem. (Tokyo), 84*:1061 (1978).
27. B. M. Dunn and I. M. Chaiken, *Biochemistry, 14*:2343 (1975).
28. I. M. Chaiken and H. C. Taylor, *J. Biol. Chem., 251*:2044 (1976).
29. H. C. Taylor and I. M. Chaiken, *J. Biol. Chem., 252*:6991 (1977).
30. J. Danner, J. E. Sommerville, J. Turner, and B. M. Dunn, *Biochemistry, 18*:3039 (1979).
31. P. J. Hogg and D. J. Winzor, *Arch. Biochem. Biophys., 234*:55 (1984).
32. P. J. Hogg and D. J. Winzor, *Biochim. Biophys. Acta, 843*:159 (1985).
33. P. J. Hogg and D. J. Winzor, *Arch. Biochem. Biophys., 240*:70 (1985).
34. R. J. Yon, *J. Chromatogr., 457*:13 (1988).
35. P. Kyprianou and R. J. Yon, *Biochem. J., 207*:549 (1982).
36. C. L. Ford and D. J. Winzor, *Anal. Biochem., 114*:146 (1981).
37. A. J. Muller and P. W. Carr, *J. Chromatogr., 284*:33 (1984).
38. D. J. Anderson and R. R. Walters, *J. Chromatogr., 333*:1 (1985).
39. D. J. Winzor, L. D. Ward, and L. W. Nichol, *J. Theor. Biol., 98*:171 (1982).
40. A. D. Riggs, H. Suzuki, and S. Bourgeois, *J. Mol. Biol., 48*:67 (1970).
41. C. P. Woodbury, Jr., and P. H. von Hippel, *Biochemistry, 22*:4730 (1983).
42. D. F. Senear, M. Brenowitz, M. A. Shea, and G. K. Ackers, *Biochemistry, 25*:7344 (1986).
43. S. T. Olson, P. E. Bock, and R. Sheffer, *Arch. Biochem. Biophys., 286*:533 (1991).
44. D. J. Winzor, P. D. Munro, and C. M. Jackson, *J. Chromatogr., 597*:57 (1992).
45. P. J. Hogg, P. E. B. Reilly, and D. J. Winzor, *Biochemistry, 26*:1867 (1987).
46. L. W. Nichol, W. H. Sawyer, and D. J. Winzor, *Biochem. J., 112*:259 (1969).
47. C. Lagercrantz, T. Larsson, and H. Carlsson, *Anal. Biochem., 99*:352 (1979).
48. R. J. Fitzgerald and H. E. Swaisgood, *Arch. Biochem. Biophys., 268*:239 (1989).
49. H. D. Jang and H. E. Swaisgood, *Arch. Biochem. Biophys., 283*:318 (1990).
50. A. Gow, D. J. Winzor, and R. Smith, *Biochemistry, 26*:982 (1987).
51. P. J. Hogg and D. J. Winzor, *Anal. Biochem., 163*:331 (1987).
52. D. Eilat and I. M. Chaiken, *Biochemistry, 18*:794 (1979).
53. B. M. Dunn and W. A. Gilbert, *Arch. Biochem. Biophys., 198*:533 (1979).
54. B. M. Dunn, J. Danner-Rabovsky, and J. S. Cambias, in *Affinity Chromatography and Biological Recognition* (I. M. Chaiken, I. Parikh, and M. Wilchek, eds.), Academic Press, Orlando, Fla., 1983, p. 93.
55. F. M. Veronese, R. Bevilacqua, and I. M. Chaiken, *Mol. Pharmacol., 15*:313 (1979).
56. I. M. Chaiken, *Anal. Biochem., 92*:302 (1979).
57. S. Angal and I. M. Chaiken, *Biochemistry, 21*:1574 (1982).
58. D. M. Abercrombie, T. Kanmera, S. Angal, H. Tamaoki, and I. M. Chaiken, *Int. J. Peptide Protein Res., 24*:218 (1984).
59. H. S. Swaisgood and I. M. Chaiken, *J. Chromatogr., 327*:193 (1985).
60. G. Fassina, H. E. Swaisgood, and I. M. Chaiken, *J. Chromatogr., 376*:87 (1986).
61. H. E. Swaisgood and I. M. Chaiken, *Biochemistry, 25*:4148 (1986).

62. G. Fassina and I. M. Chaiken, *J. Biol. Chem., 263*:13539 (1988).
63. Y. Shai, M. Flashner, and I. M. Chaiken, *Biochemistry, 26*:669 (1987).
64. J. P. Hummel and W. J. Dreyer, *Biochim. Biophys. Acta, 63*:530 (1962).
65. J. R. Cann, A. G. Appu Rao, and D. J. Winzor, *Arch. Biochem. Biophys., 270*:173 (1989).
66. L. L. So and S. J. Goldstein, *Biochim. Biophys. Acta, 165*:398 (1968).
67. A. Neuberger and R. D. Marshall, in *Glycoproteins* (A. Gottschalk, ed.), Elsevier, Amsterdam, 1966, p. 299.
68. E. Antonini, M. R. Rossi Fanelli, and E. Chiancone, in *Protein–Ligand Interactions* (H. Sund, ed.), Walter de Gruyter, Berlin, 1975, p. 45.
69. E. Antonini and M. R. Rossi Fanelli, *Methods Enzymol., 44*:538 (1976).
70. E. Antonini, G. Carrea, P. Cremonisi, P. Pasta, M. R. Rossi Fanelli, and E. Chiancone, *Anal. Biochem., 95*:89 (1979).
71. E. Chiancone and D. J. Winzor, *Anal. Biochem., 158*:211 (1986).
72. K. C. Aune, L. C. Goldsmith, and S. N. Timasheff, *Biochemistry, 10*:1617 (1971).
73. T. A. Horbett and D. C. Teller, *Biochemistry, 13*:5490 (1974).
74. T. P. Walsh, D. J. Winzor, F. M. Clarke, C. J. Masters, and D. J. Morton, *Biochem. J., 186*:89 (1980).
75. M. R. Kuter, C. J. Masters, T. P. Walsh, and D. J. Winzor, *Arch. Biochem. Biophys., 212*:306 (1981).
76. S. J. Harris and D. J. Winzor, *Arch. Biochem. Biophys., 243*:598 (1985).
77. S. J. Harris and D. J. Winzor, *Anal. Biochem., 169*:319 (1988).
78. J. E. Wilson, *Trends Biochem. Sci., 3*:124 (1978).
79. S. J. Harris and D. J. Winzor, *Biochim. Biophys. Acta, 911*:121 (1987).
80. P. J. Hogg and C. M. Jackson, *J. Biol. Chem., 265*:241 (1990).
81. P. J. Hogg and D. J. Winzor, *Arch. Biochem. Biophys., 254*:92 (1987).
82. P. J. Hogg, S. C. Johnston, M. R. Bowles, S. E. Pond, and D. J. Winzor, *Mol. Immunol., 24*:797 (1987).
83. S. C. Johnston, M. Bowles, D. J. Winzor, and S. E. Pond, *Fundam. Appl. Toxicol., 11*:261 (1988).
84. D. J. Winzor, J. A. Nagy, and H. A. Scheraga, *J. Protein Chem., 10*:629 (1992).

11

Weak Affinity Chromatography

Sten Ohlson

HyClone Laboratories, Inc., Logan, Utah

David Zopf

Neose Pharmaceuticals, Inc.,
Horsham, Pennsylvania

I. INTRODUCTION

Affinity chromatography (1) has been used routinely since the 1970s to purify biomolecules from crude mixtures in a single step. The purification factors are generally impressive, often in the range of 10- to 1000-fold. The usual format involves (bio)specific adsorption of freely soluble ligate onto an immobilized ligand under quasi-physiological buffer conditions followed by an abrupt change to nonphysiological, often harsh buffer conditions that drastically reduce binding affinity, leading to elution of the ligate. The term *affinity chromatography* [coined originally by Cuatrecasas et al. (2)] as applied to such a single-step solid-phase extraction procedure is somewhat misleading, as it is not a true chromatographic procedure where solutes that interact with a matrix are retarded differentially.

In recent years, affinity chromatography has been experiencing a major metamorphosis spurred by the introduction of a "dynamic" form of affinity chromatography. This new approach, termed *weak affinity chromatography* (WAC), uses readily reversible interactions between ligand and ligate ($K_a = 10^2$ to 10^4 M^{-1}) as the basis for separation under mild, isocratic buffer conditions (3,4). Compared with traditional affinity chromatography, WAC offers improved chromatographic resolution and recovery as well as higher specific bioactivity of purified ligates, and thus WAC overcomes many of the limitations commonly associated with affinity purification techniques. In this chapter we present the basic aspects of the theory of weak affinity chromatography as well as several practical examples to illustrate the technology. We also discuss in some detail where the technology might lead in the future.

II. CHROMATOGRAPHIC THEORY

Weak affinity chromatography is a form of interactive chromatography where a combination of weak short- and long-range physical interactions constitute the basis for separation. It can be seen as a mixed-mode chromatography system where interaction of solutes with the solid matrix may include contributions from hydrophobic, ion-exchange, hydrogen bond, and van der Waals interactions. WAC distinguishes itself from other forms of chromatography, such as reversed phase, only in the type of interactions utilized for the separation.

The various modes of separation, including that of weak affinity, can be understood in terms of basic chromatography theory. This theory is well established in a long series of publications (see, e.g., Refs. 5 and 6). Application of chromatographic theory to weak affinity systems is important (see below under simulation studies), as it gives a broad perception of the possible uses of weak affinity chromatography. In WAC, the binding of a ligate dissolved in the bulk phase to immobilized ligand can be reasonably assumed in most cases to be a second-order, reversible action:

$$\mathrm{Lt} + \mathrm{Ld} \underset{}{\overset{K_a}{\rightleftharpoons}} \mathrm{LtLd} \tag{1}$$

where Lt, Ld, and LtLd are the ligate, ligand, and ligate–ligand complex, respectively, and K_a is the association constant (M^{-1}).

The following kinetic equation can be derived:

$$q^* = \frac{Q_{\max} K_a c^*}{1 + K_a c^*} \tag{2}$$

where q^* is the concentration of adsorbed ligate at equilibrium (moles per kilogram of solid phase), c^* the concentration of free ligate at equilibrium (M), and $Q_{\max}$ the maximum accessible ligand sites (moles per kilogram of solid waste). This is the Langmuir adsorption isotherm, where $K_a c^*$ reflects the deviation of the expression from linearity. When $K_a c^* \ll 1$ (i.e., the adsorption isotherm is in the linear range), Eq. (2) becomes

$$q^* = Q_{\max} K_a c^* \tag{3}$$

Retention in chromatography is usually expressed as the capacity factor (k'), where k' is $(t_R - t_0)/t_0$. t_R is the time for the ligate to traverse the column and t_0 is the time for the mobile phase to run through the column. By definition, the capacity factor, k', is equal to n_s/n_m, where n_s is the number of total moles of ligate bound to the stationary phase and n_m is the number of the total moles in the mobile phase. Furthermore, $n_s = q^*M_s$ and $n_m = c^*V_m$, where M_s is the weight of the stationary phase (kg) and V_m is the volume of the mobile phase (liter). Accordingly, under conditions where the adsorption isotherm is linear:

$$k' = \frac{q^*M_s}{c^*V_m} = CQ_{\max}K_a \qquad (4)$$

where $C = M_s/V_m$. C is a constant (usually in the range of 0.5 to 1.0) that reflects the characteristics of the column such as void fraction, porosity, and density of stationary phase.

Equation (4) is a simple expression of retention of ligate when the adsorption isotherm is linear, a condition that is usually valid when performing weak affinity chromatography ($K_ac^* \ll 1$). Below we discuss the effect of nonlinear conditions on the performance of WAC. It is clear from Eq. (4) that significant retention and thereby specificity ($k' = 1$–10) can be achieved by high affinity (K_a) and/or high ligand load ($Q_{\max}$). In a weak affinity system as defined above, retention is achieved by increasing the ligand load. For example, to obtain a $k' = 1$ and $C = 1$ with a system of an immobilized IgG monoclonal antibody interacting with its antigen at $K_a = 10^3\ M^{-1}$, the necessary concentration of active antibody in the column bed is approximately 75 mg/mL.

III. SIMULATION OF WEAK AFFINITY CHROMATOGRAPHY

As WAC is a new technique, attempts have been made to predict its outcome under various conditions so as to promote a more thorough understanding of the fundamentals of the technology and its potential applications. For this purpose we developed a computer simulation model to study the performance of WAC under conditions of linear- and nonlinear chromatography (7). The theoretical model is based on the kinetic equations as described above, the axial dispersion of ligate in the mobile phase, and the mass transfer resistance of ligate inside the pores of each particle and through the unstirred layer of liquid surrounding each particle. Kubin (8), Kucera (5), and Carr and coworkers (9) derived the analytical solutions to this model under both linear and nonlinear conditions. For the computer simulation of WAC we selected as one model the antigen–antibody system, where immobilized monoclonal IgG antibodies weakly interact with their corresponding small carbohydrate antigens.

A. Computer Simulation of WAC: Comparison of High Versus Low Affinity

Figure 1 illustrates the difference in performance between a weak affinity system ($K_a = 10^3\ M^{-1}$) and a high-affinity system ($K_a = 10^5\ M^{-1}$). It can be seen that at high affinity, the performance is essentially lost by trying to run chromatography under isocratic conditions within a reasonable period of time. When ligate–ligand interactions are weak, however, peaks are sharpened and high performance is achieved. It is clear that high-affinity interactions are not suitable for "dynamic" affinity chromatography but rather should be amenable only to adsorption–desorption techniques.

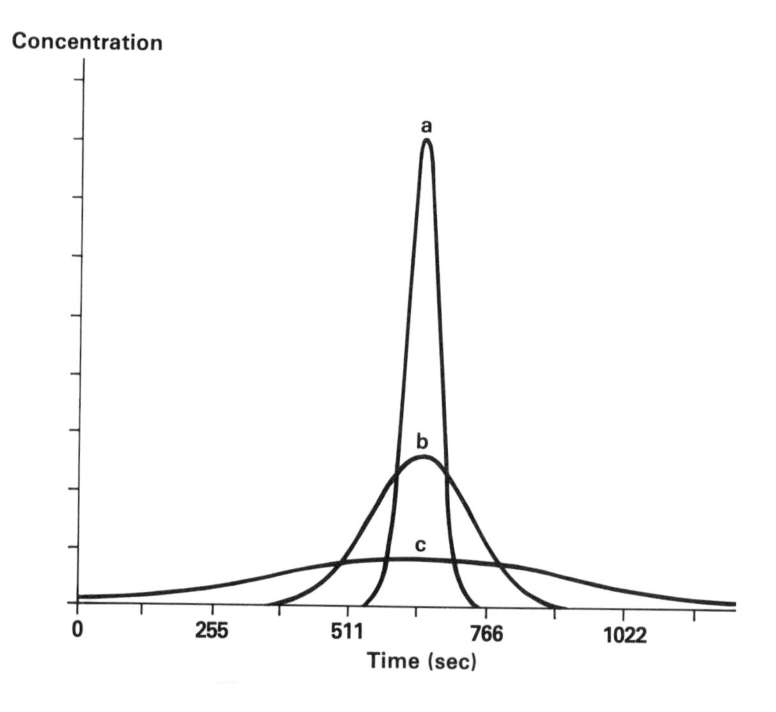

Figure 1 Computer simulation of WAC: comparison of high affinity versus low affinity. 10 μL of a carbohydrate antigen ($M_r \sim 700$, concentration = $1.4 \times 10^{-6}\ M$) was chromatographed on a 0.5 × 25 cm column (particle size, 10 μm; porosity, 0.64; void fraction, 0.42; density, 2.42 g/cm^3). Isocratic elution is performed at 1 mL/min. $K_a = 10^3$ (a), 10^4 (b), and 10^5 (c) M^{-1}. (See Ref. 7 for details.)

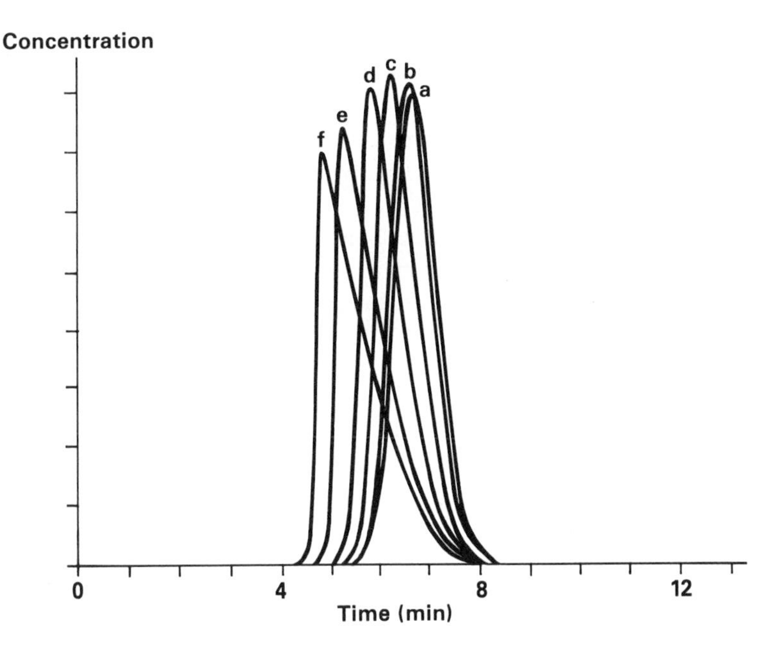

Figure 2 Computer simulation of WAC: effect of ligate concentration. The same parameters were used as described in the legend to Figure 1 except $Q_{max} = 3.9 \times 10^{-4}$ mol/kg gel, length of column = 10 cm, and $K_a = 5.2 \times 10^3\ M^{-1}$. (See Ref. 7 for details.)

B. Computer Simulation of WAC: Effect of Ligate Concentration

As indicated earlier [Eq. (4)], the value of the parameter $K_a c^*$ determines the extent of linearity in the adsorption isotherm. When the "overload" term, $K_a c^*$, is increased (i.e., > 1), the number of ligand sites will not be adequate to retard all ligate molecules properly. As a result, a significant portion will be eluted earlier and we obtain asymmetrical and less retarded peaks. Figure 2 illustrates this effect clearly where highly skewed peaks with loss in retention are obtained. Eventually, at very high values of $K_a c^*$, fractions of the ligate will be seen in the void volume and no meaningful chromatography will be attained. Fortunately, in WAC the ligate concentration can be considerably higher than in medium- or high-affinity chromatography without jeopardizing the criterion for linearity.

IV. IS WEAK BINDING SPECIFIC?

It is generally observed in chromatography that many nondesired interactions (e.g., interactions with the support itself) are of weak affinity. Although we often refer to these interactions as being nonspecific, this does not mean that weak affinity interactions are by nature nonspecific. Specific retention in a WAC system is achieved as a result of the time-averaged, cumulative action of many specific weak binding events that are created by a variety of kinds of binding forces between biomolecules. The fact that high-performance chromatographic separations within the weak affinity range are exquisitely sensitive to both affinity and ligand load virtually ensures that conditions for selective separation can be established. Even when the nondesired interactions of the ligate with some component of the solid matrix might occur in the same weak affinity range as the desired interaction, it is almost always possible to distinguish the two types of interactions and to take steps to eliminate effects of the nondesired interactions. In general, this is because desired versus nondesired interactions occur at sites with dissimilar densities in the solid matrix and are subject to differential changes in affinity in response to changes in physical–chemical conditions. The practical results that can be achieved using specific weak affinity biorecognition is obvious from analyses of a group of target molecules in crude samples such as urine and serum (presented below): target molecules are differentially retarded due to subtle differences in molecular structure but well separated from thousands of components, some of which undoubtedly interact to some minor extent with the solid phase but do not interfere with the chromatographic analyses.

V. APPLICATIONS OF WEAK AFFINITY CHROMATOGRAPHY

A. Separation of Weak Inhibitors with Enzymes

Enzymes play an important role in cellular metabolism where the trafficking of metabolites is often under a dynamic weak affinity control. The extent of enzymatic reactions is governed by the concentration of substrates and inhibitors, which is usually in the m*M* range. In a preliminary study we have selected the proteolytic enzyme, α-chymotrypsin, as a suitable model of weak enzyme–inhibitor interactions, since many of its inhibitors are known to bind with weak affinities ($K_a = 10^1$ to 10^4 M^{-1}) (10). We were able to immobilize onto aldehyde-derivatized silica (11) 44.5 mg of α-chymotrypsin per milliliter of column volume (1.8 m*M*) with a remaining enzymatic activity of approximately 50% of total bound.

As an illustration of this application, Figure 3 shows the separation of two inhibitors, *N*-acetyl-L-tryptophan and indole, in a simulated crude extract of human serum. The separation was performed on a 5-cm column under isocratic conditions. The amount of active bound ligand and the binding constant were estimated by frontal analysis (12), where the Langmuir adsorption isotherm was

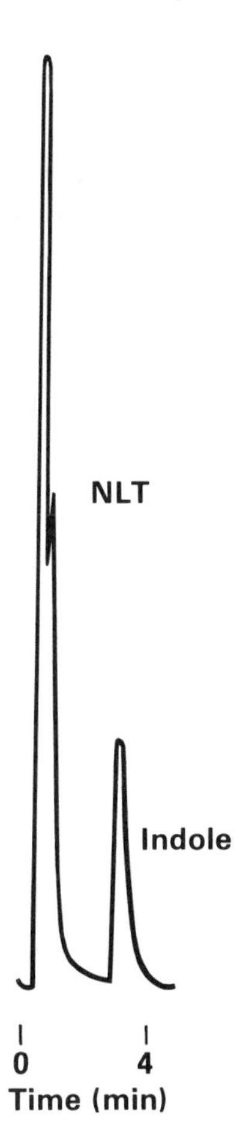

Figure 3 WAC separation of enzyme inhibitors. Separation of *N*-acetyl-*L*-tryptophan and indole (1.7 μg/mL of each in fetal bovine serum diluted 1:120 with mobile phase) on a 5 × 0.5 cm microparticulate silica column (particle size, 10 μm; porosity, 300 Å) with immobilized α-chymotrypsin. Isocratic chromatography was performed at 20°C using Tris (0.08 *M*, pH 7.8) with $CaCl_2$ (0.1 *M*) as the mobile phase (1 mL/min). Injection volume was 100 μL and emerging peaks were monitored at 274 nm.

established by determining bound ligate (inhibitor) at various ligate concentrations. The amount of immobilized ligand, α-chymotrypsin, capable of interacting with the inhibitors, was found to be 38.5 mg per milliliter of column volume (87% of total). The *N*-acetyl-L-tryptophan was slightly retarded, with an estimated $K_a = 1.5 \times 10^2\ M^{-1}$ [literature value, $K_a = 0.57 \times 10^2\ M^{-1}$ (13)], whereas the indole showed significant retention with an estimated $K_a = 1.1 \times 10^3\ M^{-1}$ [literature value, $K_a = 1.4 \times 10^3\ M^{-1}$ (13)]. The estimated binding constants show a reasonable correlation with the literature values determined for interactions in free solution. In this sample we have shown that inhibitors can be separated according to their weak but different affinities on a WAC column with immobilized enzymes.

B. Analytical and Semipreparative Separation of Antigens Using Weak Affinity Monoclonal Antibodies

1. *Analytical Separations*

Monoclonal antibodies (MoAbs) provide a versatile source of immobilized ligands for analytical WAC separations. MoAbs can be raised against a wide variety of substances, and once selected for desired specificity and affinity, can be conveniently produced as homogenous reagents in gram quantities. Weak binding antibodies ($K_a < 10^4\ M^{-1}$) are abundant among natural circulating immunoglobulins and represent the majority of clones discarded in many laboratory experiments where the goal has been to select a reagent with maximal affinity and specificity for use in classical immunoassays. The relative ease of producing weak-binding MoAbs simplifies the task of creating reagents for weak affinity chromatography, and in fact, may permit successful development of a WAC analytical method where development of high-affinity antibodies against a particular analyte has proved difficult or impossible to achieve.

For example, antibodies against carbohydrate antigens rarely exhibit $K_a > 10^7\ M^{-1}$ and most commonly range from 10^2 to $10^4\ M^{-1}$. In an attempt to produce a high-affinity MoAb reagent for development of a radioimmunoassay of a human urinary oligosaccharide, designated $(Glc)_4$ (see Table 1 for oligosaccharides structures), Lundblad et al. (14) performed six hybridoma fusion experiments and selected 16 clones that bound a polyvalent, semisynthetic $(Glc)_4$-BSA conjugate in ELISA, but found that of these antibodies, only two bound the free oligosaccharide with $K_a > 10^4\ M^{-1}$ (15). As high-affinity antibodies—prerequisite to creating a sensitive radioimmunoassay that depends upon antibody affinity for sensitivity—proved difficult to obtain, a weak affinity MoAb specific for the oligosaccharide hapten was selected for use as immobilized ligand in WAC, and high sensitivity (≤ 10 µg/L) was achieved by electrochemical detection of the analyte (16). Figure 4 illustrates WAC analysis for $(Glc)_4$ in the urine of a normal individual and in the urine and blood of a patient with acute pancreatitis.

The relationship between chromatographic performance and ligand–ligate affinity, defined in detail by Ohlson et al. (3), is graphically illustrated in Figure 5, where chromatography was repeated at increasing temperatures (10 to 50°C) that induced progressively weaker interactions ($K_a = 10^4$ to $5 \times 10^2\ M^{-1}$). Temperature-

Table 1 Structures of Oligosaccharides

Name	Structure
Glucose-tetra [(Glc)4]	Glcα1-6Glcα1-4Glcα1-4Glc
Lacto-*N*-fucopentaose II (LNF II)	Galβ1-3GlcNAcβ1-3Galβ1-4Glc 4 | Fucα1
Lacto-*N*-fucopentaose III (LNF III)	Galβ1-4GlcNAcβ1-3Galβ1-4Glc 3 | Fucα1

dependent changes in affinity between MoAbs and oligosaccharides has been observed repeatedly (15,17,18), but their physical basis has not been studied in detail. This phenomenon might relate to requirements for numerous hydrogen bonds to stabilize the secondary structures of oligosaccharides in conformations favorable for binding, and/or to entropic changes related to displacement of numerous water molecules from the peptide combining site (19). The practical consequence of this effect is that by adjusting the column to an appropriate constant temperature, it is possible to achieve efficiency of chromatographic performance similar to that of a noninteracting solute such as water. For the system shown in Figure 5, at temperatures >45°C, calculation of theoretical plate numbers indicates that the kinetic contribution of ligate–ligand binding appears to play little role, if any, in the peak-broadening process.

2. Semipreparative Separations

WAC can offer a rapid means for semipreparative recovery of a weakly bound antigen from complex mixtures that contain many chemically similar compounds. Purification of LNF II (18) (see Table 1 for oligosaccharide structures), a relatively abundant oligosaccharide in human milk, from its structural isomer LNF III and from at least 20 structurally similar oligosaccharides (20) is demonstrated in Figure 6.

For low-molecular-weight ligates such as oligosaccharides ($M_r \approx 1000$) the use of intact monoclonal antibodies (binding subunit $M_r \approx 80{,}000$) as immobilized ligands places certain practical limitations upon preparative scaleup. For example, extrapolation from experimental data for WAC of $(Glc)_4$ (see Figure 5) would predict from theory that approximately 100 μg of ligate could be separated per

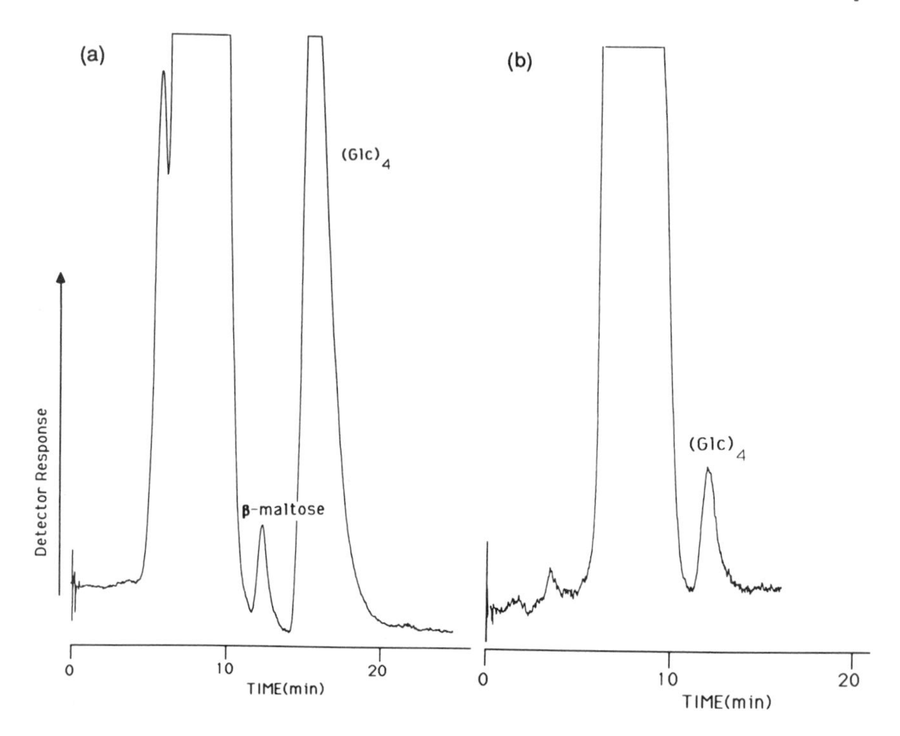

Figure 4 WAC analysis for $(Glc)_4$ in urine and serum on an affinity column containing immobilized monoclonal antibody 39.5. (a) Urine (25 μL) from a patient with acute pancreatitis was ultrafiltered (cutoff M_r = 10,000) and deionized before injection onto a 100 × 5 mm column containing 86 mg of monoclonal antibody 39.5 (IgG2b) coupled isocratically at 0.2 mL/min in 0.1 *M* Na_2SO_4 plus 0.02 *M* phosphate, pH 7.5 at a column temperature of 30°C. The column effluent was mixed with 50 m*M* NaOH prior to passage through a pulsed amperometric detector (see Ref. 16 for further details.) (b) Serum (25 μL) from a patient with acute pancreatitis was treated as described for urine in (A) and then extracted on a C-18 Bond–Elut™ column prior to injection. WAC was performed under the same conditions as described in (a), except the column temperature was maintained at 40°C. Integration of the peaks indicated "$(Glc)_4$" give 1.01 ng and 0.35 ng $(Glc)_4$ in urine and serum, respectively.

chromatographic run on a 10-mL bed volume column containing 1 g of immobilized immunoglobulin. In practice, the situation is likely to be somewhat better, since retention is not highly dependent on operating in the linear region of the ligand–ligate equilibrium isotherm.

Semipreparative WAC has been used to isolate oligosaccharide bearing chemically unknown, or incompletely known, antigenic epitopes in quantities sufficient

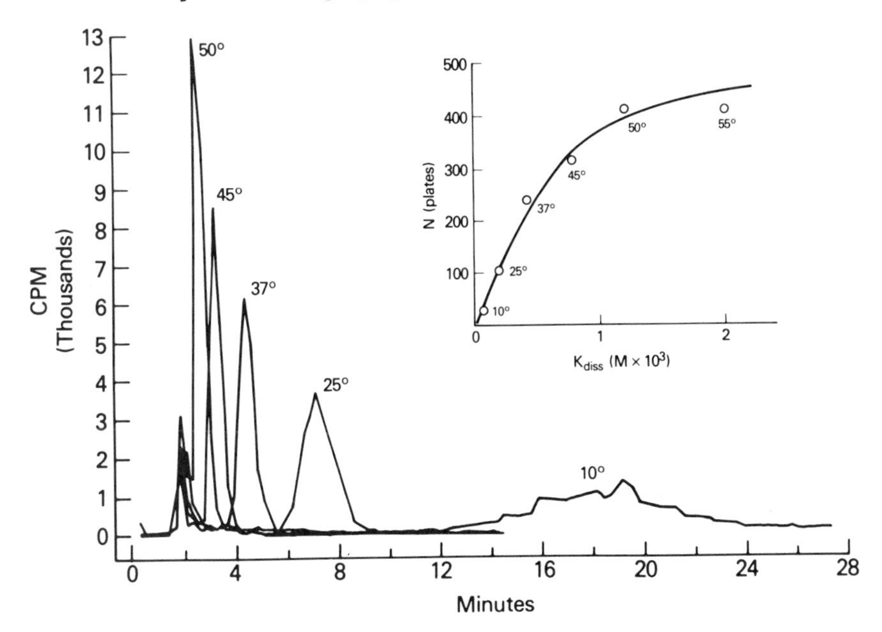

Figure 5 Affinity chromatography of $(Glc)_4$-ol as a function of temperature. The column (the same as described in the legend to Figure 4) was run isocratically at 1 mL/min using 0.2 *M* NaCl plus 0.02 *M* sodium phosphate buffer (pH 7.5). Fractions (0.23 to 0.46 mL) were collected and counted by liquid scintillation. Identical samples containing tritiated $(Glc)_4$-alditol (1.5 ng/ sp. act. = 8.4 Ci/mmol) in 10 μL of distilled water were injected and run isocratically with the system maintained at the temperature shown. The inset shows the relationship between calculated theoretical plate number and the ligand–ligate dissociation constant K_d (i.e., $1/K_a$) determined by frontal and zonal analysis (see Ref. 3 for details).

for complete structural characterization. For example, evidence from immunostaining of a thin-layer chromatogram (21) prompted Wang et al. (22) to search for a larger, more complex form of the carbohydrate antigen structure with higher binding affinity than the already known sialyl-Le^x-active hexaglycosylceramide. A MoAb, Onc-M26, originally selected for its specific ability to bind mucins secreted by breast cancer cells, was employed in WAC as immobilized ligand to isolate two sialyl-Le^x-active oligosaccharides with branched core structures from among hundreds of naturally occurring monosialylated oligosaccharides in pooled human milk (Figure 7). The difference in affinity that enabled separation of the simpler known hexasaccharide from the more complex nonassacharides was determined to be on the order of 20-fold, an amount that, in principle, could increase avidity of the Onc-M26 IgM molecule for multivalent mucin antigens by several logs.

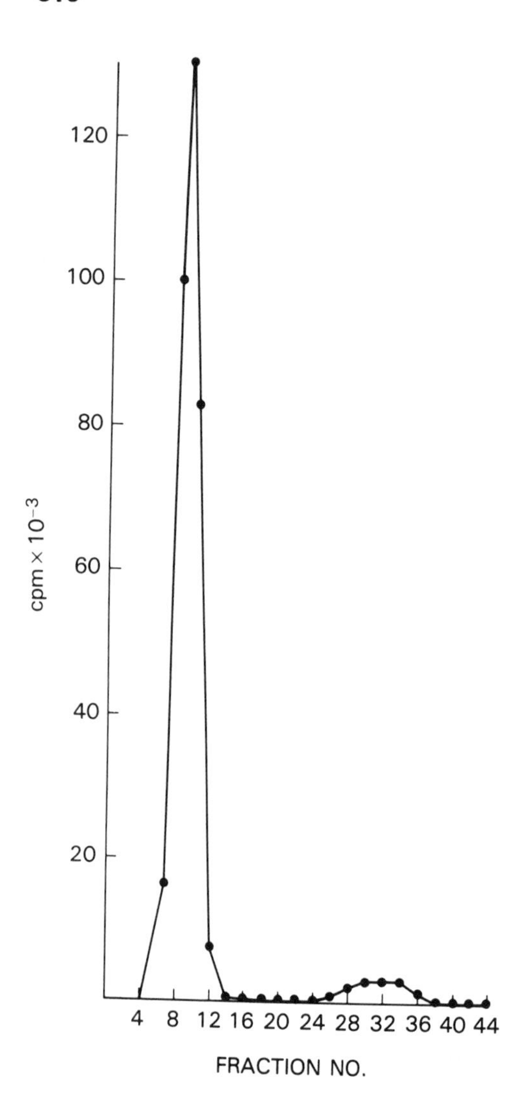

Figure 6 Affinity isolation of LNF II from a mixture of human milk oligosaccharides. A mixture of neutral di- through hexasaccharides from pooled human milk (1.2×10^{-2} nmol) was applied in 50 µL to a 0.3×10 cm glass column containing 9.5 mg of monoclonal antibody CO 514 (IgG3) noncovalently immobilized on *Staphylococcus* protein A–Sepharose. The system was run isocratically in 0.14 *M* NaCl plus 0.05 *M* Tris–HCl (pH 8.0) at 0.1 mL/min, 25°C (see Ref. 18 for details).

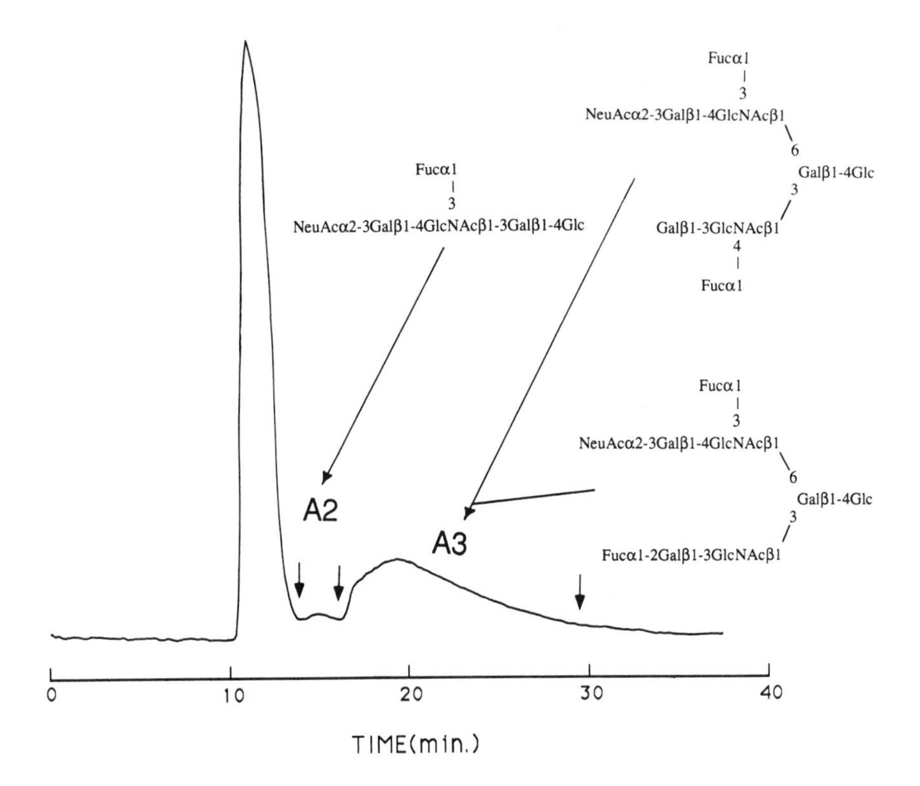

Figure 7 WAC isolation of SLe^x-active oligosaccharides. A sample containing 1 μg of partially fractionated monosialylated oligosaccharides was injected on column containing 33.7 mg of immobilized monoclonal antibody Onc-M26 (IgM) noncovalently bound to SelectiSpher-10 ConA (250 × 5 mm). The column was eluted isocratically at 0.2 mL/min with 0.1 *M* Na_2SO_4 plus 0.02 *M* phosphate (pH 7.5) at 20°C. NaOH (0.1 *M*) was added at 0.3 mL/min via a postcolumn mixing tee and oligosaccharides were detected with a pulsed amperometric detector (see Ref. 22 for details). Fractions A2 and A3 were pooled separately as indicated by arrows, and fraction A3 was further fractionated by high-performance ion-exchange chromatography, giving two major peaks that were isolated and shown by NMR to have the chemical structures indicated.

VI. LIGAND DESIGN FOR WAC AND FUTURE TRENDS

A. Why Study Weak Affinity?

Weak affinity ligand–ligate pairs abound in natural systems, often forming the subunit interactions that comprise multisite cooperative binding between ordered clusters of macromolecules: Senescent glycoproteins lacking terminal sialic acid residues on branched carbohydrate side chains are selectively removed from blood plasma via interaction between clustered galactosyl residues and a membrane-bound lectin on hepatocytes (23); hexameric Clq molecules interact with aggregated immunoglobulin heavy chains in immune complexes to activate the complement cascade (24); microbial pathogens specifically attach to cells (25) and cells specifically aggregate with each other and interact with the extracellular connective tissue matrix (26–28) via a multitude of weak affinity interactions that must be readily reversible in order to permit a dynamic environment suited to constant adaptation. Studies of these and many other phenomena will be facilitated by utilization of WAC as an analytical and/or semipreparative tool.

B. Technical Developments

Technologies are evolving rapidly for modeling, manipulating, and mimicking polypeptide-binding sites. For example, advances in genetic engineering have greatly reduced the effort required to produce bioactive recombinant proteins containing partial or modified active sites (29). Furthermore, definition of the exact positions of critical residues in protein combining sites of x-ray crystallographic studies has provided a rational basis for organic synthesis of relatively small molecules that mimic specific binding exhibited by bioactive proteins (30). These and other approaches provide a variety of tools to create appropriate "designer" ligands for WAC analysis or preparation of virtually any biomolecule.

From the information presented above, it is clear that an important feature in considering the design of a column for WAC is availability of a ligand that can be immobilized in active form at millimolar concentrations in the column bed. We have shown examples where an enzyme (chymotrypsin, $M_r \approx 25$ kD) or bivalent immunoglobulin (IgG, $M_r \approx 160$ kD) have been immobilized successfully on macroporous silica beads. In preparing high-performance liquid chromatography (HPLC) columns containing these proteins, the density of chromatographically active sites was limited by the Stokes radius of the ligand rather than the concentration of binding sites on the activated silica surface. Production of recombinant peptides, such as cloned immunoglobulin heavy-chain variable region (V_H) fragments (29), that retain specific ligate recognition, albeit at reduced affinity, may have properties advantageous for WAC in that such peptides might be loaded onto columns at densities perhaps five- to 10-fold greater than for the native protein molecule. The commonly observed 10- to 100-fold reduction in affinity of V_H fragments with respect to intact antibody or Fab fragments (31–33) will in many cases represent an advantage for WAC, permitting enhanced chromatographic performance associated with weaker affinity in the context of increased ligand load.

C. Future Applications

In principle it should be possible to immobilize small molecules (e.g., indole derivatives) at nearly maximal density on silica supports and achieve excellent analytical WAC for macromolecules (e.g., chymotrypsin). Following the same argument, mixtures of several small ligands (each present at millimolar concentrations on the column) that are targets for specific, weak binding by different proteins might be used to analyze for several bioactive macromolecules (e.g., serum enzymes) simultaneously.

The utilization of monoclonal antibodies to isolate previously unknown, weakly bound antigens provides a model for many potential applications of WAC in biological research. For example, the recent discovery that certain selectins—membrane-bound proteins that mediate leucocyte–endothelial cell recognition essential for selective cell trafficking—are C-type lectins that specifically bind complex carbohydrates (34,35), provides an interesting opportunity for utilization of WAC to identify the native carbohydrate receptors on various classes of leukocytes. Recombinant fragments of C-type lectins immobilized as active ligands could serve to isolate and resolve oligosaccharide receptors that may exhibit a range of weak affinities dependent on variations in core structures.

In summary, we are confident that WAC will find a wide variety of applications. Recombinant and synthetic techniques that create active "subsites" derived from macromolecular binding sites will provide economical sources of ligands suitable for WAC separations of important analytes.

REFERENCES

1. W. H. Scouten, *Affinity Chromatography*, Wiley, New York, 1981.
2. P. Cuatrecasas, M. Wilchek, and D. B. Anfinsen, *Proc. Natl. Acad. Sci. USA, 61*:636 (1968).
3. S. Ohlson, A. Lundblad, and D. Zopf, *Anal. Biochem., 169*:204 (1988).
4. D. Zopf and S. Ohlson, *Nature, 346*:87 (1990).
5. E. Kucera, *J. Chromatogr., 19*:237 (1965).
6. S. Goldstein, *Proc. R. Soc. London, 219*:151 (1953).
7. M. Wikstrom and S. Ohlson, *J. Chromatogr., 597.*:83 (1992).
8. M. Kubin, *Col. Czech. Chem. Commun., 30*:1104 (1965).
9. J. L. Wade, A. F. Bergold, and P. W. Carr, *Anal. Chem., 59*:1286 (1987).
10. T. E. Barman, ed., *Enzyme Handbook*, Springer-Verlag, New York, 1969.
11. P. O. Larsson, M. Glad, L. Hansson, M. O. Mansson, S. Ohlson, and K. Mosbach, in *Advances in Chromatography* (J. C. Giddings, E. Grushka, J. Cazes, and P. R. Brown, eds.), Marcel Dekker, New York, 1983, p. 41.
12. K.-I. Kasai, Y. Oda, M. Nishikata, and S.-I. Ishii, *J. Chromatogr., 376*:33 (1986).
13. P. D. Boyer, ed., *The Enzymes*, Academic Press, New York, 1970.
14. A. Lundblad, K. Schroer, and D. Zopf, *J. Immunol. Methods, 68*:217 (1984).

15. A. Lundblad, K. Schroer, and D. Zopf, *J. Immunol. Methods, 68*:227 (1984).
16. W.-T. Wang, J. Kumlien, S. Ohlson, A. Lundblad, and D. Zopf, *Anal. Biochem., 182*:48 (1989).
17. J. Dakour, A. Lundblad, and D. Zopf, *Anal. Biochem., 161*:140 (1987).
18. J. Dakour, A. Lundblad, and D. Zopf, *Arch. Biochem. Biophys., 264*:203 (1988).
19. N. Sharon and H. Lis, *Chem. Br.*, July 1990, p. 679.
20. A. Kobata, in *Methods in Enzymology* (V. Ginsburg, ed.), Vol. 28, Academic Press, New York, 1972, p. 262.
21. P. S. Linsley, J. P. Brown, J. L. Magnani, and D. Horn, Monoclonal antibody onc-M26, which detects elevated levels of mucin in sera from breast cancer patients, binds a sugar sequence in sialylated lacto-*N*-fucopentaose III, *Proc. 9th International Symposium on Glycoconjugates*, Lille, France, 1987, p. F69.
22. W.-T. Wang, T. Lundgren, F. Lindh, B. Nilsson, G. Grönberg, J. P. Brown, H. Mentzer-Dibert, and D. Zopf, *Arch. Biochem. Biophys.*, in press.
23. Y. C. Lee, in *Carbohydrate Recognition in Cellular Function*, Ciba Foundation Symposium 145 (G. Bock and S. Harnett, eds.), Wiley, New York, 1989, p. 80.
24. M. G. Colomb, G. J. Arlaud, and C. L. Villiers, *Complement, 1*:69 (1984).
25. H. Leffler and C. Svanborg-Eden, in *Microbial Lectins and Agglutinins: Properties and Biological Activity* (D. Mirelman, ed.), Wiley, New York, 1986, p. 83.
26. S. H. Barondes, *Annu. Rev. Biochem., 50*:207 (1981).
27. W. Frazier and L. Glaser, *Annu. Rev. Biochem., 48*:491 (1979).
28. N. Sharon, *FEBS Lett., 217*:145 (1987).
29. L. Sastry, M. Alting-Mees, W. D. Huse, J. M. Short, J. U. A. Sorge, B. N. Hay, K. D. Janda, S. J. Benkovic, and R. A. Lerner, *Proc. Natl. Acad. Sci. USA, 86*:5728 (1989).
30. H. U. Saragovi, D. Fitzpatrick, A. Raktabutr, H. Nakanishi, M. Kahn, and M. I. Greene, *Science, 253*:792 (1991).
31. E. Haber and F. F. Richards, *Proc. R. Soc. London Ser. B, 166*:176 (1966).
32. J.-C. Jaton, N. R. Klinman, D. Givol, and M. Sela, *Biochemistry, 7*:4185 (1968).
33. J. H. Rockey, *J. Exp. Med., 125*:249 (1967).
34. M. L. Phillips, E. Nudelman, F. C. A. Gaeta, M. Perez, A. K. Singhal, S.-I. Hakomori, and J. C. Paulson, *Science, 250*:1130 (1990).
35. E. L. Berg, M. K. Robinson, O. Mansson, E. C., Butcher, and J. L. Magnani, *J. Biol. Chem., 266*:14869 (1991).

12

Investigating Specificity via Affinity Chromatography

Lawrence M. Kauvar

Terrapin Technologies, Inc., So. San Francisco, California

I. INTRODUCTION

All chromatography depends on differential affinity of analytes for the chromatographic sorbent (1). Variation in solvent composition was long ago shown to be a way to modify the differential affinities of analytes for a limited set of sorbent types. The development of coated and bonded phases in the 1970s increased the set of sorbent types, reducing the need for complex solvent gradients to achieve acceptable resolution (2).

Affinity chromatography, which was developed in parallel over the same period (3), utilizes biospecific ligands immobilized on a sorbent base designed to be as invisible as possible with respect to binding analytes. In effect, this approach raises the goal of sorbent modification to that of devising specialized sorbents which provide large differential affinities between analytes of interest and contaminants in particular samples matrices.

Delineation of a quantitative continuum ranging from affinity chromatography to ion- and hydrophobic-exchange chromatography raises two major questions of fundamental theoretical interest. Is such a paradigm for laboratory-scale chromatography relevant to understanding comparable chemical separation phenomena in nature and in industrial processes? Can sorbents be deliberately devised to span the specificity gap between traditional chromatographic sorbents and sorbents derivitized with strictly biospecific (unitary interaction site) ligands?

The goal of this review is to demonstrate that the answer to both questions is "yes." Following a discussion of the relevance of information theory to chemical separations, data will be presented on one particular system for generating intermediate levels of sorbent specificity.

II. INFORMATION THEORY AND CHROMATOGRAPHY

In general, sorting items into bins is antientropic, and in a thermodynamic sense involves work, a property of nature summarized in the familiar metaphor of Maxwell's demon (4). Applied to chromatography, this property means that for a fixed collection of analytes, the specificity of a sorbent can be measured either in terms of the number of distinguishable bins into which it sorts the analytes, or in terms of the unevenness with which analytes are sorted into the bins. These alternative definitions capture the essential difference between the traditional chromatographic sorbent, which ideally functions as a general-purpose tool to sort into hundreds of bins the components of a single mixture, and the traditional affinity sorbent, which ideally functions to create two bins: one for the target analyte and one for everything else. For purposes of this discussion, the first type of specificity will be referred to as resolution and the second will be referred to as selectivity (Figure 1).

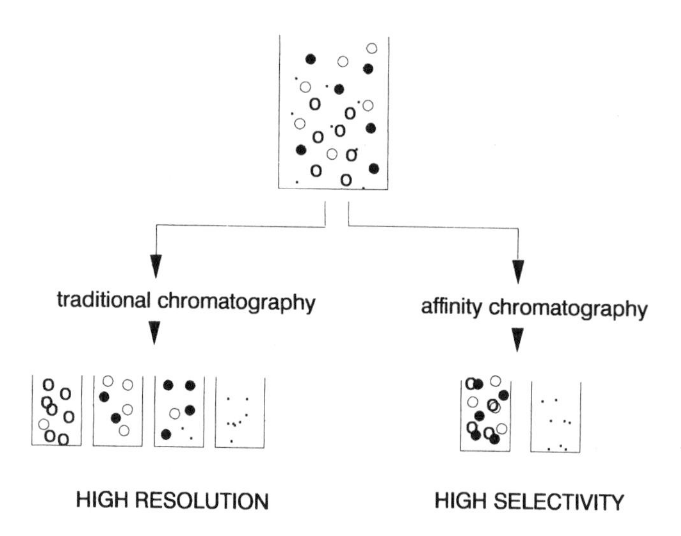

Figure 1 Sorting items is an antientropic activity, requiring work. Two forms of sorting are possible, however, here designated as resolution versus selectivity. The former distributes items across a large number of bins, while the latter distributes the items across a small number of bins.

The goal of a chromatographic separation normally determines the measure of specificity to be used. Organisms face comparable information processing problems. Two clear instances arise in the functioning of the immune system and of the olfactory system. In each case, trade-offs emerge between the need for high selectivity of individual binding agents and the necessity to keep the total number of binding agents in a reasonable range while providing resolution of a variety of analytes.

III. FAMILIES OF DEDICATED BINDING PROTEINS

In the antibody case, a two-tier system generates selectivity. The primary repertoire, encoded in the germ line, provides antibodies of moderate selectivity (5) which are fine tuned by selection following a randomizing process of somatic mutation (6) (Figure 2). Even the fine-tuned antibodies do not show absolute specificity, however. Cross-reactions to a variety of small molecules are readily found, particularly using short peptides as the probes (7).

Antibodies are the prototype of highly specific binding agents and are a prime example of ligands used to construct affinity sorbents. Even in this case, however, specificity is not an easily defined property, since it depends on context. In vivo,

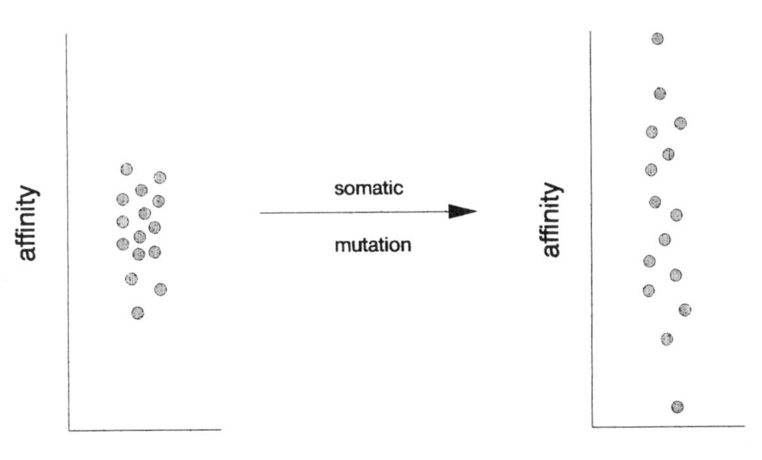

Figure 2 The initial antibody response to an antigen is attributable to preexisting clones of moderate affinity (selectivity) for a wide range of antigens (5). Later in the response, somatic mutation of the early responding antibodies leads to a wider variation in affinity, with some of the resulting antibodies showing substantially higher affinity for the antigen (6).

the most critical context for defining specificity is self/nonself. This fact presumably explains why the frequency of antibodies that can bind an analyte at reasonable affinity in vitro is typically 1/1000, while the frequency of antibody-producing cells that respond in vivo to antigenic stimulation by the analyte is much lower, typically 1/100,000 (8).

In the olfactory case, two major theories for the functioning of this highly specific chemical sensor have been discussed for almost a century (9). One model, the labeled line theory, postulates millions of exceedingly odorant-specific receptors. The other model, the combinatorial theory, postulates a much smaller number of broadly tuned receptors whose pattern of activation acts to encode the olfactory stimulus into a pattern that can be decoded by the brain. This model is almost certainly applicable to the taste system, in which all stimuli are categorized with respect to only four basic receptor types: sweet, sour, bitter, and salty.

A third model, which postulates a miniature gas chromatograph inside the nose, has not been easy to distinguish from the second model, and has not received support from anatomical studies. The clearest evidence in favor of the combinatorial model for processing the broad range of olfactory stimuli has come from analysis of the olfactory system of insects (10). Although instances of labeled lines have clearly been documented, as in the case of moth-mating pheromone detectors (11), only a small number of genes have been identified that affect more general olfactory processing, and mutations in these genes affect responsiveness to classes of chemicals, not single compounds. Recently, putative olfactory receptors have been cloned from mammalian libraries and the number of homologous genes determined to be in the hundreds, not millions (12).

IV. MODERATE AFFINITY INTERACTIONS IN BIOLOGY

The specificity trade-offs seen in families of proteins dedicated to chemical binding have deeper implications for other areas of biology. By the nature of evolution, selection for special-purpose binding agents ultimately requires reproductive advantages. Each increase in the number of such binding agents thus imposes a genetic load, making combinatorial mechanisms preferable in a population of finite size for which selective mortality needs to be kept low (13).

For example, the development of resistance to pesticides is much more rapid in the case of insects than rodents because the number of individuals that can independently mutate to resistance is so much larger. For large organisms such as mammals, trade-offs with regard to the two standards are likely to favor combinatorial mechanisms since small population sizes do not permit high mortality to function as a means of removing deleterious mutations in a large number of genes.

Analysis of the formation of detailed neural connections, for example, has revealed the existence of cell adhesion molecules that provide a modest level of specificity to control gross features of the developing nervous system (14). Observation of axonal growth also suggests a process quite similar to chromatography is taking place as pseudopodia extend in many directions before a particular direction for axon extension is taken (15). The cues for such directional guidance are totally uncharacterized at the biochemical level and provide a challenging area for applying the concepts and tools of affinity chromatography.

Intracellularly, too, there is apparently a variety of moderate-specificity chromatographic separation work taking place, for example in the distribution of cytoskeleton-associated proteins under different metabolic states (16).

V. INDUSTRIAL CHEMICAL MONITORING

The need for specificity trade-offs is also evident in the industrial applications of chromatography. In this case, economic forces act similarly to biological evolutionary forces to limit the number of sorbents. The commercial availability of only a small variety of sorbents for liquid chromatography is compensated for by the ability to mix solvent components in enormous variety. By contrast, a far larger set of sorbents is available for gas chromatography, which permits only minor variation in solvent composition. Since mixing solvents is a poorly reproducible process in different labs, which furthermore involves undesirable solvent expenses and column turnaround drawbacks for repetitive use, there is now an increasing interest in embedding specificity in the sorbent for liquid chromatography. Affinity chromatography is the logical endpoint of this trend, with a distinct sorbent for each application. How far down this path the industry can afford to go is an open question.

VI. OPTIMIZATION OF SPECIFICITY

One final general point regarding specificity in an information-processing context is that selectivity leading to a small number of bins can be a drawback for practical applications even if it is technically feasible and economically practical to achieve. For example, an important application of chromatography in the pharmaceutical industry has been in quality assurance analysis of drugs. Extending this function to the new drugs produced by recombinant DNA technology introduces a range of new problems (17). When grown in large fermentors, microorganisms undergo mutation, leading to microheterogeneity of the recombinant pharmaceutical. Purification of the drug can also introduce modifications. For characterizing these variants, an affinity column incorporating a high-selectivity monoclonal antibody against a single epitope is of limited utility. Conversely, a low-selectivity sorbent, such as a traditional ion exchanger, is insensitive to many of the changes. Intermediate levels of specificity are thus desirable.

The same logic that applies to analysis of variants produced by growth under nonselective conditions in a fermentor also applies to the selectively neutral variants that comprise the molecular clock of evolution (18). The omnipresent occurrence of such variants, together with the good match between the prevalence of amino acids in the genetic code and in proteins generally, implies that nearly every position in a protein is subject to changes which can be compensated for at a functional level by changes at other positions within the protein. The specific nucleation sites for folding, for example, apparently find each other by a process that is chromatographic in character rather than an all-or-none lock-and-key mechanism.

The observation of significant hydrophobic moments in a wide range of protein helices, often with the hydrophobic faces of two helices juxtaposed, is consistent with this view (19). Proteins would presumably get hopelessly tangled up with each other during synthesis if the specificity of this interaction were as low as on a typical hydrophobic interaction sorbent, but the folding process would be too sensitive to mutation if it were as high as that of a typical high-affinity antibody. Evidently, intermediate levels of specificity are optimal over the long range.

VII. PARALOG CHROMATOGRAPHY

In the preceding sections, a wide variety of biological and chemical engineering problems were shown to have at their core a trade-off between competing specificity objectives. To study these phenomena, in vitro model systems are needed. In particular, it should be useful to construct systems that span the gap between traditional chromatographic sorbents and the highly specific sorbents that ar normally used in affinity chromatography. In this section we present one approach to bridging the gap.

An outstanding feature of the prototypical affinity ligand, an antibody-binding site (or paratope), as compared to traditional chromatographic sorbents, is that the antibody provides a mixture of binding modes at the size scale of analytes: cation and anion exchange, hydrophobic interactions, and hydrogen bonding. The enormous variety of possible antibody-binding sites is due to the combinatorial explosion inherent in polymers composed of monomers that are diverse in multiple properties.

Comparable diversity in mixed-mode binding properties can be achieved using short peptides, or other polymers composed of monomers diverse in multiple characteristics, as chromatographic ligands; these paratope analogs are referred to as *paralogs* (20). Figure 3, for example, shows variation in the pattern of proteins from a complex yeast extract mixture that bind to an agarose support derivatized with three paralogs. The paralogs illustrated differ in two of eight positions. Many

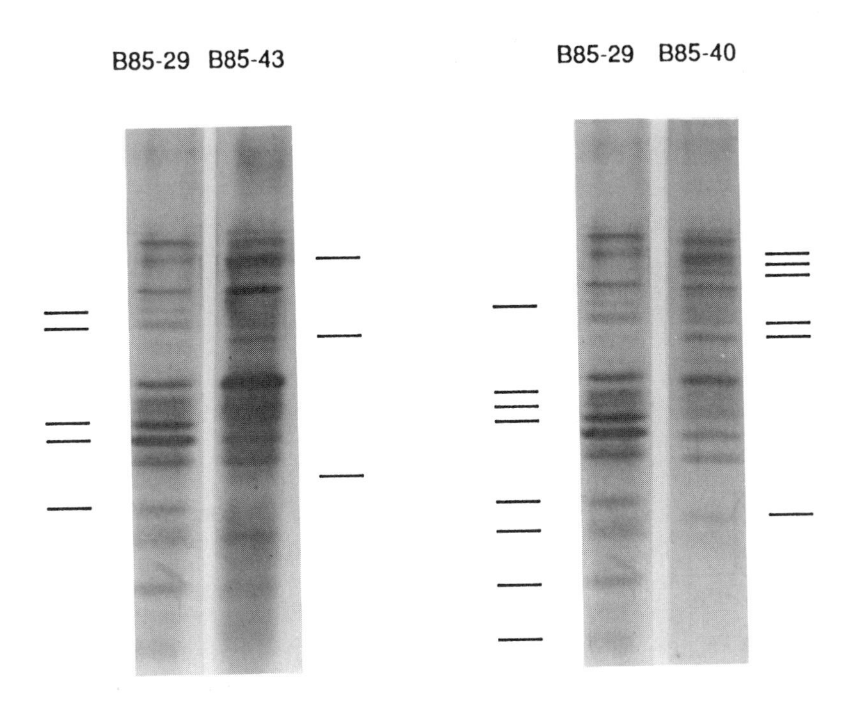

Figure 3 Paralogs are short polymer analogs of antibody binding sites (paratopes). Variation in paralog properties can be achieved by systematic design, resulting in differences in the set of proteins resolved. A DEAE salt cut of a yeast lysate was prepared (20) and loaded onto sorbents created by coupling short peptides to activated agarose. After washing, adsorbed proteins were eluted with 250 m*M* NaCl and examined by SDS gel electrophoresis and silver staining. Markings on the side of the gels indicate prominent differences in the patterns of proteins bound. B85-29 is a conformationally restricted oligocationic paralog; B85-43 differs from it in polarity, while B85-40 differs in isoelectric point.

of the proteins visualized by one-dimensional SDS gel electrophoresis bind to two of the sorbents, but few bind to all three.

As a practical matter, the ability to engineer sorbents that are partially similar is useful in protein purification. A common problem in such work is that one runs out of separation techniques before one runs out of contaminating proteins. Inevitably, the contaminants remaining are those that are most similar to the protein of interest. In the normal course of protein purification, the only solution is to refractionate the mixture more finely, at the cost of low recovery (Figure 4).

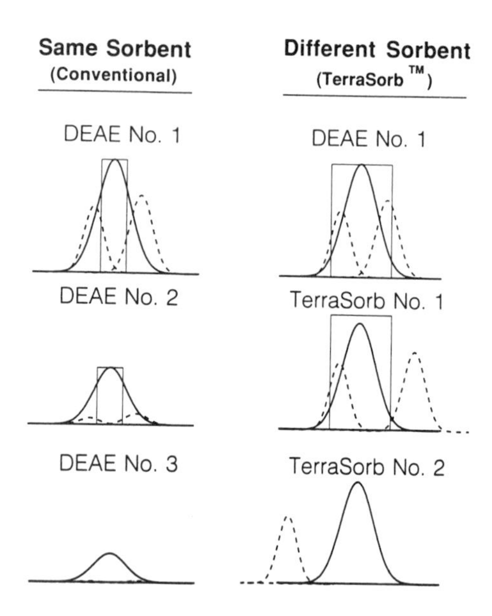

Figure 4 Purification of a target protein (solid line) from co-chromatographing contaminants (dashed lines) can often be achieved by re-chromatographing the peak fraction of the target on a higher-resolution column or with a shallower elution gradient. An alternative strategy is to re-chromatograph the full target peak on columns with differing selectivities. Paralog technology allows a large number of such selectivities to be easily prepared.

The data in Figure 5 illustrate this principle. For two of the paralog sorbents compared in Figure 3 in a single large salt step elution protocol, a series of smaller salt steps were used to elute proteins from a complex mixture. In addition to the overall difference in pattern of proteins bound, the relative binding strengths of many of the proteins evidently differ on the two sorbents, as different salt concentrations are needed to elute the same bands from different sorbents.

This figure also illustrates the moderate conditions needed for elution of proteins from paralog columns. By contrast, a well-known drawback to immunoaffinity columns is the difficulty in eluting bound analytes under conditions mild enough to retain biological activity (21). The tight binding of antibodies compared to paralogs is likely to be due in part to the rigidity of the antibody-binding site (22). In the paralog case, some of the free energy gained in binding is expended in stabilizing the many degrees of rotational freedom present in the paralog structure. Constraining the paralog structure via an intramolecular disulfide bond, for example, increases the average binding strength as measured by the increased salt

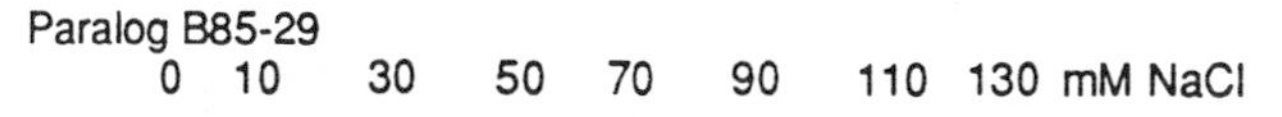

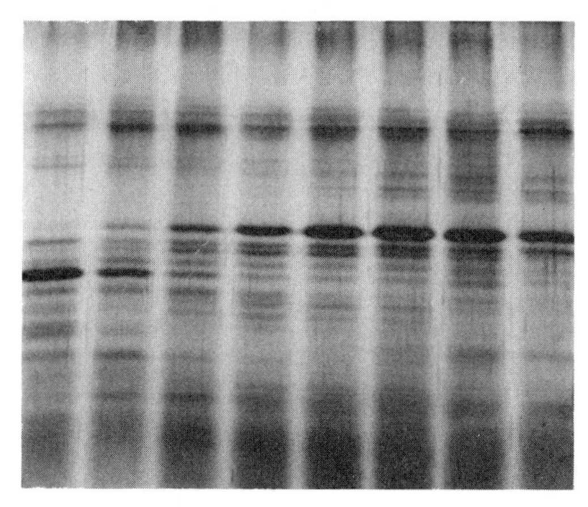

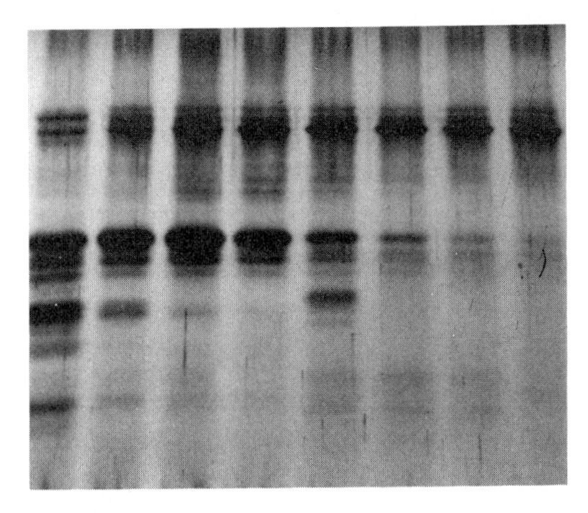

Figure 5 Even proteins that bind to the same paralog sorbent display differences in strength of binding, as evidenced by the increasing NaCl concentrations needed to elute the proteins. Experimental protocol and reagents are the same as in Figure 3, with eluates from step increases in salt concentration illustrated. Paralog sorbents can thus function in both high-selectivity and high-resolution modes.

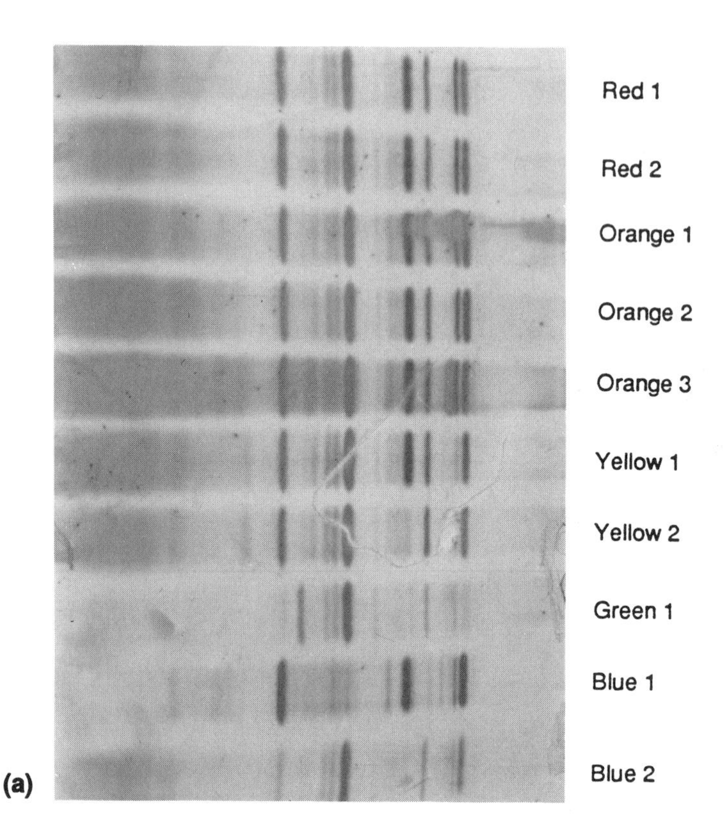

Figure 6 Diversity in binding properties of families of sorbents was evaluated by the protocol of Figure 3. Panel (a) illustrates results from a diverse set of paralog sorbents (Terrapin Technologies TerraSorb sorbents), while panel (b) illustrates results from a diverse set of textile dye sorbents (ICN Protrans sorbents). In addition to differences among the family members in the variety of proteins that are bound, the paralog and dye sorbents differ in the absolute amount of protein bound. The dye sorbents bind about 2% of the yeast extract, while the paralog sorbents bind from 2 to 25%. The paralogs thus offer higher overall resolution than the dyes but lower selectivity for these particular proteins.

concentration needed to elute the majority of the proteins (data not shown). Evidently, the paralogs combine aspects of the high-resolution traditional chromatographic style of sorting with aspects of the high selectivity style of traditional affinity chromatography.

The principle of using mixed-mode sorbents to create novel properties is not unique to paralogs, although prior efforts have used a limited variety of random copolymers rather than a single defined ligand structure for achieving the mixed

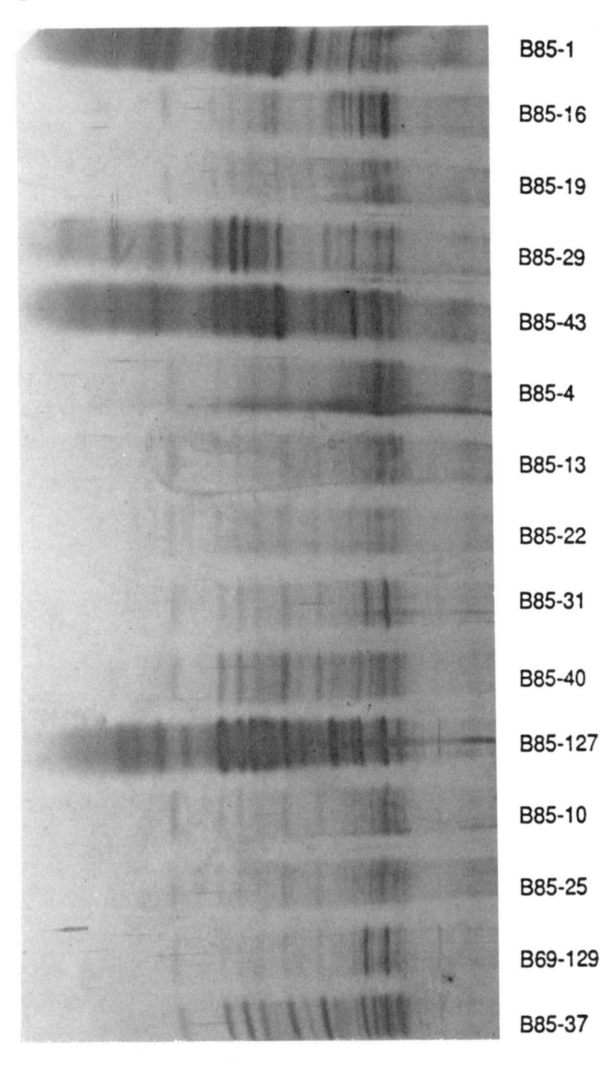

(b)

modes (23). Kiniwa (24) describes use of random copolymers of amino acids as a chromatographic matrix. Polysorb MP-3 from Interaction Chemicals, Inc. (Mountain View, California) is described as a mixed-mode polymeric sorbent derivatized with both C-18 and sulfonic acid moieties. Another approach to mixing modes operates at a much larger size scale than the molecular dimensions of paralogs. For example, Cibacron Blue dye, thought to act as an affinity ligand for NADH-utilizing enzymes, has been combined with traditional ion exchangers in a single sorbent. Several varieties of Cibacron Blue attached to DEAE or CM agarose are commercially available from multiple vendors, for example Bio-Rad (Richmond,

California). Such a column's separation characteristics are due to both the dye and the ion exchanger but are not fundamentally different from those attainable by operating two columns in series. Furthermore, the mixing of ligands does not provide the sort of diversity that the paralog approach offers, wherein the mixing of properties is at the size scale of the ligands themselves.

Still another approach to providing a mixture of binding sites has come from careful analysis of the Cibacron Blue dyes, which have been used as affinity ligands for many years. As normally synthesized by the textile dye industry, Cibacron Blue is a mixture of related molecules. Presumably, if individual members of this family were used as affinity ligands, they would show differences in binding properties. Exactly this result has been reported (25). Evidence has also been presented that the mechanism for the dye–protein interaction is more complex than simply mimicry of NADH (26). Sets of such dye-ligand sorbents are now commercially available from several sources.

A set of paralog sorbents has also been produced. The LI-KIT plate (Terrapin Technologies, So. San Francisco) is a 96-well membrane-bottomed flow through microplate with different sorbents in different wells, allowing parallel testing of numerous chromatographic sorbents, including paralog sorbents. Using the same complex yeast extract mixture of proteins as in the experiments described above, the two sets of sorbents were compared, with the results shown in Figure 6. Two significant features are revealed by this comparison. First, the dye ligands show differences in their selectivity with respect to this mixture of proteins, but the differences are more subtle than in the paralog case. Second, the fraction of proteins that bind to any of the dye sorbents is about 2%. For the paralog sorbents, the fraction bound ranges from 2 to 25%.

This experiment shows that the dye sorbents represent variations at the high-selectivity end of the specificity spectrum, while the paralogs represent variation more toward the high-resolution end, with applicability to a correspondingly larger number of proteins. Together, the two classes of novel sorbents demonstrate the feasibility of spanning the gap between traditional chromatographic sorbents and biospecific ligand derivatized affinity sorbents.

Following the antibody prototype, it is likely that paralogs of higher selectivity can be obtained by creating variations on an initial moderate affinity ligand chosen from a diversified starting panel. Application of these tools to a range of biological and chemical separations should lead to new insights into the roles played by molecular interactions of different degrees of chemical specificity.

ACKNOWLEDGMENTS

Portions of the experimental work reported here were funded by a Small Business Innovative Research Grant to LMK from the U.S. National Science Foundation

(ISI-9022271), with contributions by Richard Gomer, Karen Zabel, Kalman Benedek, and Peter Cheung. I thank Carol Topp for editorial and artistic input.

REFERENCES

1. A. Jaulmes and C. Vidal-Madjar, *Adv. Chromatogr., 28*:1 (1988).
2. O. Mikes, *HPLC of Biopolymers and Biooligomers*, Elsevier, New York, 1988.
3. I. M. Chaiken, M. Wilchek, and I. Parikh, *Affinity Chromatography and Biological Recognition*, Academic Press, New York, 1983.
4. L. K. Nash, *Chemthermo: A Statistical Approach to Classical Chemical Thermodynamics*, Addison-Wesley, Reading, Mass., 1971.
5. R. A. Goldsby, B. A. Osborne, D. Suri, J. Williams, and A. D. Mandel, *Curr. Microbiol., 2*:157 (1979).
6. C. Milstein, *Science, 231*:1261 (1986).
7. E. D. Getzoff, J. A. Tainer, R. A. Lerner, and H. M. Geysen, *Adv. Immunol., 43*:1 (1988).
8. N. R. Klinman, A. R. Pickard, N. H. Sigal, P. J. Gearhart, E. S. Metcalf, and S. K. Pierce, *Ann. Immunol., 127C*:489 (1976).
9. R. F. Schmidt, ed., *Fundamentals of Sensory Physiology*, Springer-Verlag, New York, 1978.
10. O. Siddiqi, *Trends Genet., 3*:137 (1987).
11. R. G. Vogt and L. M. Riddiford, *Nature, 293*:161 (1981).
12. L. Buck and R. Axel, *Cell, 65*:175 (1991).
13. P. W. Price, *Evolutionary Biology of Parasites*, Princeton University Press, Princeton, N.J., 1980.
14. G. M. Edelman, B. A. Cunningham, and J. P. Thiery, eds., *Morphoregulatory Molecules*, Wiley, New York, 1990.
15. J. R. Jacobs and C. S. Goodman, *J. Neurosci., 9*:2412 (1989).
16. T. T. Puck, A. Krystosek, and D. C. Chan, *Somatic Cell Mol. Genet., 16*:257 (1990).
17. F. E. Regnier, *Science, 238*:319 (1987).
18. J. L. King and T. H. Jukes, *Science, 164*:788 (1969).
19. D. Eisenberg, R. M. Weiss, T. C. Terwilliger, and W. Wilcox, *Faraday Symp. Chem. Soc., 17*:109 (1982).
20. L. M. Kauvar, P. Y. K. Cheung, R. H. Gomer, and A. A. Fleischer, *BioTechniques, 8*(2):204 (1990).
21. S. Ohlson, A. Lundblad, and D. Zopf, *Anal. Biochem., 169*:204 (1988).
22. A. Fersht, *Enzyme Structure and Mechanism*, 2nd ed., W. H. Freeman, New York, 1985.
23. D. R. Nau, *BioChromatography, 1*(2):82 (1986).
24. H. Kinawa, U.S. Patent 4,694,044, 1987.
25. S. J. Burton, S. B. McLouglin, C. V. Stead, and C. R. Lowe, *J. Chromatogr., 435*:127 (1988).
26. R. S. Beissner and F. B. Rudolph, *J. Biol. Chem., 254*:6273 (1979).

Index